OXFORD ENGLISH MONOGRAPHS

Agon

Poetry's Challenge to the Mathematatization of Reality (1920s–1960s)

ANIRUDH SRIDHAR

OXFORD
UNIVERSITY PRESS

OXFORD
UNIVERSITY PRESS

Great Clarendon Street, Oxford, OX2 6DP,
United Kingdom

Oxford University Press is a department of the University of Oxford.
It furthers the University's objective of excellence in research, scholarship,
and education by publishing worldwide. Oxford is a registered trade mark of
Oxford University Press in the UK and in certain other countries.

Published in the United States of America by Oxford University Press
198 Madison Avenue, New York, NY 10016, United States of America

British Library Cataloguing in Publication Data
Data available

Library of Congress Control Number: 2025940296

ISBN 9780198985457

DOI: 10.1093/oso/9780198985457.001.0001

Printed and bound by
CPI Group (UK) Ltd., Croydon, CR0 4YY

The manufacturer's authorised representative in the EU for product safety is
Oxford University Press España S.A. of Parque Empresarial San Fernando de Henares,
Avenida de Castilla, 2 – 28830 Madrid (www.oup.es/en or product.safety@oup.com).
OUP España S.A. also acts as importer into Spain of products made by the manufacturer.

Acknowledgements

I thank my parents for their almost unreasonable constancy of support—and Snytch, for ever delighting my memories—through all the shifting graces of time.

Michael Whitworth, my DPhil supervisor at Oxford, maintained heroic calm through my frenetic ambitions. His structure in method and erudition in subject were the reasons I completed the thesis upon which this book is based; and his scrupulous anatomization of verbal ambiguity has bettered me as a writer and critic. I would like also to thank Simon Palfrey, Nicholas Gaskill, and Matthew Bevis for reading through parts of my thesis and acknowledge their insightful ways of sending me in sensible and practical directions.

I have relied considerably on my mathematician friend Adam Jones; our long discussions on the beauty and theories of mathematics have undoubtedly bolstered the insights of this book. I am also thankful to my friend—and in some ways, comrade in crime—Mir Ali Hosseini, who has kindly lent to me his keen editorial eye. I also thank Curtis Brown, the literary estate of William Empson, for allowing me to publish his poems in full.

Last, I thank Hans—the general of my nightmares, still governing where each sally of mischief tends.

Contents

List of Figures

1
The Agon

In one of those richly cryptic aphorisms that distinguish the prophetic etchings of Blake, he says,

> Energy is the only life, and is from the Body; and Reason is the bound or outward circumference of Energy. (1790)[1]

At the violent birth of rational modernity, Blake restores 'Energy' and 'Reason' thus to their antique roles in Plato's allegory of soul: a drama raging between *Thumos* (spiritedness), *Eros* (desire), and *Logos* (reason), steeds and charioteer tossing through firmament, curvetting as one on the ridge of Heaven: mounting resistance thus warbling to fresh harmony. So 'Energy', both desire and spirit in Blake's aphorism, juts out from 'Body' in an eternalized simple present to be contained and channelled by reasoning mind.

But a closer look at its modifiers reveals a Reason less in command than is at first apparent. The word 'bound' can signify multiple and opposing senses, from 'bound with/within' to 'bound towards', as in the "winds merciless / Bound".[2] In the aphorism, the contraposed senses of 'binding' and 'bound to' are somehow reconciled, so that Reason, rather than describing an abstract boundary, becomes involved in the Energy blazing whole from body. To see how, let us take an ordinary sentence like 'the idea was so excellent she was *bound* to write about it'. Will she pen the idea because, as Hazlitt says, "it is in the nature of greatness to propagate an idea of itself, as wave impels wave, circle without circle",[3] or because she recognizes a moral duty to share the idea with the society that gave rise to it? Does the endowment of excellence compel self-perpetuation or establish a covenant with the less favoured? Is she bound to or bound by? It is some such modulation of agency that is also

[1] William Blake, *The Marriage of Heaven and Hell*, ed. Geoffrey Keynes (Oxford: Oxford University Press, 1985).

[2] From the first book of the Urizon (William Blake, *The Book of the Urizon* (London: Thames and Hudson, 1794)), [63–4].

[3] From 'The Indian Jugglers', in William Hazlitt, *Selected Essays*, ed. George Sampson (Cambridge: Cambridge University Press, 1917), p. 135.

Agon: Poetry's Challenge to the Mathematization of Reality (1920s–1960s). Anirudh Sridhar, Oxford University Press.
 DOI: 10.1093/oso/9780198985457.003.0001

effected by how the sentence structure unfolds: 'Energy' as subject explodes onto the scene with a compact declarative, 'is the only'; but 'Reason', beginning the following clause with the same declarative structure 'is the', is stunted by the awkward foregrounding of the word 'bound' (turning the voice passive) which together with 'outward' describes the semantic arc along which the *assertion* of the first clause is subtended by the *definition* of the second. Read together, the ambiguity and syntax thus undermine the presumptive hierarchy between Reason and Energy: in binding Energy, Reason is bound to it, bound with the outward process of 'body' that Blake calls 'life'.

Bearing in mind these mechanics, now, we can try and make sense of the aphorism. We might begin figuring Reason-bound-Energy as the wild spirit of body purposed by intuitions of order: a daily process of living, from which scientific man extracts "the ratio of all we have already known" (as Blake elsewhere defines abstract "Reason").[4] "Knowledge", declares Blake, "is not by deduction but Immediate by Perception or Sense. Christ addresses himself to the Man not his Reason" whereas "Plato did not bring Life [...] to light": for "Harmony and Proportions are Qualities & not Things. The Harmony & Proportion of a Horse are not the same with those of a Bull. Every Thing has its own Harmony & Proportion."[5] The 'Reason' of the aphorism is thus tortured by ambiguity and syntax until it reverts from an alienated ratiocination of mind to its medieval friendship with 'will'—recall Aquinas, for instance, who says, "to act for an end is to order one's action to an end; but this is the work of reason".[6] Blake involves reason in bodily appetite in direct rebellion against to an Age of Reason whose central precept he casts as follows: "That Energy, called Evil, is alone from the Body; and that Reason, called Good, is alone from the Soul."[7] But the division, he says, is illusory, for "Man has no Body distinct from his Soul. For that called Body is a portion of Soul discerned by the five senses".[8] He thus grants the age its principle—that reason is a function of soul—but then recasts body as phenomenal soul. Reason-bound-Energy, we might then say, figures souled body as participant in the always coming and going of order, in the cosmic agon between matter and form, through the clash of Energy and Reason—with an implicit caution that an age which

[4] William Blake, *There Is No Natural Religion* (Cambridge: Harvard College Library Department of Graphic Arts [1787] 1948), p. 1.

[5] From Blake's annotations to Bishop Berkeley's *Siris: A chain of philosophical reflexions and inquiries concerning the virtues of tar water, and divers other subjects connected together and arising from one another* (Dublin, 1744), pp. 213–14. Blake's annotated copy is in Trinity College, Cambridge.

[6] Thomas Aquinas, *Summa Theologica*, vol. 1 (London: Blackfriars, 1969), article II, question 1, objection 2.

[7] Blake, *Marriage*, p. 9. [8] Ibid.

seconds body to Hell, abstracting lofty reason from faulty flesh, will eventually amuse its patterns and ratios for the actual stuff of life.

We must preserve in the terms 'Reason' and 'Energy' a civilizational distinction once immediate to hand in order fully to recover Blake's vision. He hopes 'Energy' will sustain the moral urgency in such terms as desire and passion. With Puritan England's apparent return to Augustine, we can consider Energy in part as eros, the common form that love took for Augustine, appetite, desire for what is not possessed, the worldly things which body outwardly seeks: towards which the body instinctively bounds.[9] The reason binding energy can then be taken as a meridian circling the already motivated body, as if a longitude marking time by the crossing of other moving bodies and—to borrow a later sense of meridian—the line of ch'i along which the physicians of China found the vital elan of body channelled. The latter sense of meridian would not have been available to Blake, but the sense of a line around a planet marking space and time allows us to imagine, through magnification and contraction, the body as having a 'circumference' and as cosmic agent in the agon between order and chaos. It is important, however, that 'Reason' is defined in the geometric jargon of 'circumference' because reason becomes in the eighteenth century that which grants access to mathematical deduction; or as Dr Johnson would have it, "the power by which man deduces one proposition from another, or proceeds from premises to consequences";[10] or, as Adorno and Horkheimer rather bluntly put it, the Enlightenment "equates thought with mathematics".[11] Another equivalence implicit in *The Marriage of Heaven and Hell*,[12] and more peculiar to Blake, is that between the body and the poem, which seem in his early writings twin acts of creation thrust into being as divine and mortal jabs at the unanswerable. The poem for Blake, we may venture, registers the sensory prehension of the lived world,

[9] See Hannah Arendt, *Love and Saint Augustine*, eds. Joanna Vecchiarelli Scott and Judith Chelius Stark (Chicago: University of Chicago Press [1929] 1996), pp. 9–18.

[10] Samuel Johnson, *A Dictionary of the English Language*, eds Beth Rapp Young, Jack Lynch, William Dorner, Amy Larner Giroux, Carmen Faye Mathes, and Abigail Moreshead (London: William Strahan, [1755, 1773] 2021).

[11] Theodor Adorno and Max Horkheimer, *Dialectic of Enlightenment* trans. Edmund Jephcott (Stanford: Stanford University Press, 2003), p. 18. Oakeshott distinguishes the senses of word 'reason' before and after the emergence of science: "The 'reason' to which the Rationalists appeal is not, for example, the Reason of Hooker, which belongs still to the tradition of Stoicism and of Aquinas. It is a faculty of calculation by which men conclude one thing from another and discover fit means of attaining given ends not themselves subject to the criticism of reason, a faculty by which a world believed to be machine could be disclosed" in Michael Oakeshott, *Rationalism in Politics and Other Essays* (Carmel: Libert Fund [1962] 1991), pp. 22–3.

[12] Blake also connects the supposedly evil Energy with poetry when he says Milton, the "true poet", "wrote in [the] fetters" of circumference when "he wrote of Angels and God", of the "Son, a ratio of the five senses", but was liberated when writing "of Devils and Hell" [10–11].

beating a measure of lived time on souled body—in contradistinction to the bodied soul of Plato granted but the measure of Eternity by the numbers.

If we then take *poesis* as the bodied expression of soul and *mathesis*, as its lucent meridian of reason, we may read into the original aphorism a volatile union of antitheses: of deixis and deduction, the contrasting creative modes of poetry and mathematics, displaying in our outward works the same bounding and binding tension as in the inner agon between Energy and Reason. Given Blake's attempt to restore Reason from mental deduction to bodily intelligence, I have betokened Energy and Reason with the archaisms, *poesis* and *mathesis*, to signify the *actions* of 'bringing forth' and 'bringing into order', leaving the artefacts of those actions free to partake of both. We may consider the poem, for instance, an artefact of bodily energy but it is also an artifice of numbering and ordering—of feet and metrical periods;[13] and in the sonnet form, where lyric comes finally detached from lyre, the poem becomes an argument at times forensic in nature.[14] But the structure of Blake's aphorism even calls to mind a mathematical discovery: a feat of imaginative energy given form by proof, the strictures of deductive reason. For Blake, it was through such interpenetrations of opposites, the full capability of which he likely reserved for poetry and painting alone, that world would disclose itself to Being. Only amid the sturm und drang of body and mind, Energy and Reason, did Nature discover to her devout the exultations of her terrible aspect in the vigour of pith.

The attempted redistribution of these principles after a century in the thrall of Newton is one of Blake's lasting legacies, if not achievements: it would carry through Coleridge and Yeats well into the modernist movement. Blake addresses his *Marriage of Heaven and Hell* to what Carlyle would later call the Age of Victorious Analysis[15]—referring deridingly to the dazzling successes of French mathematicians in the science of dynamics; to the refinements of the salon, where even poets, according to Pope, "by Numbers judge a Poet's Song, And smooth or rough, with such, is right or wrong" (ll. 339).[16] Although the modernists wrote into a world almost unrecognizable to the

[13] So much so that the word 'number', from referring to feet, metrical periods, and lines, came in the nineteenth century simply to mean 'poem' (*OED* 4.b.).

[14] Michael Spiller discusses the origins of the sonnet among the learned lawyers of Italy (*The Development of the Sonnet* (London: Routledge, 1992)).

[15] Thomas Carlyle, *French Revolution, Book 2* in *The Collected Works, vol. 3* (London: Chapman and Hall, 1864), p. 26.

[16] Alexander Pope, 'An Essay on Criticism', in *The Works of the Poets of Great Britain and Ireland with Prefaces, Biographical and Critical, vol. 6*, ed. Samuel Johnson (Dublin: Pat Wogan [1711] 1804), p. 21.

Augustan, it had come to resemble, in many ways, the dystopia of Blake's prophetic visions: a human world bound by the ironclad rule of number and ratio. The modernist movement, some of whose central aspects this book will attempt to detail, can be seen as successor to Blake insofar as it, too, sought to unite Energy with Reason—a binary variously and varyingly distinguished in the history of poetics as sense and intellect, emotion and wit, and feeling and thought. For Blake, as we have seen, such a marriage is forged and governed by resistance. The poets we will study likewise attempted to show that a world built on Reason alone, as they felt theirs to be, is an outline without substance, a circumference without circle that would give into itself apart from the resisting Energy of body. The abstract symbols through which modern man conceived the world must make way for a fuller picture of reality comprehensible only to the sensuous intelligence aroused in encounter with poetic form.

It is this dance of opposites that gives name to this book. The term 'agon' will recall, I hope, that spirit of competition so central to Dionysian Greece. Donning their buskin of insubordination, modernist poets,[17] in a latter age of science, vied not with each other but their mathematical peers. Considering poets and mathematicians as peers, or poetry and mathematics as sisterly arts in diverging quests for the unanswerable, we might have titled this story 'paragone'. But unlike the Renaissance paragone—the rivalry between poetry and painting or painting and sculpture—the battle in the twentieth century would not be an aristocratic game adjudicated by the degree to which one art is able to entreat and exploit the excellence of man; by whether poetry required more learning or if painting disclosed diviner creative powers.[18] Jusepe de Ribera's blind aesthete fingering a marble head with an unseen painting sprawled at his side, for instance, frames a sororal jibe at the art of painting judged against the art of sculpture by a painter in a painting. But the rivalry between poetry and mathematics was not such a game: it was exigent and political: for theirs was a contest for authority, discursive and intellectual, to reckon and represent reality in their own frames. It can in some ways be seen as a contest in the tradition of Kant's *Conflict of the Faculties*: of lower faculties challenging the authority of their supposed superiors.[19]

[17] I use the term 'modernist poet' generally in the introduction to paint in broad strokes the mood among poets in their period (demarcated in this work from the 1920s to the early 1960s) as regards science. The idea of a common modernist attitude to science will be considered in the following section. Subtleties of inclination peculiar to individual poets will be explored in their respective chapters.

[18] See Chapter 1 of Sarah J. Lippert, *The Paragone in Nineteenth-Century Art* (Abingdon: Routledge, 2019).

[19] Immanuel Kant, *The Conflict of the Faculties*, trans. Mary J. Gregor (Lincoln: University of Nebraska Press, 1979).

The frame of the 'agon' will also secure this narrative, I hope, from the implication of isolation that attends the phrase, overused in literary criticism, of 'two cultures':[20] for the modernist did not attempt siege from the confines of the ivory tower or pastoral estate: to them, resistance to what Auden called the "Kingdom of Number" would involve all the clamour of assembled engagement.[21] Rather than renouncing altogether the imperious dictates of mathematical reason, they engaged, contended, and reconciled. Geoffrey Hill likens agon to a "spiritual wrestling match with the forces of darkness" that is "worked through stylistically" in the syntax and diction of poetry.[22] The grappling and contortion of wrestling perhaps best suggests the style of the modernist agon against Blake's forces of darkness, or Auden's infernal mathematics, for "could not one almost say that the / Cold serpent on the poisonous tree / Was l'espirit de géométrie"?[23]

1.1 The Protagonists

The American critic Louis Menand chanced on one of those pregnant summaries of an age, as when Vasari condensed two hundred years of Florentine genius to a rebirth of the golden age: "modernism is a reaction against the modern".[24] It is difficult to rank among the features of modernity that stirred this rebellion the importance of mathematization: the antagonist with which this narrative is chiefly concerned. But this much can be proffered: that although primers on modernism will cast in its gallery of villains such commonplaces as the death of God, the triumph of Mammon, the automation of daily life, urbanization, utilitarianism, and suchlike,[25] what we are dealing with is of unique significance to poetry. The aforementioned and suchlike are

[20] The phrase nowadays implies that the sciences and the arts are two separate and diverging cultures and was popularized by C.P. Snow (*The Two Cultures* (Cambridge: Cambridge University Press [1959] 1998)).

[21] W.H. Auden, *Nones: 1947–1950* (New York: Random House, 1951), p. 54.

[22] Geoffrey Hill, 'The Violent Bear It Away' (Lecture delivered at Cambridge University on the occasion of the 400th anniversary of the publication of the King James Bible, 2011).

[23] W.H. Auden, *New Year Letter* (London: Faber and Faber, 1941), p. 34.

[24] Louis Menand, 'T. S. Eliot and Modernity', *The New England Quarterly*, 69.4 (1996), 554–79, p. 554.

[25] We can see these in Horace M. Kallen, 'Modernism', in *The Encyclopaedia of Social Science*, eds Edwin R.A. Seligman and Alvin Johnson (New York: The Macmillan Company [1933], 1962), and four Peters to boot: Peter Nicholls, *Modernisms: A Literary Guide* (Basingstoke: Palgrave Macmillan, 2009), p. 13; Peter Gay, *Modernism: The Lure of Heresy from Baudelaire to Beckett and Beyond* (New York: Vintage Books, 2007), p. 28; Peter Childs, *Modernism* (London: Routledge, 2000); Peter Howarth, *Cambridge Introduction to Modernist Poetry* (Cambridge: Cambridge University Press, 2012), p. 97.

experiences common enough to all working modernists but are not such as would affect all alike. One imagines the Harlem poets to have experienced the effects of urbanization rather differently than the Southern Agrarians; so too, the decline of Christianity among secular versus religious poets. But reaction against the dominance of the mathematical world-picture was less adventitious than these, and, in ways we will anon see, more innate.

The resistance in this age took, broadly speaking, two forms. Some poets recognized the language of mathematics as integral to twentieth-century culture and so, did not attempt usurpation. They critiqued the mathematical world-picture's overreach while recognizing that modernity was too much shaped by it for lordly disdain. The poetry of William Empson and Michael Roberts, the first two poets we seriously consider, exhibits this style of defence. Both treat the language of mathematics and science as part of the language of modern culture and therefore, the diction of poetry. They attempted, in a more thoroughgoing fashion than any other poets of the twentieth century, to take the discursive empire of mathematics seriously and make poetry work in a world reshaped by it but all the while against its sovereignty.

The other mode of resistance recognized that poetry would never regain the authority to tell truths while accommodating with the kingdom of number. These poets faced the antagonist with a determined fatalism that led them to withdraw poetry altogether from the competition to *describe* the world. The most extreme partisans of this view were Laura Riding and Charles Olson—and, when he meditated on the nature of poetry, W.B. Yeats. Riding and Olson attempted to baffle a culture in the thrall of mathematical description by introducing entirely alien standards of truth. They challenged the very division of the world into description and descriptum by rendering the poetic word ingredient in the realization of reality—the fourth chapter will elaborate on this rather cryptic summary. Theirs was a more metaphysical challenge to the mathematical world-picture than Empson's and Roberts's attempts to imbricate mathematical notions into the diction of poetry and subvert thereby.

We must note now that although all the poets we discuss grappled seriously with mathematics, not all enjoyed an equal grasp of mathematics—this is not an entirely fair standard of judgement, given Empson may have been a mathematical prodigy before he was turned to poetry. The poets we discuss may at times make loose claims about the significance of mathematical ideas that would tend to undermine what is being asserted: this is particularly true of Olson and Ezra Pound (whose poetics we explore in the fifth chapter). We will attempt in such instances to identify the inaccuracies and speculate on what the misinterpretations about mathematics are nevertheless meant to signify in the poetics. Usually, we find that misprisions (what Bloom regards as defensive distortions[26]) also tell us something interesting about cultural and artistic views of mathematics; they are counterparts, in some ways, to seventeenth-century natural philosophers miscasting poetic metaphor (discussed later in this chapter) to fashion their polemic against poetry: we are after all narrating an agon, and rhetorical manoeuvres, in whatever degree mistaken or mendacious, are of interest.

Our protagonists, Empson, Roberts, Riding, Olson, and Yeats, all contend with the mathematical world-picture in a manner that shapes their poetics. With Empson and Roberts, this is unsurprising. They emerged from the epicentre of mathematical activity, the University of Cambridge in the 1920s, where the most advanced debates in the philosophy of mathematics were regularly being staged. Riding and Olson, though of opposite dispositions, are decidedly American in that Puritan way of presuming to start history afresh, which in their case, led them to attempt remaking poetry *ex nihilo* in the wake of scientific civilization. Yeats, however, cannot be said to represent anyone but himself: he is both liminal and pivotal to the history of poetry, bringing to close a certain lyric mode begun with Sidney, while, with Blake, being out of place in the world into which he wrote; and like Blake, the opposition of mathematical reason and aesthetic energy seems to have been central to his strange poetics and 'theory', so to put it, of art.

We have used phrases so far like 'world-picture', 'kingdom of number', 'discursive empire' and will continue recurring to them in the course of framing the agon. We will thus pause awhile, in the next section, to clarify their meaning. The following is a theoretical exercise undertaken not to establish a theory for the explanation of poetry but to elucidate the concepts that will allow us to better narrate events within the great fact of poetry in an age of science.

1.2 World-Pictures

We will begin with a single, though not uncontroversial, assumption that what is known about the world is secured chiefly through representation: in words but also in pictures, musical notes, numbers, and the like. As we

[26] Bloom is discussing poets productively misreading other poets. (Harold Bloom, *The Anxiety of Influence: A Theory of Poetry* (New York: Oxford University Press, 1973), p. 23.)

hinted, our poets, especially Riding and Olson, will challenge the notion that poetry re-presents rather than being part of or an event in the world itself. Erich Auerbach, in his famous anatomy of *Mimesis*, notes how even the thing being represented does not always remain passive: an allegorical telling or foretelling of a historical event, for instance, "establishes a connection between two events or persons in such a way that the first signifies not only itself but also the second, while the second involves or fulfils the first".[27] In this book, we will gradually cultivate a more dynamic sense of representation than is generally meant. But let us begin with 'representation' as the conception of something in the world that issues from a particular mode of speaking about it. What is represented is thus dependent on what is doing the representing; in this way, it is somewhat like a plaintiff's case, her picture of the facts, that depend for their framing on her representing counsel. We bring in the legal sense of representation because the traditional sense of 'imitation' typically elides the political dimension where representers are involved with one another in a competitive relationship; and in turn involved with the represented in a hierarchical one. Closer inspection may show that the represented, the so-called 'world', is itself a corporate entity assembled in imagination from a variety of worlds: for, as in the legal analogy, competing representations will issue different pictures of one notionally known entity or event. Now unless one mode of representation can incorporate a competing mode entirely into its own mechanism, there is no reasonable way to assess the value (a word we will use in this book to mean the claim to or purchase on reality)[28] of the two accounts outside the criteria for correctness established by each individual convention. For instance, we cannot reason whether Henry Mark Anthony's picture of the landscape at Tintern Abbey agrees more with the 'real' abbey than Wordsworth's. Anthony's painting is not simply "clad in one green hue" but it also fails to "connect / The landscape with the quiet of the sky".[29] What are the grounds on which to decide which omission is the more significant; is there any objective standard to adjudicate this paragone?

The problem is starker when two world-pictures contain assertions about one notional thing that contradict one another. In 1859, Charles Darwin

[27] Erich Auerbach, *Mimesis: Representation of Reality in Western Thought*, trans. Willard B. Trask (Princeton: Princeton University Press [1953] 2003), p. 142.

[28] This is a common enough practice in twentieth-century philosophy and rests on the notion that the value of representation increases with its purchase on that which it seeks to represent.

[29] William Wordsworth, *The Poems of William Wordsworth: Collected Reading Texts from the Cornell Wordsworth series vol II*, ed. Jared Curtis (Tirril: Humanities-Ebooks, 2011).

synthesized his observations of organic life into a theory of "universal warfare".[30] One year hence, John Ruskin published a competing theory:

> [I]n a plant, the taking away of any one part [...] injure[s] the rest. Hurt or remove any portion of the sap, bark, or pith, the rest is injured. If any part enters into a state in which it no more assists the rest, and has thus become 'helpless', we call it 'dead'. The power which causes the several portions of the plant to help each other, we call life. Much more is this so in an animal.[31]

The theories of "universal warfare" and "help" are opposite accounts of the same thing, namely, the fundamental disposition of life in nature. But they begin with different frames of reference—the biologist observes from the individual's perspective against the collective (not accidental, we might venture, in the first age of Capital) while the art critic studies the relations of the parts to the whole; and thereby, the meaning they draw from their observations cannot be refuted from each other's or a neutral point of view from first principles.

We reach such impasses due to mutual exclusivity among claims to truth. Nelson Goodman, for instance, asks us to

> Consider, to begin with; the statements 'The sun always moves' and 'The sun never moves' which, though equally true, are at odds with each other. Shall we say, then, that they describe different worlds, and indeed that there are as many different worlds as there are such mutually exclusive truths?[32]

Extending Goodman's argument from frames of reference to discursive modes (which he does himself), if it is true that two representational systems cannot be reduced to one without losing essential aspects of meaning created by either, and it is also true that we only 'know' the world through systems of representation, we may consequently ask if they in fact represent the self-same world—say (recalling Wordsworth and Anthony), an ultimate, hypostatic Tintern Abbey beneath the many fabricated Abbeys. But whether or not there are multiple worlds corresponding to mutually exclusive truths, one world apart from representation, or no world knowable beneath representation is a question for metaphysics that does not hinder our progress

[30] Charles Darwin, *On the Origin of Species by Means of Natural Selection or The Preservation of Favoured Races in the Struggle for Life* (London: John Murray, 1859), p. 65.

[31] John Ruskin, *Modern Painters vol. V* (London: George Allen [1860] 1888), p. 159.

[32] Nelson Goodman, *Ways of Worldmaking* (Indianapolis: Hackett Pub., 1978), p. 2. The two propositions are 'equally true' seen from the galactic and solar systemic perspective, respectively.

here.[33] What is, however, a central premise of the agon we are about to characterize is the fact that two systems of representation do not reduce, or factorize, to a master mode. We will call the effects of such mutually exclusive modes 'world-pictures'.

World-pictures do not fully correspond chiefly because their modes of signification are distinct. The theorist of literature may have already been put off by groupings of this kind. After all, it is not clear that poetry as such generates a world-picture about which anything can profitably be said. Luckily, though, what we have said about world-pictures does not require us to proffer any constant attributes of world-pictures—the poetic or mathematical—which may or may not vary from base to appearance in each iterative essay. Despite the potential for variability of attributes in every poem or painting, the concept of a world-picture may yet be well defined: as the result of a unique mode. That is, a poetic world-picture is poetic in that it is the result of a poetic mode of representation. But even so, the literary professional might point to the fact that the history of literary criticism shows a great deal of variation in characterizations of poetic mode. During the modernist period, a popular opinion was that poetry disclosed the world chiefly through metaphor, by revealing connections between a vehicle, and its intended destination;[34] say, between a lion and Achilles, there was transferred courage, insolence, caprice, and fierceness but perhaps not the tail (though in my personal imagination, the mane is an essential): and thus, by articulating various connections, metaphor would bring the density of the poetic world-picture into focus from and through its interstices. Empson and his followers viewed the poetic mode chiefly as working through and organizing ambiguity in words. My own sense of Empsonian ambiguity pictures words as the atoms or elements of a world-picture that poetry, through orbital interactions, exchanges of energy, discloses by mutually enlightening semantic haloes. There was also an American train of thought which numbered Wallace Stevens, William Carlos Williams, John Crowe Ransom, and Charles Olson among its members that described the poetic mode as primarily perlocutionary:[35] that diction and syntax made things happen within and

[33] See Ernst Cassirer, *The Philosophy of Symbolic Forms* (New Haven: Yale University Press, 1955); Ernst Gombrich, *Art and Illusion: A Study in the Psychology of Pictorial Representation* (New York: Phaidon Press, 1960); Nelson Goodman, *The Structure of Appearance* (Cambridge: Harvard University Press, 1951).

[34] I.A. Richards is perhaps the flag bearer of this view.

[35] See William C. Williams, *Selected Essays* (New York: Random House, 1954). p. 287; Wallace Stevens, *Collected Poetry and Prose*, eds Frank Kermode and Joan Richardson (New York: Library of America, 1997); and Charles Olson, *Collected Prose*, eds. Donald Allen and Benjamin Friedlander (Berkeley: University of California Press, 1997), pp. 123, 243.

without a poem and it was these happenings or events that formed poetic reality; to this school, the poetic world-picture was entirely composed of how poems mean and the question of what a poem means was of lesser moment. I defend the above somewhat idiosyncratic characterizations and attempt to reconcile some of these conceptions of poetic mode in my essay in *The Work of Reading*.[36] The term 'represent' with which we began may thus be too passive to catch all the modes of showing or disclosing: but it retains I hope enough of 'display' from its French root (*representer*) and 'presence' from its Latin root (*praesens*) for us to persist with the word.

We use the term 'mode' somewhat in the manner in which Michael Oakeshott describes a 'civil association' as that which results from "a mode of intelligent relationship". "And by a mode of relationship I mean a categorically distinct manner of being related which, while it may subsist in conjunction with other modes of relationship, cannot be reduced to any other."[37] A civil association might share a great many outward characteristics with a constitutional monarchy, social democracy, or even dictatorship. And conversely, European states that approximate to civil associations have many characteristics they do not share with one another. But a civil association yet remains a unique, yet abstract, object of philosophical interest because results from a unique mode of human relationship characterized by a specific set of attitudes—the poetic world-picture is an analogous object of interest to us. Despite great variety in conceptions of poetic mode, the concept of a poetic mode can yet be well defined within the family of modes, each member of which each member of the family of world-pictures is a result. Our assumption here is that the set of characteristics of each individual mode will be distinct. That is not to say that there must be a complete disjuncture between any two modes, but that the union of their sets of attributes will be greater than each individual set or that the intersection will be lesser. If, for instance, the set of characteristics of the poetic mode is = {metaphorical, ambiguous, perlocutive, reflexive} and that of the scientific mode is = {inductive, deductive, mensural, reflexive}, we see that their union, {metaphorical, ambiguous, perlocutive, reflexive, inductive, deductive, mensural} is greater than either individually, and their intersection, {reflexive}, is lesser. A mode is thus well defined when it has a unique identity in the family of modes.

[36] Anirudh Sridhar, 'Poem as a Field, Canon as Crystal: Geoffrey Hill's Historical Semantics', in *The Work of Reading: Literary Criticism in the 21st Century*, eds Anirudh Sridhar, Mir Ali Hosseini, and Derek Attridge (New York: Palgrave Macmillan, 2020), pp. 224–30.

[37] Michael Oakeshott, *On Human Conduct* (Oxford: Oxford University Press, 1975), p. 112.

We could also argue for the fact that the world-pictures birthed by unique modes will be as or more distinct as were their parents. To do so, we will have to restructure our analogy so that the domain set consists of supposed real-world objects being represented in the co-domain. The operations of modes would then be functions that translate items from a 'real-world' domain set to a 'representation' co-domain set: and so, a world-picture would be composed of the range of items that manifests within the co-domain. It becomes the task of propagandists to then insist that these functions with respect to their own world-picture are 'invertible': implying that the picture generated by their mode uniquely corresponds back to the real-world objects that it is designed to represent. But I am starting to feel that an elucidation of these hints will result in over-theorizing—it may be more prudent to appeal to the common-sense intuition that the more different the discursive mode, the more different will be the resulting world-picture.

Modes are meant to precede world-pictures ontologically, but they are not hidden from view, as background labour behind the picture encountered, as in, say, the packaging of candy. Goodman says,

> Worlds are made not only by what is said literally but also by what is said metaphorically, and not only by what is said either literally or metaphorically but also by what is exemplified and expressed—by what is shown as well as by what is said. In a scientific treatise, literal truth counts most; but in a poem or novel, metaphorical or allegorical truth may matter more, for even a literally false statement may be metaphorically true and may mark or make new associations and discriminations, change emphases, effect exclusions and additions.[38]

Mode thus actively communes with and conditions reception. We take what is said based on how it is said, where it is said, and what is said around it. If we look only at what is said in the following lines, "these structures, like the combination of amino acid sequences into plant proteins, constitute the phase map for any given species within which the period-functions are continuous with the epoch",[39] we might think it intended as a factual statement, the register and diction bidding us to assume the context of a biology paper. But if we look at the next line, "The conclusion to Professor Lichen's somewhat leathery discourse is greeted with polite, photosynthetic applause", we

[38] Goodman, *Worldmaking*, p. 18.
[39] J.H. Prynne, *Poems* (Freemantle: Freemantle Arts Centre Press, 2005), p. 235.

might guess instead at an attempt at high Dickensian parody. But when we see that the title has the phrase "plant time manifold", which appears in a collection of poems, we focus less on the narrative humour and more on the juxtaposition of register, 'epoch' with 'period-functions', 'applause' with 'photosynthetic', and wonder if some mutual accommodation or appropriation of scientific and common diction is not being attempted through concentration in verse. A historicist might even argue that Jeremy Prynne, the author of the poem, came under the tutelage of Joseph Needham[40] and might thereby be attempting to revive an organicist or vitalist view of plant-life against the physicist's reduction of reality to a 'space-time manifold'. Of course, attention to illocution—the waggish play with register, say—will, as usual, show the historicist's reading to be skewed: but it nevertheless complicates Goodman's argument about the sovereignty of frames of reference by noticing frames of reading or reception—the latter of which introduces the political dimension to the agon.

Oakeshott, in an essay on the modern university, theorizes the relationship of discursive modes with each other as urbanely as he does the fictional citizens or *cives* of his civil association: "modes of thinking [are] related to one another conversationally", he says, through "oblique recognition and accommodation".[41] This might describe the behaviour of some Oxford dons with others but it is a fact of history that modes of representation have rarely, if ever, coexisted peacefully. Communities of world-pictures are constantly being forced into varying degrees of discursive imperialism.[42] Seen conversely, world-pictures are typically involved in a process of conquest. Eduard Jan Dijksterhuis, for instance, calls his history of science *The Mechanisation of the World Picture,* not 'the mechanical world picture' because he narrates the long process by which one mode of investigating and representing reality came to establish for all *the* world picture.

Discursive imperialism occurs whenever one or more world-pictures that do not correspond are nonetheless forced to be assessed under uniform standards of correctness or purchase upon the posited entity being described. We customarily leave it to the institutes of peer-review, criticism, taste, and suchlike, to determine the value of a work for a community of admirers. But

[40] J.H. Prynne and Keston Sutherland, 'Introduction to Prynne's Poems in Chinese', *The Cambridge Quarterly*, 41.1 (2012), 197–207, p. 199.

[41] Oakeshott, *Rationalism*, p. 195.

[42] The word 'discursive' is used here not to suggest an elaborate style but as the adjectival form of discourse. And we mean the word discourse as a system of thought, encompassing both its formal and conceptual aspects.

even so, epochs come to be distinguished by the master-lens through which their world was seen—our own, for instance, sees reality through a fog of numbers, ratios, and units. On the other hand, it is the visual image that discloses to us the *lebenswelt* of the Renaissance. And in the eighteenth and nineteenth centuries, there is a branch of philosophy, which can perhaps be said to begin with Lessing, that attempts to judge all the arts against the 'pure' art of music, culminating in Pater's declaration that "All art constantly aspires towards the condition of music":[43] because, it was argued, painting and sculpture are constantly doing away with their subject-matter to display pure form, assuming that in symphony, form (note patterns) is all there is. The shifting reigns of world-pictures—barring faith in the benevolence of Hegel's Geist—depends to a great degree on what we have called frames of reception; a broader lay or even intellectual community that is concerned less with the value of individual works and comes through historical and political forces to judge the value of world-pictures against others.

F.M. Cornford, describing the attacks on the old religions of Greece by Ionian science, says, "henceforth natural science annexed to its province all that went on 'aloft' in the sky or 'under the earth' ".[44] Although the scientific world-picture of the Ionians did not succeed in sweeping away all others as our own post-Newtonian mathematical one seems to have, Cornford's observation is a reminder of the always totalizing ambitions of discursive empires. That is, discursive empire, we are assuming, is the ambition not of science alone but the natural tendency of discursive modes. In the agon between scholasticism and humanism in the late Middle Ages, for instance, Quentin Skinner observes "the growing confidence with which the humanists were in fact willing to invade the scholastic fields of study, to denounce their rivals for continuing to follow benighted methods, and so to insist with increasing imperialism on the needs for the special techniques of humanism to be put to use throughout the entire spectrum of the intellectual disciplines".[45] Inheriting the same ambition, the mathematical world-picture that emerges from the Renaissance steadily tends to a view that all elements of all experience should be represented, if not explained, in its own terms: "for enlightenment", says

[43] Walter Pater, *The Renaissance: Studies in Art and Poetry* (New York: The Modern Library, 1873), p. 134.

[44] F.M. Cornford, *Before and After Socrates* (ebook: Pickle Partners Publishing [1932] 2016), p. 18.

[45] Quentin Skinner, *The Foundations of Modern Political Thought: Volume 1 The Renaissance* (Cambridge: Cambridge University Press, 1978), p. 105.

Adorno, in describing the mathematical world-picture, "is totalitarian as only a system can be".[46]

But, notes Goodman, there is a problem that the totalitarian ambition must overcome.

> if all right versions could somehow be reduced to one and only one, that one might with some semblance of plausibility be regarded as the only truth about the only world. But the evidence for such reducibility is negligible [...] How do you go about reducing Constable's or James Joyce's world-view to physics?[47]

Given the irreducibility of world-versions, the ruling picture will thus dispense with comparisons of the crude sort we conducted—between Wordsworth and Anthony or Ruskin and Darwin—as thought-experiments and appeal instead to the intuitive notion of a hypostatic world beneath the many world-pictures that is not, as it were, a 'picture' but the pure given, a fundamental reality, *ousia*. The ruling picture will typically find ways of establishing how it is the one that corresponds most closely to the nature of this fundamental reality, to the *physis* of *ousia*. The cultural, political, and intellectual acquiescence to such claims is what I will term discursive authority.

Discursive authority is successfully secured to the degree in which a world-picture is not scrutinized by its purchase on the notional entity or phenomenon being described but is itself taken as synonymous with an underlying reality. This godlike authority comes from the mode's guise as *author* of what is to be deemed real. The process might begin with the claim that one mode captures the nature of reality best or most or both. But soon enough, we will find that reality itself is being recognized by its resemblance or correspondence with a mode of description. Such authority is won, we might venture, in eroding the perceived ontological distance between description and descriptum.

This move can be observed across the works of René Descartes, a pioneering architect of the mathematical world-picture. In *The Discourse on Method* (1637), Descartes argues that because the only form of knowledge that man is competent to possess is deductive, given "the mathematicians alone have been able to find any demonstrations", the only way to gain knowledge of reality is

[46] Adorno and Horkheimer, *Dialectic*, p. 18.

[47] Goodman, *Worldmaking*, pp. 4–5.

by reducing it to "the various relations or proportions subsisting among those objects":[48] to those properties that are amenable to mathematical treatment.[49] In his *Meditations* (1641), however, Descartes moves on from method to assert an ontology so severe that it would abash any dogmatic divine in the annals of religion. Rather than proffering mathematics as a tool for securing true knowledge, he seems to suggest that mathematical properties are all that can be said to exist. He presents us with the following reasoning. "For it is highly evident that all that is true is something [truth being identical with existence]; and I have already fully shown the truth of the principle that whatever is clearly and distinctly known is true."[50] To borrow Oakeshott's conclusion on Descartes, "there are no grades in knowledge, what is not certain is mere nescience".[51] A thing is true if it can be clearly conceived, and if it is true, it exists. Descartes finds he can only conceive mathematical truths clearly: "I reckoned among the number of the most certain truths those I clearly conceived relating to figures, numbers, and other matters that pertain to arithmetic, and geometry, and in general to the pure mathematics."[52] This much Plato could have said. But what Descartes insists—in great part to fulfil his higher purpose of asserting the existence of God—is that clear conception translates to corporeal existence. I can acquire knowledge of "corporeal nature in so far as it is the object of pure mathematics".[53] Corporeal nature can be said to exist only insofar as it can be rendered in mathematical terms. "I at least know with certainty that [a] thing may exist, in as far as [it] constitutes the object of pure mathematics."[54] All things are thus subject to this rigorous check at customs before granted the passport to existence. He takes a few pieces of wax, and upon burning or otherwise deforming them, finds "but little in them that is clearly and distinctly perceived". What remains is only "magnitude or extension in length, breadth, and depth", "figure", and "motion": in fine, all that is compliant to "number". The wax is retrenched of all other properties such as "colours", "odours", and "tastes", which makes him doubt "whether or not the ideas I have of these qualities are in truth the ideas of real objects".[55] We find therefore that the mathematical account is not judged by its

[48] René Descartes, *The Discourse on Method*, trans. John Veitch (London: Everyman's Library [1637] 1912), p. 16.

[49] Descartes does question whether mathematics can ever be the sole means of inquiry (Descartes, *Method*, p. vi.); but, says Oakeshott, "the lesson his successors believed themselves to have learned from Descartes was the sovereignty of technique and not his doubtfulness about the possibility of an infallible method" (Oakeshott, *Rationalism*, p. 22).

[50] René Descartes, *Meditations on First Philosophy*, trans. (John Veitch ([1641] 1901), p. 63.

[51] Oakeshott, *Rationalism*, p. 21.

[52] Descartes, *Meditations*, p. 62.

[53] Descartes, *Meditations*, p. 64.

[54] Descartes, *Meditations*, p. 66.

[55] Descartes, *Meditations*, pp. 28–9.

purchase on the posited reality but that the latter's claim to existence is granted on account of its mathematical properties.

Isaiah Berlin says that the success of a discursive empire is determined by the impression of such propaganda on cultural imagination:

> What happens as a rule is that some subject gains the ascendancy—say physics, or chemistry—and, as a result of the enormous hold which it has upon the imagination of its generation, it is applied in other spheres as well. This happened to sociology in the nineteenth century, it happened to psychology in our own.[56]

Physics was able to fix mathematical theory as a sure sign of a world-picture's correspondence with reality by establishing mathematizability as a sign of the object's correspondence with reality; or as Adorno and Horkheimer put it, "in the preemptive identification of the thoroughly mathematized world with truth".[57] That mathematization became a culturally recognizable sign of veracity attests to the fact that the scientific world-picture had in the course of a few hundred years instituted for all other world-pictures a universal standard of assessment. In the centuries following Newton, many areas of experience not traditionally amenable to mathematical representation would forcibly be repainted in numbers—sociology and psychometry are Berlin's examples. It is against the imposition of such a standard that the poets in this book will be shown to revolt.

The main challenge to poetry was that the world which it represented or created no longer under the empire of science bore the cultural stamp of verity. The contours of reality had been drawn by mathematics and were more or less widely observed by a doting culture. Its boundaries are crisply clarified by Lord Kelvin in his lecture on "Electrical Units of Measurement" (1883):

> I often say that when you can measure what you are speaking about, and express it in numbers, you know something about it; but when you cannot measure it, when you cannot express it in numbers, your knowledge is of a meagre and unsatisfactory kind; it may be the beginning of knowledge, but you have scarcely, in your thoughts, advanced to the stage of *science*, whatever the matter may be.[58]

[56] Isaiah Berlin, *Roots of Romanticism*, ed. Henry Hardy (Princeton: Princeton University Press, 1999), p. 6.

[57] Adorno and Horkheimer, *Dialectic*, p. 18; We will explore the significance of such a reversal in discussing Empson's poem, "Doctrinal Point", in Chapter 2.

[58] Published in *Popular Lectures and Addresses, vol. 1: Constitution of Matter* (Cambridge: Cambridge University Press [1889] 2011), p. 73.

Following an age of the steam engine and The Great Exhibition, of grand scientific spectacle, it seemed to twentieth century modernists that the only culturally acceptable way to think and speak about the world was as tiny abstract units moving according to mathematical laws, observed by agents following the same formal laws of interaction. In order to posit an alternate world-picture, they had to begin from or at least mimic the sound of these beliefs.[59]

Bastian van Fraassen says that

> Acceptance involves not only belief but a certain commitment. Even for those of us who are not working scientists, the acceptance involves a commitment to confront any future phenomena by means of the conceptual resources of this theory. It determines the terms in which we shall seek explanations. If the acceptance is at all strong, it is exhibited in the person's assumption of the role of explainer, in his willingness to answer questions ex cathedra. Even if you do not accept a theory, you can engage in discourse in a context in which language use is guided by that theory—but acceptance produces such contexts.[60]

Mathematics, we have noted, had under the empire of science become the sign of reference to and identification with reality. Fraassen concludes from this that even dissent, or a challenge to the scientific world-picture, had to be couched in the master-picture's symbolic mantle. Indeed, that is how modernist poets differed in their antagonism to mathematical reason from the Romantics. By bringing mathematical concepts into the lines of verse, the modernists were able to cast their hats into the arena of accepted discourse while forcing the reader subversively to see the concept through the poetic frame, and to thereby, in some measure, retard the unidirectional process of discursive imperialism. We could thus regard the mathematical theories evoked in modernist poems as bids for cultural authority: to gain the poetic mode discursive authority by mere resemblance to the master world-picture—though the task they set for themselves is to then intrigue and undermine, as it were, from within the court of authority.

It is in this way that we are able in this book to speak of a general 'modernist' reaction to mathematization. Poets of the period were writing in a climate of uncertainty and their continued ability to talk seriously of a shared reality

[59] We will observe in Chapter 5 how Ezra Pound and Louis Zukofsky attempted to define poetic meaning in mathematical terms.

[60] Bastian van Fraassen, *The Scientific Image* (Oxford: Clarendon Press, 1980), p. 12.

was endangered. Prior to concerns about the outward manifestations of modernity loomed in the modernist conscience a consuming anxiety about poetry's public authority to speak truths.

1.3 Mathematics and Reality

On its face, the description of mathematics as a world-picture might seem to the reader mistaken. Mathematics operates from its own criteria and does not depend on the world, physical or human, to ratify its internal correctness. Pure mathematics, as we see it, is indifferent to whether reality corresponds to it: so, it is not, in this sense, a world-picture. But that does not preclude the existence of a mathematical world-picture. Auden coined the phrase 'Kingdom of Number' to signify the whole domain of the mathematical, from the abstract to the physical world, ever extending, in his time, to the social and psychological. Such is the scope we intend by talking of the world-picture of mathematics. To the twenty-first century ear, however, such composite denominations of disciplines seem outdated, and certainly, mathematics and science today conjure diverging images: a formal art of aethereal symmetries on the one hand, a mundane pursuit of substantiate fact on the other. To the modernists, however, it had been but half a century since mathematics was but the language of physics: seemingly originating in a common desire for discovering order in Nature, and conspiring in the process by which observed events in flux are abstracted to stable law, science and mathematics were still recognizable as one—or two counterparts, at least, of a greater world-picture. The individuality of their characters, though by the modernist period sharply defined (in the arcana of philosophy), was not yet deeply rooted nor widely diffused.[61] Although this book will, when the occasion of the poem demands, consider pure mathematics distinctly from science, the titular 'mathematization' is intended to be understood in a broader sense, as a grand social instinct, peculiar to one civilization in one epoch—steadily intensifying in Europe from the seventeenth to the nineteenth century—to order all knowledge in mathematical form—a phenomenon that would have seemed more

[61] Graeme Gooday and Daniel Jon Mitchell, 'Rethinking "Classical Physics"', in *The Oxford Handbook of the History of Physics*, eds Jed Z. Buchwald and Robert Fox (Oxford: Oxford Academic, 2013), pp. 732–43; Andrew Warwick discusses deep connections between physics and 'applied' mathematics in *Masters of Theory: Cambridge and the Rise of Mathematical Physics* (Chicago: University of Chicago Press, 2003), p. 367; Dominique Pestre, *Physique et Physiciens en France 1918–1940* (Paris: Éditions des Archives Contemporaines, 1984), pp. 104–8.

familiar to modernist imagination than the thousand subfields of STEM fractalizing in the modern university.

Literary scholarship, however, has been eager to winnow out strains within the 'other culture' to produce new sub-fields that seem to grow more esoteric by year—'quantum modernisms', declares a recent title, and 'modernist relativities'.[62] Talk of a literature-and-mathematics sub-field distinct from literature-and-science began around 2018, when a series of works began to appear that dealt exclusively with the influence of pure mathematics on literature.[63] Andrea Henderson's *Algebraic Art: Mathematical Formalism and Victorian Culture*[64] was followed by Nina Engelhardt's *Modernism and Mathematics: Modernist Interrelations in Fiction*, and Baylee Brits's *Literary Infinities: Number and Narrative in Modern Fiction.*[65] In 2019, Joseph Jarrett wrote about the influence of mathematics on late Elizabethan drama;[66] Tim Armstrong then published an essay, "'A Transfinite Syntax': Modernism and Mathematics", to which our fourth chapter owes a great deal.[67] In 2020, a collected volume, *The Palgrave Handbook of Literature and Mathematics*, was published, which explores a long history of interaction between mathematics and literature since the Middle Ages, and seems to assert itself as a new domain of interdisciplinary study.

This book will notice certain limitations in the prevailing critical approach of influence tracing in these interdisciplinary endeavours, which can be classified under what Derek Attridge in *The Work of Reading* calls empirico-historicism,[68] a trend that may be summarized as the search for explanation

[62] Rachel Fountain Eames, *Physics and the Modernist Avant-Garde: Quantum Modernisms and Modernist Relativities* (London: Bloomsbury, 2014).

[63] There were one-off instalments in the preceding decades, such as Miranda Hickman's *The Geometry of Modernism: The Vorticist Idiom in Lewis, Pound, H.D., and Yeats* (Austin: University of Texas Press, 2005) and Robert Tubbs's *Mathematics in Twentieth Century Literature and Art: Content, Form and Meaning* (Baltimore, Johns Hopkins University Press, 2014), which I am not regarding as indicating the emergence of a subfield. I am also not counting works in other disciplines that discuss, however marginally, the relationship between literature and mathematics, such as Linda D. Henderson, *The Fourth Dimension and Non-Euclidean Geometry in Modern Art* (Cambridge: MIT Press [1985] 2013) and Amir R. Alexander, *Duel at Dawn: Heroes, Martyrs, and the Rise of Modern Mathematics* (Cambridge: Harvard University Press, 2010).

[64] Andrea K. Henderson, *Algebraic Art: Mathematical Formalism and Victorian Culture* (Oxford: Oxford University Press, 2018).

[65] Nina Engelhardt, *Modernism, Fiction and Mathematics* (Edinburgh: Edinburgh University Press, 2018); Baylee Brits, *Literary Infinities: Number and Narrative in Modern Fiction* (London: Bloomsbury, 2018).

[66] Joseph Jarrett, *Mathematics and Late Elizabethan Drama* (Cham: Palgrave, 2019).

[67] Tim Armstrong, '"A Transfinite Syntax": Modernism and Mathemtatics', *Affirmations of the Modern*, 6.1 (August 2019), 1–29.

[68] Derek Attridge, 'Introduction', in *The Work of Reading: Literary Criticism in the 21st Century*, eds Anirudh, Sridhar, Mir Ali Hosseini, and Derek Attridge (New York Palgrave Macmillan, 2020), p. 4.

through fact-collection. The summary of "mathematics, modernism, and literature" offered by the editors of the *Literature and Mathematics* might prove illustrative.

> The rise of mathematical formalism in the nineteenth century [...] plays out broader tensions in the period. [...] The historian of mathematics Jeremy Gray summarizes: "the mathematics of the nineteenth century is marked by a growing appreciation of error leading to a note of anxiety, hesitant at first but persistent by 1900" (Gray 2004, p. 23). The period from the 1880s to around 1930 is accordingly called the "foundational crisis" of mathematics, and the questions and anxieties surrounding mathematics also occupied non-professionals and appear in literary texts.[69]

The notion of formalism, as the information-rich chapters of Jocelyn Rodal, Arkady Plotnitsky, and Nina Engelhardt show, seems to have been shared, or at least been recognizably similar, in the literature and mathematics of the time—and this hermeticism could perhaps indicate to us something about the decadence of the age. But does the exercise of drawing influence from mathematics to literature explain why modernist theories of formalism coincided with their mathematical counterparts? Looking at the interaction of faculties apolitically, that is, without an agonistic perspective, the best explanations of cross-application we have are the trendiness of mathematics and such ideas as happened to be in the air, in the zeitgeist. But through an agonistic frame, given literature's fast-waning authority under the discursive empire of mathematics, we may read literary formalism as a political gesture—both affecting the tenor of the master-discourse and securing itself from the daunting litmus test of naturalistic world correspondence. On the other hand, as we will see, the more dominant rhetorical move of modernism, particularly of modernist poetry, was to firmly establish the correspondence between word and world: to show the poetic world-picture to more accurately comprehend the world as held out to experience than that of science.

The limitations of the influence-tracing protocol might become clearer if we further scrutinize the notion of formal autonomy which motivates many works on modernism and mathematics. As suggested by the precis we quoted, what seems to have triggered interest in the topic is the mathematical historian Jeremy Gray's now-famous text in literary studies, *Plato's Ghost: The*

[69] Nina Engelhardt and Robert Tubbs, 'Introduction: Relationships and Connections between Literature and Mathematics', in *Palgrave Handbook of Literature and Mathematics*, eds Robert Tubbs, Alice Jenkins, and Nina Engelhardt (Cham: Palgrave Macmillan, 2022), pp. 6–7.

Modernist Transformation of Mathematics (2008).[70] Gray draws his main theme of modernism in mathematics from the German historian of mathematics Herbert Mehrtens' book *Moderne—Sprache—Mathematik* (1990).[71] Mehrtens and Gray pose the tantalizing idea that mathematics, through new developments in analysis, set theory, modern algebra, and logic underwent a modernism that can be spoken of in terms similar to modernism in the arts.[72]

Before Mehrtens, this connection would likely have seemed to literary scholars fantastic. And for many literary critics in the anglophone world, the same is true before Gray, given Mehrtens was not translated to English. According to Michael Whitworth, whose *Einstein's Wake* (2001) includes a chapter on non-Euclidean geometry, mathematics seems "removed from the conventional concerns of literature, allowing the fewest possibilities for metaphorical exchange".[73] Mehrtens and Gray have thus proven key in sparking the interest of literary scholars in the apparent connections between literary and mathematical modernity, including, it must be confessed, that of this author. I have since, however, come to regard many of these connections to be specious and limiting. Most critical works in the field dwell on the infinities of Georg Cantor, the formalism of David Hilbert, and the various geometries of non-Euclidean bent—all presented by Gray as seminal in the growing modernism of mathematics.[74] Although we are in the present work invested in modern developments in mathematics, we go where the poets send us:

[70] Jeremy Gray, *Plato's Ghost: The Modernist Transformation of Mathematics* (Princeton: Princeton University Press, 2008); I have ascertained this from conversations with scholars in the field; the meeting and general discussion about the importance of Jeremy Gray to the new field took place at the conference, 'Mathematics and Modern Literature' (Manchester University, 3–4 May 2018).

[71] See Herbert Mehrtens, 'Modernism vs Counter-Modernism, Nationalism vs Internationalism: Style and Politics in Mathematics, 1900–1950', in *L'Europe mathématique: Histoires, mythes, identités*, eds Catherine Goldstein, Jeremy Gray, and Jim Ritter (Paris: Editions de la Maison des Sciences de l'homme, 1996), p. 521.

[72] There are also a few lesser known authors in between who concur with this notion. See William R. Everdell, *The First Moderns: Profiles in the Origins of Twentieth-Century Thought* (Chicago: University of Chicago Press, 1997).

[73] Michael H. Whitworth, *Einstein's Wake: Relativity, Metaphor, and Modernist Literature* (Oxford: Oxford University Press, 2001), p. 198.

[74] Gray, *Plato's Ghost*, pp. 5–8; See Peter Johnson, '"Presences of the Infinite": J.M. Coetzee and Mathematics' (PhD dissertation: Royal Holloway, University of London, 2013); Roberto Natalini, 'David Foster Wallace and the Mathematics of Infinity', in *A Companion to David Foster Wallace Studies*, eds Marshall Boswell and Stephen J. Burn (New York: Palgrave, 2013), pp. 43–58; Loveday Kempthorne's PhD thesis, 'Relations between Modern Mathematics and Poetry: Czesław Miłosz; Zbigniew Herbert; Ion Barbu/Dan Barbilian' (PhD Dissertation: Victoria University of Wellington, 2015). Brits and Armstrong focus on Cantor's theories of infinity, and Jocelyn Rodal, in her upcoming book, focuses on mathematical formalism, particularly the influence of Hilbert on Eliot and Pound. Engelhardt surveys a wider range of ideas yet focuses mainly on whatever Gray regards as 'modernist'. The chapters of Howard Pollack-Milgate and Chris Ackerley in *Literature and Mathematics* explore older theories of mathematics, such as that of Leibniz, but only as extended tracings of influence on modern theories of formalism and infinity.

Pythagoras, Plato, and Euclid, when the mood is ideal, calculus, matrices, and quaternions when the tone is factual. The frame of the agon allows us to conjure a many-headed antagonist 'mathematization'—a civilizational force rather than index of intellectual history.

Nevertheless, the importance of Gray to the field is such that it would do to dwell awhile on his story.

> You treat world history as a mathematician does mathematics, in which nothing but laws and formulas exist, no reality, no good and evil, no time, no yesterday, no tomorrow, nothing but an eternal, shallow, mathematical present.[75]

Gray says Hermann Hesse, in *The Glass Bead Game*, "seems to get so much right about modern mathematics".[76] Hesse seems to register here what amounted to a revolution, during the late nineteenth and early twentieth centuries, in the philosophy of mathematics. But we must pedal back to the seventeenth to understand its significance. Although the relationship of mathematics to the physical world has, at least since the Ionians first posed the question, been liminal, a great paradox began to materialize in seventeenth-century Europe. We will let the venerable Whitehead state it for us:

> Nothing is more impressive than the fact that as mathematics withdrew increasingly into the upper regions of ever greater extremes of abstract thought, it returned back to earth with a corresponding growth of importance for the analysis of concrete fact. The history of the seventeenth century science reads as though it were some vivid dream of Plato or Pythagoras.[77]

The irresistible march of modern science, acting out some Hegelian reconciliation of medieval aristotelianism and Renaissance platonism, had by the end of the eighteenth century come to be identified almost entirely with the power of mathematics.

> Mathematical research in calculus and differential equations went hand in hand with work in the mathematical sciences of mechanics, astronomy, fluid mechanics, acoustics, and others. There was no such thing at the time as

[75] Hermann Hesse, *The Glass Bead Game* (London: Pan Books & Cape [1949] 1987), p. 168.

[76] Qtd from personal email correspondence (Jeremy Gray, interviewed by Anirudh Sridhar, 23/04/2016).

[77] A.N. Whitehead, *Science and the Modern World* (New York: Open Road [1925] 2021), p. 32.

> separate fields of pure and applied mathematics, although distinct areas of mathematics did start to separate from physics as the century wore on.[78]

The schism between mathematics and physics that succeeded their two-hundred-year harmony has been named, by a number of mathematical historians now, 'mathematical modernism', the most active period—1890 and 1930—of which coincides with high modernism in the arts.[79]

Gray's thesis is as follows:

> Taken together, all the changes in mathematics (during this period) and the connections to other intellectual disciplines that were then animated constitute a development that cannot be described adequately as progress in this or that branch of mathematics (logic and philosophy) but must be seen as a single cultural shift, which I call mathematical modernism.[80]

The mathematical developments of this period, as we discussed, were first styled 'modernist' by Herbert Mehrtens, whose rationale was as follows: "The two common traits of the various modernisms that I identify as central are, first, the autonomy of cultural production and, second, the departure from the vision of an immediate representation of the world of experience."[81] The first trait offers a simple historical explanation for the phenomenon, which is the growing autonomy of the academic profession of mathematics in the nineteenth century.[82] Gray, however, says "the professional situation of mathematicians, in particular their relative autonomy from scientists, did not cause modernism to happen, but it enabled it and it promoted it".[83] He concentrates mainly on the second trait: that mathematicians of the 'modernist' period began to insist that their subject, that is to say, the operations of and on mathematical objects, exceeded mere application to physical objects; that at a time when physics was becoming increasingly mathematical, new developments in mathematics were arising independently from concerns of the former.

[78] Calvin Jongsma, 'Mathematization and Modern Science', in *Mathematics in a Postmodern Age: A Christian perspective*, ed. Russell W. Howell and James Bradley (Grand Rapids: Wm. B. Eerdmans, 2001), p. 184.

[79] Gray, *Plato's Ghost*, p. 2.

[80] Ibid., p. 4.

[81] Herbert Mehrtens, 'Modernism vs Counter-Modernism, Nationalism vs Internationalism: Style and Politics in Mathematics, 1900–1950', in *L'Europe mathématique*, p. 521.

[82] For an account of the professional separation between mathematics and physics in Germany, see Jeremy Gray, 'Anxiety and Abstraction in Nineteenth-Century Mathematics', *Science in Context*, 17.1–2 (2004), 23–47.

[83] Gray, *Plato's Ghost*, p. 5.

This formalist retreat can, for instance, be seen in the definition of the real numbers. The gaps in the number line between rational numbers (numbers that could be represented as fractions) were not simply filled in by the mathematician as centimetres are infested with millimetres—this would have been recognizable to the physicist who is constrained by "senses and instruments".[84] Instead, the real numbers are defined as limits of Cauchy sequences, or the limits of ever-reducing rational numbers.[85] Poincaré observes that mathematicians are less concerned with well-rounded discrete objects like the integers and have increasingly come to understand things purely from their formal relations to other objects: '"Matter does not engage their attention, they are interested by form alone." '[86] Continuous mathematics, once the number line was completed, would betray properties that defeat intuition.[87] "The mathematical continuum has the property that any small piece of it can be magnified to resemble the whole, and to ensure this it is necessary to intercalate infinite sets of new elements."[88] Analysis thus draws mathematics further away from the sensible, empirical world. As Poincaré observes, "instead of endeavouring to reconcile intuition and analysis, we are content to sacrifice one of them, and as analysis must be flawless, intuition must go to the wall".[89] To Mehrtens, however, this departure from representing or even resembling the world of experience described but one strain of modernist mathematics; modernism also had in it a *gegenmodern* or counter-modern streak, of mathematicians who "favoured instead the *concrete*, and *intuition* or the given";[90] of which hereafter.

It is the notion of ontological independence, however, that has come to dominate literary critical imagination. There has been a tendency in the mathematics-and-literature field to accept the use of 'modernist', by Gray, to mean 'autonomous' or independent from the world. Joan Richards, in *Mathematical Visions*, narrates late Victorian mathematicians, Christian and idealist, repudiating the ascientific mathematics of vogue for undermining the union of number and fact that had obtained in the two-hundred-year

[84] Ibid., p. 130.

[85] Terrence Tao, *Analysis I* (Delhi: Hindustan Book Agency [2006] 2022), p. 105.

[86] Henri Poincaré, *La grandeur mathématique et l'expérience*, trans., 'Mathematical magnitude and experiment', Chapter 2 in *Science and Hypothesis* (London: Walter Scott Publishing, 1905): 17–34, p. 17.

[87] We will explore the reception of this idea in poetry in our section on Olson.

[88] Gray, *Plato's Ghost*, p. 130.

[89] Poincaré, 'Mathematical', p. 31.

[90] José Ferreirós, 'Paradise Recovered? Some Thoughts on *Mengenlehre* and Modernism', in *Science as Cultural Practice, Vol. II: Modernism in the Sciences, ca. 1900–1940*, eds Moritz Epple and Falk Müller (Berlin: Akademie Verlag, forthcoming); see also Herbert Mehrtens, *Moderne—Sprache—Mathematik: Eine Geschichte des Streits um die Grundlagen der Disziplin und des Subjekts formaler Systeme* (Frankfurt am Main: Suhrkamp, 1990), ch. 3.

wake of Newton. But in the epilogue, she argues that a gradual fin-de-siècle acquiescence to the new geometries led inevitably to a 'modernist' surrender to formalism.[91] Rodal similarly hopes to demonstrate that formalism is a trait shared by mathematics and poetry in the early twentieth century. She compares literary formalism with Hilbert's system, which swept aside the ontology of mathematical objects to make the proof of correctness, internal consistency, its chief end.[92] And Andrea Henderson, in *Algebraic Art*, argues that the formalist trend in literary culture grew out of the autonomizing tendencies of nineteenth-century mathematics.[93] Although Henderson is more interested in Victorian ideas of autonomy and art for art's sake, her book is certainly a specimen of a tendency to view mathematics and art as communing predominantly as formal and abstracted symmetries.

In favour of a supposed inclination to formalism, Rodal points to Eliot's essay on 'Ben Jonson',[94] wherein the world created by Jonson is compared to a non-Euclidean plane, "because they have a logic of their own"; but the hint of formalism is promptly correlated in the following line to the objective: "this logic illuminates the actual world".[95] That said, if we look to the most visible manifestations of the mathematical in modernist art, we would be hard-pressed to deny, given the abandonment of mimesis by Mondrian, say, or the diatonic scale by Schoenberg, that there is in modernism a profound yearning for representational freedom. But this freedom, which is sought to turn the quotidian, to present the everyday in strange and unusual ways, is habitually miscast to indicate a retreat into formalism. A closer study of illocution I think will show art in the modernist period, while gradually retrenched of naturalist illusion, all the while involved in the agon for representational authority, even if only the memory of real authority remained with them.

Let us take the emblematic case of a modernist departure from mimesis, namely, cubism, whose mathematical interactions and supposed formalist consequences were made popular in critical circles by Linda Henderson's *The Fourth Dimension and Non-Euclidean Geometry in Modern Art* (1983). The attempt to decompose the figure as traditionally depicted in illusionist

[91] Joan L. Richards, *Mathematical Visions: The Pursuit of Geometry in Victorian England* (Boston: Academic Press, 1988), p. 231.

[92] This is based on her presentation of her upcoming book at the 'Mathematics in Modern Literature' conference in 2018 (https://mathematicsandmodernliterature.files.wordpress.com/2018/05/programme-for-mathematics-and-modern-literature-2018-conference-pack.pdf).

[93] Henderson, *Algebraic Art*.

[94] See Jocelyn Rodal, '"A Few Sheets of Paper Covered with Arbitrary Symbols": Formalism, Modernism, Mathematics'. PhD dissertation: University of California, Berkeley, p. 85.

[95] T.S. Eliot, 'Ben Jonson', *TLS* (November 1919), pp. 637–8.

painting into angular geometric shards to then display on the canvas various aspects of the figure available to conception that are and are not to perception has by theorists of cubism like Apollonaire (whose writings Gray borrows to ground his concept of modernist autonomy[96]) and Metzinger been framed as a move from the three-dimensional space of perspective to a so-called fourth dimension. The fourth dimension is rather loosely defined by Apollonaire in *Les Peintres Cubistes* (1913) as follows:

> The fourth dimension appears to be engendered by the three known dimensions: it represents the immensity of space eternalizing itself in all directions at any given moment. It is space itself, the dimension of the infinite; the fourth dimension endows objects with plasticity.[97]

A sober redaction of the passage might simply say that cubism offers a vantage where all aspects of an object are on simultaneous display unbeholden to a point of view in time. Henderson presents this fact as suggesting a cubist longing for freedom from representation.[98] She quotes Olivier-Hourcade as celebrating that, with the advent of cubism, "art is free!" (1912),[99] and Gleizes and Metzinger as proclaiming in *Du Cubisme* (1912) that "There is nothing real outside ourselves."[100] But the technical evolution of cubism indicates that it was precisely such misreadings of its style, the assumption that cubist art is rendered for pure mind divorced from embodied perspective, that agitated the turn from analytic to synthetic cubism. Even before the use of collage techniques, newspaper, grains of sand, and so forth—what are often used to mark the split between synthetic and analytical cubism (around 1912)—we find tokenistic gestures by painters to preserve cubism as an art of representation and illusion. The graphic tack with shadow atop Braque's otherwise flat *Still Life with Violin and Pitcher* (1910) or the three-dimensional tassel-and-stud on the top-left margins of his otherwise flattened *Man with a Guitar* (1911) evince the desire to separate the depicted flatness of the figure within the painting and the literal flatness of the canvas 'outside' the painting.

96 Gray, *Plato's Ghost*, pp. 1–2.

97 Guillaume Apollinaire, 'Aesthetic Meditations: On Painting, the Cubist Painters', trans. Mrs Charles Knob. Launch for the Société Anonyme, *The Little Review*, vi (Spring 1922) from Guillaume Apollinaire, *Les Peintres Cubistes: Méditations esthétiques* (Paris: Eugène Figuière, 1913), pp. 15–17.

98 Henderson, *Fourth Dimension*, p. 180.

99 Olivier-Hourcade, 'La Tendancede al peinture contemporaine (Notes pour une causerie sur l'art contemporain)', *Revue de France etudes Pays Français*, Feb. 1912, p. 36; trans. in Edward Fry, *Cubism*, trans. Johnathan Griffin (London: Thames and Hudson Ltd, 1966), p. 74.

100 Albert Gleizes and Jean Metzinger, *Du Cubisme*, trans. R. Girard & Cie for La Compagnie Française des Arts Graphiques, Paris, 1947, p. 30.

Clement Greenberg reads these awkward additions of trompe-l'oeil—awkward, given they do not interact with the picture in any substantial way—as an attempt to keep the depicted plane "separate enough from literal flatness to permit a minimal illusion of three-dimensional space to survive between the two".[101] And he proceeds to observe that literally flat objects like newspapers were later in synthetic cubism used to identify the canvas over which it would lay adhered; to create a sense of depth in the painting that terminated at the literal canvas; which would, somewhat like *repoussoir* guiding the eye into the illusion of the painting (though working in reverse in cubism, from the background forward), bring the two-dimensional plane of the picture into relief: "Sealed between two parallel flatnesses—the depicted Cubist flatness and the literal flatness of the paint surface—the illusion is made a little more present."[102] We might say that the cubists' apparent abandonment of illusionism was actually a more complex mode of representation: one that allowed painting to continue vying, if only from past inertia, for representational authority over the human world in an age of social statistics and the photographic camera. "Painting had to spell out", says Greenberg, "rather than pretend to deny, the physical fact that it was flat, even though at the same time it had to overcome this proclaimed flatness as an aesthetic fact and continue to report nature."[103] In this manner, one often finds upon closer inspection that the realism that modernist art seeks to escape is merely the literal style it at times took in the nineteenth century. Auerbach argues that the nineteenth century is the odd one out in the history of mimesis for having rejected different "levels of literary representation" and "distinct levels of style",[104] insisting on literal imitation. Far from a mathematical retreat into the internal symmetries of art from a society become vulgar under the rule of science and technology, modernist art is often involved in a deeper engagement with the world as held out to experience.

The formalist readings growing popular in the influence-tracing exercises of the literature-and-mathematics field betray, strangely, a kinship with the mid-century mysticism of English departments that regarded poems as formal units—as well wrought urns with an internal logic of consistency. The connection between mathematical and literary autonomy was even explicitly posited by Northrop Frye: "The poet, like the pure mathematician, depends,

[101] Clement Greenberg, 'Collage', in *Art and Culture: Critical Essays* (Boston: Beacon Press, 1961), p. 71.
[102] Ibid., p. 72. [103] Ibid., p. 72. [104] Auerbach, *Mimesis*, p. 792.

not on descriptive truth, but on conformity to his hypothetical postulates."[105] Frye explains this alleged process of abstraction as follows:

> Just as in mathematics we have to go from three apples to three, and from a square field to a square, so in reading a novel we have to go from literature as reflection of life to literature as autonomous language.[106]

But this much could be said of any symbolic convention. To borrow Arthur Eddington's striking example, in a physics problem, we go from talking about an elephant falling down a hill to the weight of the elephant and the angle of the hill and continue in this vein of abstracting the actual to the symbolic units of the mode until the elephant and hill entirely disappear from the equation, as it were.[107] But this does not prove that physics is no longer engaged in the agon for representational authority. The main tendency of world-pictures is to compete over whose symbolic convention is better equipped to represent the world, we have said: and to say that literature is ensnared in its special literariness and literary devices does not show it to be less interested in conquering the real. This entire book will in some ways prove an elaboration of this simple point. But for now, if we simply recall the most remembered concepts of high modernism, the poetic image and objective correlative, we see that they evince a governing desire to ground poetry in the reality of sensual experience.[108]

Mehrtens' counter-modern strain in modernist mathematics may in fact be a more fertile comparison to the modernism of art; or Mehrtens' anti-modern strain, to Menand's 'reaction against the modern'.[109] Mehrtens relies mainly on the example of intuitionism to characterize the counter-modern.[110] The founder of intuitionism, the morose Brouwer, attempted assiduously to ground the concepts of mathematics in sense-intuitions.[111] Brouwer's school sprang from Leopold Kronecker's nineteenth-century theory of finitism,

[105] Northrop Frye, *Anatomy of Criticism* (Princeton: Princeton University Press, 1957), p. 74.

[106] Frye, *Anatomy*, p. 350.

[107] Arthur S. Eddington, *The Nature of the Physical World* (Cambridge: Cambridge University Press, 1928), p. 127.

[108] We will explore Pound's poetic image and its analogy to analytic geometry in Chapter 5.

[109] In a draft of his essay in *Modernism in the Sciences*, Leo Corry similarly questions the application of preconceived modernist tropes to mathematics. Leo Corry, 'How Useful Is the Term "Modernism" for Understanding the History of Early Twentieth-Century Mathematics?' in *Cultural Practice*, p. 394.; See also Solomon Feferman, 'Modernism in Mathematics', *American Scientist*, 97.5 (2009), 417–20, p. 420, who is largely in sympathy with Corry's stance.

[110] See Mehrtens, *Moderne—Sprache—Mathematik*, ch. 3.

[111] See Luc Bergmans and Teun Koetsier, eds, *Mathematics and the Divine: A Historical Study* (Amsterdam: Elsevier, 2005), ch. 30.

which countered the philosophical problems posed by autonomous mathematical concepts with the rather easy move of labelling them fantastic: or only accepting as real the more tractable objects of mathematics. Much of what Gray regards as modernist mathematics, infinite sets varying in size, for instance, was rejected by this school as idealist figments of mathematical dreams. But even Kronecker was forced to compromise at some stage. He admits “God made the integers”, though “all else is the work of man”.[112] But this does not testify to an inevitable departure of modern mathematics from the mundane. It is not peculiarly modern to concede the divinity of numbers.

But we are not in this book interested in tracing the tenets of literary modernism to a different set of mathematical ideas, for the agon is a battle for representational sovereignty. Our excursus into Jeremy Gray was, among other things, to defend our departure from certain dogmas of interdisciplinary research in literary studies today, chief among which is the empirico-historicist one of influence picking; and to state that in considering the ‘mathematization of reality’, we are no longer bound to the divisions of concepts that the individual histories of physics or mathematics might engross unto themselves today.

1.4 Competition Begins

Art has not always played the part of rebel. When the poetic world-picture enjoyed the kind of authority in the seventeenth century that the mathematical one did at the turn of the twentieth, we find it is science plotting coups to unseat the ruling power. Votaries of the *Novum Organum* in the seventeenth century began by presenting mathematical plainness as a way of clearing up the metaphorical obfuscations of Renaissance poetry. The attack of puritan science on the effusive profanities of Elizabethan verse and prose gave to the modernist narrative of modernity a Tacitean myth that a once noble union of human faculties had been dissolved and corrupted by the empire of number. The Renaissance mind, argues Eliot, as Roberts after him, was one that could hold reason and feeling in one breath: they could “feel their thought as immediately as the odour of a rose”.[113] This sense the modernists inherited from Blake but also felt directly from the heady sensuality of Renaissance verse, its passionate mechanical images and sensuous astronomical conceits. “Donne”,

[112] Found in Gray, *Plato's Ghost*, p. 153.
[113] T.S. Eliot, ‘The Metaphysical Poets’, *TLS*, (1921), 669–70, p. 670.

says Roberts, "was living on the edge of the Elizabethan age, and knew the prose of Nashe, a prose which spoke not only to the eye and brain, but to the senses of touch, taste and hearing. For Donne, prose was nearer to music and to painting or to the texture of leaf or carpet than to mathematics"[114]—the observation applied not just to prose but to language itself. A significant group of modernist poets, who would later be credited with a 'metaphysical revival',[115] narrated the rise of science as in the main an unnatural austerity imposed on the poetic richness of meaning: science, they argued, to move the metaphorical soul of man, had erected on the ochreous debris of his livid babel the marble tomb of mathematical purity.

Bacon introduced the new philosophy identifying the independent life of words—their "living powers", which Coleridge later said was released by poetry[116]—as "the most troublesome of all" obstacles to its success:

> Idols [...] have crept into the understanding through the alliances of words and names. For men believe that their reason governs words; but it is also true that words react on the understanding; and this it is that has rendered philosophy and the sciences sophistical and inactive. Now words, being commonly framed and applied according to the capacity of the vulgar, follow those lines of division which are most obvious to the vulgar understanding. And whenever an understanding of greater acuteness or a more diligent observation would alter those lines to suit the true divisions of nature, words stand in the way and resist the change.[117]

Density of literary signification had thus to give way to the 'greater acuteness' of science's 'diligent observation'. Hobbes deemed it an abuse of language "when they [poets] use words metaphorically; that is, in other sense than that they are ordained for; and thereby deceive".[118] Falseness and deception are not new charges against poetry. Whether by aged Plato, in defence of philosophy, or the adolescent Augustine, in defence of grammar,[119] poetry has always

[114] Michael Roberts, *Elizabethan Prose* (London: Jonathan Cape, 1933), p. 2.

[115] Our topic for Chapter 3.

[116] S.T. Coleridge, *Aids to Reflection and the Confessions of an Inquiring Spirit* (London: George Bell & Sons, 1884).

[117] Francis Bacon, *Novum Organum*, ed. Joseph Devey (New York: P.F. Collier & Son, 1902), Aphorism LIX.

[118] Thomas Hobbes, *Leviathan or The Matter Forme, and Power of a Commonwealth Ecclesiastical and Civil*, ann. G.W.F. Hegel (Amazon Digital Services, 2019), p. 13.

[119] Plato, *The Republic of Plato*, trans. Benjamin Jowett (Oxford: Clarendon Press, 1888), Book X; That is, Augustine claims in later life to have developed his views of poetry as an adolescent, St Augustine, *The Confessions*, trans. by E.B. Pusey (London: Aldine Press [1907] 1957), Book I: both target the seductive lure of Homer's stories.

been accused by its antagonists for what it openly displays: its fiction. However trite the general accusation, the attacks on verbal ambiguity and metaphor were novel and designed specifically to counterpose the poetic mode against mathematical directness. The seventeenth century's sustained battle against various aspects of the literary soon received a royal charter. The Royal Society, one of the restored Charles's olive branches to the disappointed radicals of the revolution, promptly declared in 1664 a proscription of "luxury and redundance of speech, [the] trick of Metaphors" in favour of "a close, naked, natural way of speaking [...] bringing all things as near the mathematical plainness as they can"[120]—perhaps assuming, even as blind Milton was committing *Paradise Lost* to ink, that the muse favours only the royalist. This programmatic censure of the literary carried on to the end of the century, when Locke wanted the language purged of all ambiguity, which seemed to him but "noise and sounds".[121] The purification of semantic density sought by votaries of the new philosophy, while favourable to 'neat and syllogistic science', diminished the licensed range of poetry, which creates meaning precisely through the concentration of coexisting senses.

The mode of speaking about reality, as we have discussed, determines the picture of reality that issues. Whitehead states how this chastening of the Renaissance world-picture by the straitening of its language was designed to wrest reality from the poet's grip.

> Objects in nature were perceived as with qualities which in reality do not belong to them, qualities which in fact are purely the offspring of the mind. Thus nature gets credit which should in truth be reserved for ourselves: the rose for its scent: the nightingale for his song: and the sun for his radiance. The poets are entirely mistaken. They should address their lyrics to themselves, and should turn them into odes of self-congratulation on the excellency of the human mind. Nature is a dull affair, soundless, scentless, colourless; merely the hurrying of material, endlessly, meaninglessly.[122]

The romantic, and later modernist, retelling of scientific history as the effort to erase significant aspects of sensory and lived experience from the world-picture to render it in mathematical terms, thus mirrored, in many ways, the

[120] Thomas Spratt, *The History of the Royal Society of London for the Improving of Natural Knowledge* (London: T.R. for J. Martyn at the Bell, 1667), p. 111.

[121] John Locke, *An Essay Concerning Human Understanding* (London: T. Tegg and Son [1689] 1836), p. 350.

[122] Whitehead, *Modern World*, p. 54.

early rhetorical strategies of the natural philosophy against the prevailing idiom of writing and speech in the Renaissance.

Science's "improvements", to borrow a Royal Society phrase, to "the English tongue" led,[123] according to Roberts, to a schism between reason and imagination in the eighteenth century, where none had existed in the Renaissance. While in the early 1700s, "'reason' had meant the capacity to judge all things wisely", implying "sensibility and good conduct as well as good logic", 'imagination' had "meant no more than the power of forming mental pictures".[124] Since then, as was implied in Blake's aphorism, the

> [H]istory of English and French philosophy in the eighteenth century is, in the main, the history of the gradual restriction of the concept of reason. And at the same time, because some name must be used for that which is felt to exist, it is the history of the growth and widening of the concept of 'imagination'.[125]

As the tenor of these concepts diverged, "the contrast between the precision and objectivity possible in the discussion of 'material' things and the vagueness and relative privacy of meaning possible in other fields was enhanced"[126]—thus allowing poetry, at least until psychology was mathematized to psychometry, sway over a vague and bloated inner life, while empirical reality was left safely to adult care. The fissure rent by the seventeenth century agon seemed to the modernists to have ensured a lasting separation for the parting tribes; as Eliot phrased it, "in the seventeenth century a dissociation of sensibility set in, from which we have never recovered":[127] precise description was possible in mathematics, emotional expression in poetry.

The mathematical world-picture began the agon impugning poetry's modes, ambiguity, and metaphor, but upon the prolonged entrenchment of these dogmas of objectivity, by the twentieth century, it is able to question the bodily sensorium itself, upon whose testimony art depends. Here again is the totalizing tendency of discursive empires: said Reichenbach in 1938:

> We see the world in the scale of our sense capacities: we see houses, trees, men, [...] waves, fields, woods [...] This perspective, however, is not only one-sided; it is false, in a certain sense. Even the things which we believe we

[123] Spratt, *Royal Society*, p. 111.
[124] *The Modern Mind* (London: Faber & Faber, 1937), pp. 121–2.
[125] Ibid., p. 123.
[126] Ibid., p. 124.
[127] Eliot, 'Metaphysical Poets', p. 670.

> see as they are, are objectively of shapes other than we see them. [...] We see the polished surface of our table as a smooth plane; but we know that it is a network of atoms with interstices much larger than the mass particles [...] We hear the voice coming from the mouth of a signing girl as a soft and continuous tone, but we know that this sound is composed of hundreds of impacts a second bombarding our ears [...] The [objects] as we see them have as much similarity to the objects as they are as [...] the lion seen in the cinema has to the dark and bright spots on the screen [...] we see a substitute world – not the world as it is, objectively speaking.[128]

The posterity of Plato's cave-bound prisoners is thus in the twentieth century hypnotized before a film and Plato's philosopher has become Reichenbach's scientist-with-instrument. The curation of laboratory conditions, the fortuitous deliverance of measuring instruments, the sifting of data therefrom, the cocktail of pointer-readings selected for the final mathematical theory, and the distance between mathematical theory and such verbal metaphors as 'networks', 'interstices', and 'bombardment', are all elided in the clean narrative of truth that the propaganda arm of science is able to present in the modernist period: a result of discursive authority accrued since the seventeenth century.[129]

1.5 The Resistance

Though conditions for a rebellion were ripe in the twentieth century, its target that we have identified, the supposed 'mathematization of reality', begs a further question: why among the various aspects of what some may have termed the scientific or mechanical world-picture do we focus on mathematization?

We framed our agon by stating that it is in the deepest nature of faculties to compete over their modes of discovery and definitions of quiddity. A theologian, for instance, might argue that the true nature of things is established by revelation where a philosopher, by reason. Kant, for instance, laments in the *Conflict of the Faculties* the sovereign sanction enjoyed by the divinations of theology at a time when the unbowed pursuits of philosophy were afforded

[128] Hans Reichenbach, *Experience and Prediction: An Analysis of the Foundations and the Structure of Knowledge* (Chicago: Chicago University Press, 1938), pp. 219–20.

[129] Myles Burnyeat makes a similar argument in 'Plato on why Mathematics is Good for the Soul' (1995) in *Explorations in Ancient and Modern Philosophy*, vol 3, eds C. Atack, M. Schofield, and D. Sedley (Cambridge: Cambridge University Press, 2022), p. 13.

but uneasy tolerance. But in the century that separates the modernists from Kant, "everything which manifests itself as real", said Husserl (while coining the phrase 'mathematization of reality'), was gradually required to "have its mathematical index".[130] Mathematics seemed to have inherited from theology the awesome power of drawing for all the correct shape of reality: a reality obstinately material yet abstractly devised, perceivable through instruments though beyond sensuous grasp, to be assembled from parts but not to an organic entirety.

When we identify our poets' antagonist in the pregnant phrase, 'mathematization', we hope, thereby, to hold in place such vast structures of thought as would fracture waywardly to the fragments of intellectual history, to the planes of Descartes and Gauss and the laws of Newton and Maxwell, to the principle of least action and the continuity of real numbers, to a non-Euclidean curvature of spacetime beyond the grasp of sense-perception, and uncertainties within the atom that confound the materialist foundation. Although the modernists scrutinized these various fragments with varying intensities, the genus 'mathematization' will serve to underscore the fact that their critiques extended insofar as any individual doctrine—materialism, say, or empiricism—led to the redrawing of reality in a mathematized world-picture.

At the beginning of the agon, Hobbes had declared in no uncertain terms that "the universe [...] is Corporeall, that is to say, 'Body' and hath the dimensions of Magnitude, namely, Length, Bredth, and Depth [...] and that which is not Body, is no part of the Universe: And because the Universe is All, that which is no part of it, is Nothing".[131] Michael Roberts notes how "this emphasis on material 'things', experienced by outward senses—touch, sight, taste, smell, hearing—could not give rise to mathematical science unless impressions of weight, size and number were taken as primary and all other impressions were correlated with them".[132] Thus is established a hierarchy within our sensory experience of the material world. Galileo, in *The Assayer*, said "I think that tastes, odours, colours, and so on are no more than mere names so far as the object in which we locate them are concerned, and that they reside in consciousness."[133] Of these phantoms of subjective consciousness, Descartes

[130] Edmund Husserl, *The Crisis of European Sciences and Transcendental Phenomenology* (Evanston: Northwestern University Press, 1970), p. 37.

[131] Hobbes, *Leviathan*, p. 273.

[132] Roberts, *Modern Mind*, p. 65.

[133] Galileo Galilei, *Discoveries and Opinions of Galileo,* ed. Garden City (NY: Doubleday, 1957), p. 274.

said, "we are not aware of their being anything other than various arrangements of the size, figure, and motions of the parts of these objects which make it possible for our nerves to move in various ways".[134] And thus, a reality already diminished—by the strained marriage between materialism and empiricism—to the interaction between matter and sense-perception was further attenuated to the measurable aspects of that interaction, to wit, our discernment of extension and motion. These were then appointed 'primary qualities', while the remainder, the 'secondary qualities', became their derivatives. Next, mathematics would conquer not only everything present but what was and will be: says Laplace:

> We may regard the present state of the universe as the effect of its past and the cause of its future. An intellect which at a certain moment would know all forces that set nature in motion, and all positions of all items of which nature is composed, if this intellect were also vast enough to submit these data to analysis, it would embrace in a single formula the movements of the greatest bodies of the universe and those of the tiniest atom; for such an intellect nothing would be uncertain and the future just like the past could be present before its eyes.[135]

With the faith that present conditions can be mathematized whole, and that reality obeys a mathematically determinable course, the mathematical world-picture would one day comprehend all that is above, below, before and after.

Although the birth of modern science was attended by a revolt against reason, in favour of the testimony of the senses, it was widely appreciated in the modernist period with what completeness mathematics had in the intervening three centuries come to dominate science's picture of reality. In the beginning, science posed as a means to set in order our immediate perceptions and provisional hypotheses about the workings of the surrounding environment. Seventeenth-century science was based, at least in theory, on fundamental intuitions of space and time shared by all; and in Bacon's *Novum Organum*, it becomes almost a doctrine of organized common sense for the practical advancement of civilization. But it was in the twentieth century dawning on observers just how outmoded, inadequate, and deceptive was this conception of science. The two great volleys on the Newtonian world-picture, relativity

[134] René Descartes, *Principles of Philosophy*, trans. Valentine Rodger Miller and Reese P. Miller (London: Kluwer Academic Publishers, 1982), p. 282.

[135] Pierre Simon Laplace, *A Philosophical Essay on Probabilities*, trans. Truscott, F.W. and Emory, F.L., Dover Publications (New York, 1951) p. 4.

and quantum theory, shattered the popular notion of science. It was no longer viable to maintain that spacetime and the quantum world are clarified pictures of ordinary experience. Reckoning the lapse of time to be different for observers on Earth and Neptune baffles common sense, which is tested further by the notion of objects existing in two places at once. Modernity had thus re-entered a world of ghosts and miracles—perhaps it is this very uncanniness of modern physics that recommended its latest doctrines so enthusiastically to the monotony of modern life. Indeed, despite the narrative framing of this book, it has to be noted that a kind of devilish fascination stole over the modernist imagination as regards the new physical theories, which found their way into much modernist verse of the time in the form of scientific images and metaphors[136]—nowhere is this more evident than in Empson's verse. Poets were not inclined to waste the evocative potential of relativity and quantum mechanics, which had perhaps defamiliarized the world more effectively than their fractured syntaxes and jarring semantics. Nevertheless, we will see, modernist poets were not of the devil's party, and they knew it.

Terrific though the new theories were, their popular expositions presented them as the surreal artefacts not of heightened common sense but of occult mathematical divination. The hitherto arcane theory of a German Jew was introduced to the English-speaking world in just this way. In a series of lectures to popularize Einstein's theory in England, Arthur Eddington said,

> The progress of the relativity theory has been largely due to the development of a powerful mathematical calculus for dealing compendiously with an infinite scheme of pointer readings, and the technical term tensor used so largely in treatises on Einstein's theory may be translated schedule of pointer readings. It is part of the aesthetic appeal of the mathematical theory of relativity that the mathematics is so closely adapted to the physical conceptions.[137]

A physicist and philosopher, N.R. Campbell, who attempted to bring physics to the working classes levies a similar charge against the quantum theory:

> [T]he precise analogy [used to describe quantum events] which is chosen out of the many possible alternatives is determined largely by mathematical

[136] See Whitworth's *Einstein's Wake*; Rachel Crossland, *Modernist Physics: Waves, Particles, and Relativities in the Writings of Virginia Woolf and D. H. Lawrence* (Oxford: Oxford University Press, 2018); and Alan J. Friedman and Carol C. Donley, *Einstein as Myth and Muse* (Cambridge: Cambridge University Press, 1985).

[137] Eddington, *Physical World*, p. 130.

> simplicity. About this there is nothing phenomenal at all; it is a characteristic of judgements of the internal, not of the external, world. I do not say that the framing of a theory in accordance with mathematical simplicity is any less justifiable than framing it in accordance with a mechanical analogy; but it is certainly less 'purely phenomenal'.[138]

As Adorno and Horkheimer would later put it, "Before and after quantum theory, nature is what can be grasped mathematically; even what cannot be assimilated, insoluble and irrational, is fenced in by mathematical theorems."[139] Through the popularization of the new theories, it was slowly impressed upon a modernist culture that mathematical formalization—the desire to overcome sordid fact to attain ideal theory—was not a repercussion of the scientific method but its central aim. "In a sense", concludes Whitehead, "Plato and Pythagoras stand nearer to modern physical science than does Aristotle."[140]

Modernists began to speak of surrender to pure mathematics not as the slow corruption of the old mechanical philosophy but as science's original sin. Most of the metaphysical postulates of science against which previous generations of poets from Blake on had rebelled thus gave way to the unholy scientific Assumption of *hyle* to *eidos*. The doctrine of materialism, for instance, that only the concrete part of existence—what has definite location in space and time—may be dignified as real, has seldom in history—seldom, lest we forget Lucretius and Dryden—found easy consort with the fleeting intangibles of poetic sensibility—and the modernist case is no exception. But the latter, remarkably, saw the doctrine of materialism, or what of it remained in the twentieth century, as merely science's device for the successful mathematization of reality. By fading the insubstantial pageant of experience; by filtering the *lebenswelt* through measuring instruments into a substrate evenly divisible to measurable units, science is able successfully to replace the singularity of experience, the unique particularity of being, with its hermetic and uniform ratios—so argued Yeats in *A Vision*, Lawrence in 'On Democracy', Lewis, in *Time and Western Man*. And so indeed Empson, with all the delightful bias of a budding mathematician: modern science was "forced into adopting the materialist scheme because it was the only one which ordered [reality] adequately" for mathematicians[141]—science is but "a closed system of pointer-readings about what is measurable"—simply put, what can be

[138] N. R. Campbell, *Physics: The Elements* (Cambridge: Cambridge University Press, 1920), p. 153.
[139] Adorno and Horkheimer, *Dialectic*, p. 18. [140] Whitehead, *Modern World*, p. 28.
[141] By 'modern science' is generally meant the form that science took from the seventeenth century onwards.

mathematized.[142] In the *Dialectic of Enlightenment*, Adorno and Horkheimer would mark the difference:

> Its untruth does not lie in the analytical method, the reduction to elements, the decomposition through reflection, as its Romantic enemies had maintained from the first, but in its assumption that the trial is pre-judged. When in mathematics the unknown becomes the unknown quantity in an equation, it is made into something long familiar before any value has been assigned.[143]

The mathematical world-picture's primary aim, it was thus noticed, was not to organize experience, but to reduce the mystery of phenomena to the already ordered variables of algebra, and thereby to annex the unknown and unknowable to the mathematical imperium. As the evolution of scientific thought ate away at its own foundations, materialism and empiricism falling prey to new instruments, data, and theories, the abiding aspect of mathematization thus took in the modernist period central stage. The real tyranny became neither the concept of matter, the routine of measurement, nor the habits of observation: instead, "High on some mountain shelf / Huddle the pitiless abstractions bald about the neck".[144]

Philosophers argued in the modernist period that mathematization cannot account for those facts of phenomena that impress our senses most plainly. F.H. Bradley, who would significantly influence modernist poets, argued that

> extension cannot be presented, or thought of, except as one with quality that is secondary. It is by itself a mere abstraction, for some purposes necessary, but ridiculous when taken as an existing thing. Yet the materialist, from defect of nature or of education, or probably both, worships without justification this thin product of his untutored fancy.[145]

142 William Empson, *Argufying: Essays on Literature and Culture*, ed. John Haffenden (London: Chatto & Windus, 1987), pp. 528–9; see Susan Stebbing, *Philosophy and the Physicists* (London: Methuen [1937] 2018), part II, for a history and analysis of the term 'pointer-reading', especially its various uses in Eddington.

143 Adorno and Horkheimer, *Dialectic*, p. 18. See also Matthew Handelman, *The Mathematical Imagination: On the Origins and Promise of Critical Theory* (New York: Fordham University Press, 2019), ch. 1.

144 W.B. Yeats, 'Preface', in *The Ten Principal Upanishads*, trans. Shree Purohit Swami and W.B. Yeats (1937), quoted in W.B. Yeats, *Later Essays*, ed. William H. O'Donnell (New York: Charles Scribner's sons, 1994), p. 174.

145 F.H. Bradley, *Appearance and Reality: A Metaphysical Essay* (London: Swan Sonnenschein and co. [1893] 1908), p. 17.

We can neither observe nor conceive extension without colour and other attributes deemed secondary. The promotion of the one and demotion of others seems therefore contrived. Other philosophers took Bradley's argument further to attempt a reversal of Locke's hierarchy. The physical account of colour, for instance, is contained in the fluctuating wavelengths of electromagnetic radiation and the pathway of electrical impulses through the optic nerve. J.W. Dunne finds nowhere in this narrative his most obvious impressions of colour explained.

> Now, redness may not be a thing—but it is very certainly a *fact*. Look around you. It is one of the most staring facts in existence. It challenges you everywhere, demanding, clamouring to be accounted for. *And the language of physics is fundamentally unadapted to the task of rendering that account.*[146]

The great residue of experience discharged by the industry of measurement, the specks of experience, thus found eager costermongers among the modernist philosophers of science.

> But these remnants are the most obtrusive things in our universe. [...] they produce the effect of a vast external world of flaming lights and colours, pungent scents, and clamorous, tumultuous sounds. Collectively, they bulk into a most amazing tempest of sharply-differentiated phenomena. And it is a tempest which remains to be considered *after physics* has completed its say.[147]

It is for these obtrusive facts of existence that modernist poets sought a depth and accuracy of expression to rival what differential calculus had given to motion. Styling poets "scientists of humanity",[148] the modernist, or so this book will argue, challenged the mathematical world-picture in its purchase of reality by bringing to bear a "richness of meaning which the scientist had carefully taken away in order to state his neat and syllogistic science".[149] While modernists resisted the centuries-old faith in mathematization, they displayed greater regard for the scientific *temperament* than the romantic, whose sublime ego could not be bridled to a passive study of the actual. The empirical discipline to hard fact, the careful observation and faithful recording of verit-

146 J.W. Dunne, *An Experiment with Time* (London: A. and C. Black, 1927), pp. 5–6.
147 Dunne, *Experiment*, p. 7.
148 The phrase was coined by John Middleton Murry in 'Science and Nonsense', *Adelphi*, 7.3 (1933).
149 Michael Roberts, "The Two Coleridges" *The London Mercury*, 32 (1935), 292–3, p. 292.

able experience, we will see, were virtues the modernists deemed worthy of emulation more than resistance. As we began saying, the object of the resistance was synthesis more than antithesis.

1.6 A Note on Method

The agonistic model through which we narrate modernism is more a framework than definition of critical method. In terms of method, I have attempted to find a way of dealing with works that is adequate to their purported agenda. I have also sought consciously to resist the pressures of historicism. Attending to poetic works that seek alternate ways of speaking and thinking about reality than science's in the empirico-historicist mode of criticism dominant in the humanities today would create a rift in the how and what of this book. I leave unsaid in my readings three important premises: that, assuming the agon of faculties to be the natural condition of any knowledge produced in the academy, critical method is not a politically neutral gesture; that its present phase of empirico-historicism is a symptom of the successes in the twentieth century of science's discursive empire; and that a way of resisting the lapsing authority of the literary is to regard as the main vocation of the literary critic interpretation, even appreciation, taking as principal the work of art.

The story is by now well-rehearsed that Empson, and Richards, Riding, and Graves before him, changed their discipline to reflect what is perhaps their main belief that literary criticism, the recovery and communication of literary meaning, provides some new and unique way of thinking about the world. This work is invested in their project to the extent that it believes literary criticism can create knowledge that cannot be had through other disciplinary methods: and that this knowledge is obtained by bringing to the fore the doings of words peculiar to a poem. I have, wherever possible, tried to follow Empson's advice that "you must rely on each particular poem to show you the way in which it is trying to be good".[150] For instance, the study of Roberts's poems seemed best facilitated by dividing his lyric corpus into semantic clusters; Empson's own dense poems demanded a non-linear, proleptic, and retrodictive tailing of interactions between words; Riding's philosophical poems required inferences and deductions of a syllogistic form; whereas it was through melopoetic readings that Olson's projective verse revealed its intentions. These methods will be defended in the respective

[150] William Empson, *Seven Types of Ambiguity* (London: Chatto and Windus [1930] 1949), p. 7.

chapters, but they can loosely be summed as close reading or verbal analysis. The term 'close reading' has undergone many changes since the days of Richards, as Joseph North's *Literary Criticism: A Concise Political History* assiduously traces.[151] North argues that the New Critical adoption of close reading to make of a poem an unbreachable aesthetic whole has tinged it with nostalgic idealism, preventing a return to its progressive uses by Richards and Empson, to what North calls the "heroic age of literary criticism".[152]

After the fever of Theory lapsed at the turn of the millennium, literary criticism adopted the deceptively neutral posture of historicism; and since then, the basic assumptions of positivism have gradually yet firmly conquered the study of literature. There has, however, been some resistance recently to what Derek Attridge calls "the strait-jacket of history" in literature departments.[153] Tom Eyers, with the tenor of fin-de siècle resistance to positivism, says "[H]istory, instead of being a question to be answered, has threatened to become a catch-all explanans to be passively assumed, bringing with it an obfuscation of what makes literature, literature".[154] North similarly notes that the "central logic that has dictated so much of the last three decades of literary study" is "the rejection of the project of criticism—aesthetic education for something resembling, in aspiration if not in fact, a general audience—and the embrace of the project of scholarship—the production of cultural and historical knowledge for an audience of specialists".[155] History is not necessarily at odds with close reading; Empson's *Using Biography* argues against the New Critical proscription of historical contextualization and authorial intention.[156] History and biography can illuminate a poem and unveil meanings otherwise unavailable to formal analysis. Christopher Wood in his introduction to Erwin Panofsky's *Perspective as Symbolic Form* approximates some of the assumptions guiding our work: Panofsky's "strategy was to isolate the work temporarily in order to grasp more clearly its deep structural principles, and then ultimately to reinsert the work into its primordial environment on more legitimate grounds".[157] While we do not proceed so systematically—in that history enters the analysis as and when illocution so demands—we are interested

[151] Joseph North, *Literary Criticism: A Concise Political History* (Cambridge: Harvard University Press, 2017).

[152] Ibid., p. 3.

[153] Attridge, 'Introduction', *Work of Reading*, p. 6.

[154] Tom Eyers, 'The Revenge of Form: Review of C. Levine's *Forms: Whole, Rhythm, Hierarchy, Network*', *Boundary*, 2, May 9, https://www.boundary2.org/2018/05/ tom-eyers-the-revenge-of-form-review-of-c-levines-forms-whole-rhythm-hierarchy-network/, p. 351.

[155] North, *Literary Criticism*, p. 115.

[156] William Empson, *Using Biography* (London: Chatto & Windus, 1984), p. vii and 104.

[157] Christopher S. Wood, 'Introduction', in Erwin Panofsky, *Perspective as Symbolic Form* (New York: Zone Books [1937] 2012), p. 8.

in what the voice of the work adds to the agon of the time. But we are above all after the meanings of poems under study and what can be learnt from their modes of working; what they evince about their historical period will, however interesting, be derivative information. As Hayden White argues, historical narratives are deliberately woven, much as works of literature are, by stitching together expedient details from a relatively neutral registry of available facts.[158] Attempts in literary studies to rinse historical fabric in factual, if not factitious dye, must also be subject to White's observations. In this book, we have attempted to throw our lot in with recent challenges to historicism and, with Aristotle, have taken poetry more seriously than history.

Poems, when they are composed, are intended as whole artefacts. Our readings will therefore proceed with Empson's assumption that there is a correct interpretation of a poem—a right constellation of ambiguity that emerges from seeing how the parts interact. This operational assumption not only directs analysis of verbal mechanics but creates a context for further argument rather than frustrating subsequent exegesis. But the assumption is also made to repel what Henry Staten calls the "blossom-visiting"[159] style of free association in classrooms that pre-empts interpretive labour, answering instead to the patronizing demands of "accessibility" under the predatory rule of Commodity. As verbal units in a poem are isolated from contexture for classroom entertainment, the work of criticism in turn simply becomes the heavy accumulation of historical detail "relevant"—to use Commodity's other term of art—to meandering streams of consciousness; and conversely, the serendipity of historical finding becomes the driver of further association (and, indeed, accumulation) of meaning. For our part, we will always weigh mereological reasons more than, say, letters from an archive. History will yet prove essential to drawing boundaries to the discursive arena of a poem—poems are, after all, written into a particular society, however universal their purport may be. But the establishment of a historical fact, say, that Olson was acquainted with quantum mechanics, will not be granted any privileged role in the interpretation of his mathematical metaphor, which may—if reference to quantum mechanics can indeed be argued for within its field of meaning—adapt, or even distort the science, to further a poetic agenda.

Chronicling overall an agon, the individual chapters of this book will mainly compare poetry and mathematics on the basis of how each yields

[158] Hayden White, 'The Historical Text as Literary Artefact', in *Tropics of Discourse: Essays in Cultural Criticism* (Baltimore: Johns Hopkins University Press, 1981), pp. 81–100.

[159] From Ransom's image of bees gathering honey from several blossoms without noticing the bush that supports the network of flowers in Staten, 'Is the Author Still Dead?', in *Work of Reading*, p. 26.

knowledge or understanding of the world. The next chapter is therefore concerned with epistemology. We will ask how, according to Empson, do poetry and mathematics varyingly involve the mind in its unquenchable thirst for knowledge. Chapter 3 turns to the fate of love and sensuality under the rule of the mathematical world-picture. We explore Roberts's lamentation about the lost role of poetry in giving accurate representation to the lifeworld of the senses. While Chapters 2 and 3 characterize poetry as modes by which obscured truths about the lived world may be accessed, the fourth chapter is interested in how poetry itself becomes a form of truth in the writings of Riding and Olson. We end in the last chapter by noticing crosscurrents in the agon; by charting the more complex artistic and philosophical attitudes to pure mathematics, as opposed to the mathematical world-picture. We will test the fertility of agonistic paradigm by contending with the contrary poetics of Pound and Marinetti; and end with a sustained discussion of the role of mathematics in Yeats's late doctrine of aesthetics.

2
Epistemology

The agon of the faculties retailed in the previous chapter presents poets in the twentieth century challenging mathematics over the nature and form of reality. This was in response to mathematical reason steadily over the course of two centuries engrossing unto itself all that is factual in phenomena. This chapter and the two following will compare poetry and mathematics as competing modes of world-building; in this chapter, as modes of knowing: for when seeking representational authority, a mode must first establish how through it, knowledge of the world is uniquely acquired. We therefore begin by contrasting the procedures by which the poetic and mathematical modes transport mind to understanding. Empson is the principal character to study in this regard because his poetry is concerned mainly with the limits to knowledge and the difference between knowing and acting.

But before entering the strange world of Empson's poems, we will dwell awhile on the intellectual atmosphere that set racing this remote imagination. The foundations of scientific knowledge have perhaps never been more thoroughly probed than at Cambridge in the early parts of the twentieth century.[1]

2.1 Realism at Cambridge

Among philosophers, there was in the early decades of the twentieth century a steady loss of faith in the objective truth of the scientific world-picture. There are many strains from which to trace this process. In the context of biochemistry, we see the influence of Bergson on fin de siècle philosophy giving rise to an 'organicism', a wish to, as Wyndham Lewis puts it, posit "something *alive* in place of mechanism", "organism in place of matter".[2] The German biologist Hans Driesch—whose work was Englished by C.K. Ogden, a collaborator of Empson's supervisor I.A. Richards—argues that

[1] Empson and Roberts studied mathematics at Cambridge in the 1920s (John Haffenden, *William Empson: Among the Mandarins* (Oxford: Oxford University Press, 2008), p. 104; Michael Roberts, *Selected Poems and Prose*, ed. Frederick Grubb (Manchester: Carcanet Press, 1980), p. 1).

[2] Wyndham Lewis, *Time and Western Man* (London: Chatto and Windus, 1919), p. 166.

Agon: Poetry's Challenge to the Mathematization of Reality (1920s–1960s). Anirudh Sridhar, Oxford University Press.
 DOI: 10.1093/oso/9780198985457.003.0002

> Bertrand Russell and his followers believe they can do without such concepts as becoming and change. This may be true for mere "motion" of a physical point, though even here the introduction of a particular function in the place of motion as a something sui genesis does not afford a real understanding. But such a mathematical reduction of becoming (if I may say so) is quite impossible in the realms of biology, history, etc.; and it neglects the truly original meaning of becoming.[3]

J.B.S. Haldane, in a lecture to the Heretics society, announced that a time is fast approaching "when physiology will invade and destroy mathematical physics, as the latter have destroyed geometry. The basic metaphysical working hypothesis of science and practical life will then, I think, be something like Bergsonian activism."[4] Joseph Needham's *The Sceptical Biologist* is a thoroughgoing critique of the mechanical conception of life.[5] Empson would write a favourable review of this book;[6] and Needham would go on to influence other Cambridge poets, like Jeremy Prynne.[7]

In this discussion of science and philosophy at Cambridge, however, we will restrict ourselves to mathematical physics: to the debate surrounding what has latterly been dubbed 'scientific realism', as it strikes at the very heart of the agon between the poetic and mathematical world-pictures. We do not offer here an exhaustive account of the debate; we will relate just enough to orient ourselves to the stances taken in modernist poems.[8] Alberto Coffa says, "few topics have elicited more heat and less light among philosophers during the past two centuries than the subject of realism".[9] Barry Gower notes a "bewildering variety of scientific realisms and anti-realisms confronting us".[10] To not ourselves be bewildered, we will restrict ourselves to the issue of

[3] Hans Driesch, *The History and Theory of Vitalism* (London: Macmillan & Co., 1914), p. 191.

[4] J.B.S. Haldane, *Daedalus: or Science and the Future* (London: E.P. Dutton and Co., 1924), p. 5.

[5] Joseph Needham, *The Sceptical Biologist* (London: Chatto & Windus, 1929).

[6] William Empson, 'The Sceptical Biologist', in *Argufying: Essays on Literature and Culture*, ed. John Haffenden (London: Chatto & Windus, 1987), p. 58.

[7] J.H. Prynne and Keston Sutherland, 'Introduction to Prynne's Poems in Chinese', *The Cambridge Quarterly* 41.1 (2012), 197–207, p. 199.

[8] I have relied heavily on select recent works, principal among which is Paul Dicken's *A Critical Introduction to Scientific Realism* (London: Bloomsbury, 2016). This is the clearest exposition on the causes of and responses to the debate on realism. An excellent compendium of extracts from primary texts seminal to the development of realism is *Philosophy of Science: An Anthology*, which comes with a series of quirky introductions by the editor, Marc Lange (Marc Lange, *Philosophy of Science: An Anthology* (Malden: Blackwell, 2007)).

[9] Alberto Coffa, *The Semantic Tradition from Kant to Carnap: To the Vienna Station*, ed. Linda Wessels (Cambridge: Cambridge University Press, 1991), p. 94.

[10] Barry Gower, 'Cassirer, Schlick and "Structural" Realism: The Philosophy of the Exact Sciences in the Background to Early Logical Empiricism', *British Journal for the History of Philosophy*, 8.1 (2000), 71–106, p. 71.

correspondence between mathematical description and experienced reality—for if, in agreement with some Cartesian dream, the mathematical mode could establish a one-to-one correspondence between its world-picture and all that is salient to experience, it cannot seriously be rivalled.

To the contrary, however, early-twentieth-century physics had begun to seem a series of hermetic interrelations increasingly remote from the world of appearance. Dicken states "the issue was ultimately a problem of coordination between the abstract mathematical language of our scientific theories, and the concrete physical world they attempted to describe".[11] In other words, how can the formulae of the page capture the forms of phenomena? It was Kant who first contended with this question in the *Metaphysical Foundations of Natural Science*.[12] If a consensus emerges amid the 'bewildering variety' of contemporary scholarship, it is that the debate on correspondence begins with Kant.[13] Of course, Kant questioned the theories of his day, action at a distance, the dark magic of Newtonian gravity.[14] Newton himself elides the question elegantly—or perhaps professes a profounder humility about the limits of human understanding—by only claiming that, given the available measurements, his theories seem to account for observation well enough.

> Hitherto we have explained the phaenomena of the heavens and of our sea, by the power of Gravity, but have not yet assign'd a cause of this power [...] hitherto I have not been able to discover the cause of those properties of gravity from phaenomena, and I frame no hypotheses.[15]

Although Newton becomes the figurehead of realism in the following two centuries, his confession of ignorance about causation gains the attention and esteem of sceptics, from Hume on. A century later, Kant abandons the hope of closing the gap between formula and phenomena; he even surrenders to Hume on causation, adopting the latter's terminology of 'ground (the original state)' and 'con *sequence* (what seems to regularly follow)' instead of 'cause'

[11] Dicken, *Realism*, p. 32.

[12] Immanuel Kant, *Kant's Prolegomena and Metaphysical Foundations of Natural Science*, ed. Ernest Belfort Bax (London: George Bell [1786] 1883).

[13] It is difficult to cite a consensus. But we will note his importance in the fact that Kant's claim to have inaugurated a Copernican Revolution in philosophy by grounding Newtonian physics in psychology is generally accepted till today (Michael Friedman, *Kant's Construction of Nature* (Cambridge: Cambridge University Press, 2015), p. 41).

[14] Dicken, *Realism*, p. 10.

[15] Isaac Newton, *Newton's Principia, The Mathematical Principles of Natural Philosophy* (New York: Daniel Adee, 1846), p. 392.

and 'effect'.[16] He argues that if Newton's laws follow instead from the basic structures of mind and perception, they still bring us nearer the nature of reality than we are otherwise likely to get.[17]

> He argued that some branches of our knowledge—geometry, mathematics and the fundamental principles of Newtonian Mechanics—are simply descriptions of the way in which we structure the objects of our knowledge, and so can be known with certainty merely through the introspection of our own cognitive faculties.[18]

In so doing, however, Kant binds his lighter model of correspondence to Euclidean and Newtonian precepts. And so, the idea that mind structures and mirrors reality became founded on a set of laws that were to face their demise in the coming century. With the advent of non-Euclidean geometry and relativity, Kant's compromised realism would remain no longer tenable.[19] And it was difficult to adapt the Kantian idea—that is, to maintain the correspondence between the structures of mind and mathematical physics—to a non-Euclidean and relativistic world:[20] we do not perceive space as curved, for instance.

The discovery of non-Euclidean geometry also brought to the fore the artificial aspects of the Newtonian world-picture, demanding further surgery to sustain the connection between experience and mathematical law. Philosophers of science began the process in the nineteenth century by attempting to limit the purely abstract elements of the picture. Ernst Mach, for instance, tried to do away with Absolute Space:[21] the supposed stage upon which matter rehearsed Newton's laws.[22] He attempted to circumscribe

[16] Kant, *Foundations*, p. 4. The terminology refers back to Hume's *Treatise Concerning Human Understanding*, which argues that we can never possess knowledge of causation.

[17] See Michael Friedman, who identifies the revolutionary step as being Kant's insight that "the synthetic a priori representation of space, along with the synthetic a priori science of geometry, plays a crucial role in making experience or *empirical* knowledge first possible" (*Kant's Construction*, p. 4).

[18] Dicken, *Realism*, p. 32.

[19] Clark Glymour, 'Realism and the Nature of Theories' in *Introduction to the Philosophy of Science* (Indianapolis: Hackett Publishing [1992] 1999), p. 114.

[20] Lange, *Philosophy*, p. 4.

[21] Absolute space was a hypothetical entity conceived by Newton; it was a space that "remains always similar and immovable" (Newton, *Principia*, p. 77), on this imagined plane, all his mathematical laws would remain absolutely true (see John D. Norton, 'Philosophy of Space and Time', in *Introduction to the Philosophy of Science*, pp. 180–2).

[22] Dicken, *Realism*, p. 44; For a discussion of Relativity in Mach, see Gereon Walters, 'Phenomenalism, Relativity and Atoms: Rehabilitating Ernst Mach's Philosophy of Science', in J.E. Fenstad, I.T. Frolov, and R. Hilpinen, eds, *Logic, Methodology and Philosophy of Science VIII*, (Proceedings of the Eighth International Congress of Logic, Methodology and Philosophy of Science, Moscow) (Amsterdam: North Holland, 1987), pp. 641–60.

science to shorthands and elegant summaries of the empirically observable, leaving grander 'explanations' to the metaphysicians; this attempt has been dubbed the 'descriptionist' school.[23]

But the effort to maintain correspondence between the picture and experience of reality faced new problems in the twentieth century with the arrival of relativity and quantum mechanics—and these would become recurring subjects of modernist critique. Scientific descriptions now began to hang on a greater proportion of 'unobservables': to know of events far into the void or deep within the atom required a surrender of faith to the instruments of science: instruments through whose prism we gather information about purported entities well beyond the ken of experience.[24] This information—what Eddington and Empson call 'pointer-readings'—despite its mediated acquisition, formed an indispensable part of the new scientific theories. Although the world described by the mathematical picture has always been a rarefied object of art, where facts approach truths only in artificial conditions,[25] the fiction of a world governed by sempiternal mathematical laws has since Pythagoras had its lure—and could be maintained when mathematical laws were abetted by narrative description, like a theology developing around Gospel. But once physical theories began to be largely mathematical interrelations corresponding with registers of advanced instruments disclosing a world that seemed to imagination as science fiction, theology came to another Copernican juncture.[26] The old scientific models, billiard balls for gasses, say, became a relic of Victorian science.[27] If the Newtonian laws presented an idealized picture of ordinary experience, twentieth-century physics told experience it was unfit to perceive the world as revealed to its new instruments. Eddington notes how "the pure mathematician, at first called in as servant, presently likes to assert himself as master".[28] This was a problem not for the business of doing science but for a culture in the thrall of that science—modernism was thus well poised to loudly notice the distance between theory and the sensuous grasp of world.

[23] J.L. Heilbron, 'Fin-de-Siecle Physics', in Carl Gustaf Bernhard, Elisabeth Crawford, and Per Sorbom, eds, *Science, Technology and Society in the Time of Alfred Nobel*, Nobel Symposium, 52 (Oxford: Pergamon, 1982), pp. 51–73.

[24] Lange, *Philosophy*, pp. 4–5.

[25] P.W. Bridgman, *The Nature of Thermodynamics* (Cambridge: Harvard University Press, 1941), pp. 2–3.

[26] See Dicken, *Realism*, pp. 28, 32.

[27] Mary Hesse expressed the anxiety about the growing obsolescence of mechanical models through a fictionalized persona of the real physicist N.R. Campbell in *Models and Analogies in Science* (Notre Dame: University of Notre Dame Press [1966] 1970), pp. 7–56.

[28] Arthur S. Eddington, *The Nature of the Physical World*. 1928 (Cambridge: Cambridge University Press, 1948), p. 52.

The purely mathematical theories of the twentieth century had thus reopened the problem of correspondence which had seemed briefly solved by Kant. Though there were various attempts to adapt and reassert, we will focus on the thinkers at Cambridge that were most influential on our poets, Frank Ramsey and Bertrand Russell. Both were popular at the time Empson and Roberts were undergraduates at Cambridge, and when Riding was developing her model of close reading, which, it has been argued, was influenced by contemporary work in logic.[29] Ramsey and Russell helmed two schools of the philosophy of science that have since been dubbed logical empiricism and structural realism.

The logical empiricist argued that scientific theories were shaped less by the structures of mind, pace Kant, and more by the structures of formal language.[30] Scientific explanations were true to the extent that the semantic field of their vocabulary comprehended the objects of shared experience. Scientific truth was thus brought into the realm of a kind of linguistic analysis—a development of interest for poets.

The logical empiricist, rather than getting entangled with the Kantians in the snarl of mind, saw the philosopher's task as simply making theoretical vocabulary intelligible to observation. Dicken says,

> The idea is that while some of the terms of our scientific theories are well understood and grounded in our everyday experience, others are introduced in the process of our scientific investigations and need to be somehow connected with our everyday experiences in order for us to make sense of them.[31]

Ramsey, and some in the continent like Rudolf Carnap, held that the theoretical elements of science, though expedient to the construction of new theories, ought ultimately to be grounded in past experiment. The job of the philosopher, they argued, was the archaeology of the observational origins of each theoretical premise in a newly ratified scientific theory, until an unwieldy version of the initial theory replete with experimental, and thereby experiential, sources has been wrought.[32] By thus tinkering with language, philosophy would allow mathematical science to pretend to a form of detail-laden realism not unrecognizable to a Stendhal or Balzac.

[29] See Rebecca Porte, 'An Agreement with Reality: The Poetry of Logical Modernism' (PhD Dissertation: University of Michigan, 2014), pp. 31–3.

[30] Glymour, 'Realism', p. 117.

[31] Dicken, *Realism*, p. 36.

[32] Ibid., p. 61; and Coffa, *Carnap*, p. 234.

As for the concern that scientific theories were becoming purely mathematical, one answer was that of structural realism. Its thesis, after a Kantian fashion, was that science tells nothing of things-in-themselves;[33] and so, there can be no knowledge of causation.[34] The "unreasonable effectiveness of mathematics in natural science"[35] disclosed to the structural realist but one aspect of reality, its structure. Dicken says, "For Russell, [...] the nature of our scientific knowledge is best understood as a claim about the abstract, logico-mathematical relationships that hold between the objects of the external world, rather than specific claims about those objects themselves".[36] Russell said we must be satisfied in the knowledge of abstract second-order relations between objects in the world without burdening science with higher alchemical ambitions, of the search for cause and substance. All we can gather from science is that some mathematical relations obtain in the universe. The structuralist thus drew an epistemic limit to scientific description, binding reason to the very surface of phenomena.

The logical empiricist and structural realist described different boundaries around science's episteme. But both could enhance the artist's share in the description of reality and gave poets philosophical grounds to resist science's empire. The logical empiricist underscored the extent to which the deliverances of physics were divorced from experience and dependent on measuring apparatuses. The structural realist limited scientific knowledge to the mathematical relations among things, leaving the remaining parts of reality open to alternate modes of representation.

That modes of representation determine the extent and nature of knowledge secured is a central theme in the early poetry of Empson. But instead of appending these philosophical debates as annotations to poems, we have separated the discussion of context so as to not derail the workings of poems through the alluring terminus of external reference. In the next section, we will identify the main epistemological themes in Empson's writing before moving on to a discussion of his poems, ending with our book's first full-poem close reading, of "Doctrinal Point", a poem of logical paradoxes, on the limits of scientific knowledge.

[33] Gower, *Cassirer*, p. 74.

[34] Dicken, *Realism*, pp. 159–60; On causation, see also Anjan Chakravarty, 'The Structuralist Conception of Objects', *Philosophy of Science*, 70 (2003), 867–78.

[35] Phrase borrowed from Eugene Wigner, 'The Unreasonable Effectiveness of Mathematics in the Natural Sciences', *Communications* in *Pure and Applied Mathematics*, 13.1 (February 1960), 1–14.

[36] Dicken, *Realism*, p. 144.

2.2 Empson's Value Is in Activity

F.R. Leavis once said of Empson that "he has the courage of his lack of convictions".[37] When reading him, one constantly feels that more is meant than what is said, and more is felt than what is meant. In the *Seven Types of Ambiguity*, Empson argues, when Marlowe's Faustus says "Ugly Hell gape not: come not, Lucifer", he really means to accept both Hell and Lucifer into his heart; Faustus "is willing to abandon his learning" for he is "going to a world where knowledge is immediate; [...] he has abandoned his effort to organise his preferences, and is falling to the devil like a tired child".[38] We will see repeated in Empson's works this theme of an immediate, sensuous, infernal knowledge; and its Blakean opposition to a wistful and abstract learning. In defence of this reading of Marlowe's line, in the second edition of *Seven Types*, Empson remarks in a footnote,

> A critic said that my interpretation here is wrong because the actor is meant to scream with horror not sound like a tired child. [...] But the more the actor screams the stressed words the less the audience hears the unstressed words 'not' 'not'.[39]

In his rebuttal, Empson believes he has clinched his first argument, that given the stresses of the metre, Faustus's actual meaning is quite opposite to his seemingly apotropaic words. But Empson seems to have foregone his reading of tone—that of a tired child—and almost promiscuously admitted the critic's view into his own for a quick repartee. One can already see Empson lauding this style of urbane argufying in an early review in *Granta* (1927) when discussing Russell's often fractured syllogisms:

> He [Russell] [...] is subtle only where it seems interesting, and is not pained by crudity elsewhere. This is the English way of thinking which seems so unscrupulous to the Continental; it has great virtues; it gives great resilience to the thinker, never blurs a point by too wide a focus, is itself a confession of how much always must be left undealt with, and is beautifully free from verbiage.[40]

[37] F.R. Leavis, 'Cambridge Poetry', *Cambridge Review* (1929), 317–18, p. 318.
[38] William Empson, *Seven Types of Ambiguity*. 1930 (London: Chatto and Windus, 1949), p. 206.
[39] Ibid., p. 206.
[40] William Empson, *The Book, Film & Theatre Reviews of William Empson Originally Printed in the Cambridge Magazine Granta, 1927–1929* (Kent: Foundling Press, 1993), p. 60; Empson began work on *Seven Types of Ambiguity* in 1928 (Haffenden, *Mandarins*, p. xviii).

Empson admires how Russell seems immune to the panic, when criticized, to rebut every minor challenge.[41] He sees incompleteness as a refinement—like Reynolds telling his gentlemen painters not to fuss over the finer details;[42] but also, as a confessional act, and it seems to have inspired Empson's own critical style. His arguments are replete, one might even say resplendent, with half-rebuttals. He often enters a dispute somewhere between a line of premises, produces a caustic remark by way of refutation, and moves on as if unwedded to an initial stance. And as regards freedom from verbiage, if he knows a thing cannot be articulated, he is never compelled by custom: he manages, for instance, to write an entire treatise on pastoral poetry,[43] "acknowledged to be the most brilliant study of its subject", without proffering even a cursory definition of the pastoral.[44] This mix of complete detachment, an unusual genius for accuracy, a desultory concern for minutia, and haughty disregard for completeness, distinguishes the works of Empson—more so than any doctrine he may have briefly seemed to espouse.

Mathew Bevis begins the introduction to *Some Versions of Empson* with an extended paean to Empson's elusiveness, warning the reader not to place him in any school of thought:

> Empson frequently refuses to co-operate: left-wing yet posing serious obstacles for 'any Marxist appropriation of his work'; a keen reader of Freud and of unconscious intentions but wary of psychoanalytic criticism; a precursor of certain forms of deconstruction, yet at odds with what he perceived to be a negative and suspicious hermeneutics; theoretically inclined yet opposed to some aspects of the rise of Theory. It is hard to place him, but then, 'It is not human to feel safely placed', and the longevity of his criticism owes much to the fact that he rarely plumped for the safe option.[45]

[41] One can hear this style live during Russell's famous BBC debate against Frederick Copleston on the existence of God in 1948 (printed in Bertrand Russell, *Why I Am Not a Christian* (London: George Allen & Unwin, 1957)).

[42] Joshua Reynolds, *Discourses on Art* (London: George Routledge [1778] 1909), pp. 162, 168.

[43] His *Some Versions of Pastoral* (London: Chatto & Windus, 1935).

[44] Paul Alpers, 'Empson on Pastoral', *New Literary History*, 10.1 (1978), 101–23, p. 101.

[45] Mathew Bevis, 'Introduction: Empson in the Round', in *Some Versions of Empson*, ed. Mathew Bevis (Oxford: Oxford University Press, 2007), p. 3; References are to Terry Eagleton, *Against the Grain: Essays 1975–1985* (London: Verso, 1986), p. 165; Paul de Man, *Blindness and Insight: Essays in the Logic of Contemporary Criticism, 2nd ed.* (Minneapolis: University of Minnesota Press, 1983), pp. 234–41; Christopher Norris and Nigel Mapp, eds, *William Empson: The Critical Achievement* (Cambridge: Cambridge University Press, 1993); and William Empson, *The Complete Poems*, ed. John Haffenden (London: Penguin, 2000), p. 85.

Bevis's caution here doesn't quite express the depths to which the whirl and flux of Empson's ceaseless denials ran—this we will attempt to plumb now. Given it is a rather tricky task to characterize Empson's commitment to ignorance without making ignorance sound in itself a doctrine, we will approach the question obliquely, through other ideas through which he passed, which manage to leave a trace, and modify the mosaic of attitudes that constituted Empson.

Let us begin with Buddhism. The first published version of the collected poems (1935) opens with Empson's translation of the Fire Sermon of the Pali Canon. In it, Bhikkhus, like Faustus, becomes disillusioned with the world of phenomena; the wise man "becomes weary of the knowledge of the eye, he becomes weary of the visible, he becomes weary of the knowledge of the visible, [...] he becomes weary of the ear [...] weary of the body [...] weary of the mind". Unbidden by any feature of being he can conscript to a "self", he becomes "empty of desire. When he is empty of desire, he becomes free."[46] The dramatic abnegation of these lines is discernible, more subtly, in Empson's writing—a kind of stoic rejection of all comfort in knowing. Seeing the germ in the Fire Sermon of Empson's poems, many critics suggested that Empson had taken it for his text.[47] This view spread to the extent that Empson had to resist in print the notion that he was proselytizing for Buddhism.

> It [The Fire Sermon] is said to be one of the earliest sermons of the Buddha, and carries the unearthliness of his system as far as is conceivable. One should realise that it denounces not only all existence on earth but all existence recognisable as such, even in the highest heaven.[48]

Empson did not draw from the recognition that the world is Maya—and, as the Buddha maintained, overwhelmingly sorrow[49]—a Manichean hostility to existence. Faced with the same sense of futility in knowing that enveloped Faustus and Bhikkhus, Empson responded with a mixture of intense curiosity and dark humour, what the Gardners call "a courageous pessimism".[50] While

[46] Empson, *Poems*, p. 3.

[47] See, for instance, A.E. Rodway, 'The Structure of Complex Verse: Review of *Collected Poems*', *Essays in Criticism* 6.2 (1956), 232–40; and David Simpson, 'Everything, Beggars, Is on Fire', *Arrows*, New Year Edition (1957), 5–6.

[48] William Empson, 'Everything, Beggars, Is on Fire', *Arrows* (1957), p. 5.

[49] T.W. Rhys Davids, trans. from the Pâli, *Dialogues of the Buddha (The Dīgha-Nikâya)* (London: Oxford University Press, 1899), p. 94.

[50] Philip Gardner and Averil Gardner, *The God Approached: Commentary on the Poems of William Empson* (London: Chatto & Windus, 1987), p. 44.

he was sympathetic to the Buddhist commitment to doubt, Empson could not follow through on its conclusion that sought to extinguish the innate value of earthly life. Empson thought the Sermon ultimately wrong but "fascinating [...] as one extreme of the range of human thought".[51]

The other extreme fascinated him equally. He says the Buddhist cast of mind is "as important as the belief in the value of life and the love of activity for its own sake which is common among Europeans".[52] As much as the discursive presence of Empson's works resoundingly announce the artful detachment of the author, he seems to admire the spirit of his fellow Englishman who is "satisfied with some such rule of existence as having a good time, or playing the game, or doing his duty".[53] This strangely Yeatsian feeling for the salvific powers of activity can be sensed throughout Empson's poems.

Empson's biographer Haffenden said that he "heeded throughout his career" the modified Benthamism of I.A. Richards, namely, his 'Theory of Value'.[54] Although Empson's works never "heed" any doctrine, the value he places on activity can be traced to what Richards made of the nineteenth-century term of aesthetics, 'impulse'. Richards replaced 'pleasure' in his version of the utilitarian calculus with "appetency", or positive impulse.[55] He believed the good life consisted in "the systematisation of our impulses",[56] which was best done by poetry—and the poem was in turn somewhat circularly defined as the artefact of a systematization of impulses.[57] The rather crude Benthamite addition of pleasure points is replaced by a notion of order among impulses, such that even pain, when in the service of aesthetic experience, finds place in the system. Reviewing Richards's *The Foundations of Aesthetics,* Empson says, "I am never sure that I have not missed a divine revelation lying about somewhere".[58] Beneath the sly double negative, it seems that Empson was forced to take Richards's theory more seriously than he had expected.

When invoking the value theory, however, Empson always draws it nearer his interests in activity and avoids what he must have considered the vulgarity of calculation. The theory of value can, in some ways, be seen as an

[51] William Empson, 'Mr Empson and the Fire Sermon', *Essays in Criticism,* 6.4 (1956), 481–82, p. 481.

[52] William Empson, *The Face of the Buddha*, ed. Rupert Arrowsmith (Oxford: Oxford University Press, 2016), p. 109.

[53] Empson, *Buddha*, p. 109.

[54] Haffenden, *Poems,* p. 143.

[55] I. A. Richards, *Principles of Literary Criticism* (New York: Routledge [1924] 2001), p. 43.

[56] Ibid., p. 222.

[57] Richards, *Science and Poetry* (London: K. Paul, Trench, Trubner & Co, 1926), pp. 15–16.

[58] William Empson, 'Chronicles: A Doctrine of Aesthetics', *Hudson Review,* 2.1 (1949), pp. 94–7.

adaptation of Pater's dictum in *The Renaissance*—of packing via art as many sensations as possible into a finite life—to a utilitarian scheme, and it was not the utilitarianism that interested Empson. Reviewing John Laird's *An Enquiry into Moral Notions*, Empson stated, "whether or not the values open to us are measurable, we cannot measure them, and it is of much value merely to stand up between the forces to which we are exposed".[59] To Empson, the theory was, among other things, a means to resist the tide of scientific positivism looming over much of modern thought. He states quite plainly that "the usefulness of the thing is chiefly to show that the scientific picture of the world is not necessarily at loggerheads with the aesthetic one".[60]

While with the Buddhist, Empson wanted to rout the illusion of certitude, he did not, we have said, disdain worldly matters: indeed, a lack of certitude would invest reality with fresh allure and feeling. He argues that the emotional account of an event in the world should be regarded as equally real as the factual. For instance, there is the factual "belief that Julius Caesar was stabbed" and the attendant moral lessons of hubris, say, that it has historically inspired.[61] Empson draws on Leibniz's idea that somehow, 'crossing the Rubicon' is analytically contained in the concept 'Caesar', almost in the way 'unmarried man' is contained in 'bachelor'[62]—and so, connotation, what lies beyond the denotation, of fact becomes concretely real. Empson concludes that all these beliefs—including the factual—and beliefs about the beliefs, ultimately have the same claim to our attention:

> The more you let my moral feelings form a coherent logical structure, the closer you draw the parallel to the structure of my feelings about historical fact. By the time you have put in the necessary reservations, it seems to me, the 'emotive' account of ethical judgements is on exactly the same footing as the logical-positivist or behaviourist accounts of beliefs about the external world.[63]

The arrangement of attitudes that forms a complex moral or emotive judgement, something more considered than the cry of an impulse, Empson argues, represents equally a state of affairs in the world as the facts in empirical space

[59] William Empson, 'Three Ethics', *Spectator* (November 1935), p. 912.
[60] Empson, 'Aesthetics', p. 95.
[61] William Empson, *Structure of Complex Words* (London: Chatto & Windus, 1951), p. 416.
[62] Gottfried Wilhelm Leibniz, *Discourse on Metaphysics* trans. George R. Montgomery (Chicago: The Open Court Publishing [1686] 1918), §13.
[63] Ibid., p. 416.

retailed in the narration of science. This equivalence would appreciate the value of "balancing impulses" that poetry alone, for Richards, brings about.[64]

Empson saw Richards's value theory in tandem with Buddhism, each working, in its own way, to destabilize metaphysical confidence in knowledge: Buddhism through rejection of world and the value theory through its affirmation. He calls Buddhism "the great historical antagonist" of "the Richards Theory of Value".[65] The value theory in itself, though making no claims absolute, was not a complete answer: "what satisfied the most impulses might turn out to be the same as what was to the glory of God or even as what tended to Nirvana".[66] The value-theory's lassitude may come to accommodate (and in turn, yield to) other, more dogmatic doctrines; in other words, "any serious attempt at establishing a relativity turns out to establish an absolute; in the case of Einstein [it is] the velocity of light".[67] This is not entirely fair to Richards, but the circularity of a seeming opposite bending to its antithesis is a deep mischief at the heart of the human search for knowledge in which Empson would have surely delighted.

In his verbal analyses, Empson's intense meditation on the truest meaning accessible to wit amid the ultimate meaninglessness he acutely experiences is perplexing. For Pater or Richards, one must be actively disposed to the world because value ultimately resides in beautiful experience. But it is very difficult to identify an absolute value or purpose in the writings of Empson. Such conviction in ignorance, I believe, ultimately shows Empson for a fool, or what the Elizabethans made of the fool: for "the wise man is modest and cowardly, the fool tries".[68] Empson was in awe of Shakespeare's doings with the word "fool" in *King Lear* and the clown principle embodied earlier in Falstaff.[69] It is not inconceivable to think that Empson may have been consciously or unconsciously inspired to favour the gamble that "by being [...] a complete fool", one becomes "in some mystical way superlatively wise".[70] There is a striking image in Empson's translation of his friend, Chiyoko Hatakeyama's poem, "The Fool" (1940): Empson's version introduces the lines, "Wisdom's the charger mounts him above shade, / Hanged by suspense and eternally delayed".[71] Wisdom, like Death,[72] mounts the dumb charger above the

[64] Richards, *Science*, p. 4. [65] Ibid., pp. 424–5. [66] Ibid., p. 425.
[67] Empson, 'Aesthetics', p. 95. [68] Empson, *Complex Words*, p. 106.
[69] William Empson, *Essays on Shakespeare* (Cambridge: Cambridge University Press, 1986), p. 72.
[70] Empson, *Complex Words*, p. 157.
[71] Empson, *Poems*, p. 71; see Peter Robinson, 'C. Hatekeyama and W.E.', in *Some Versions of Empson*, pp. 64–8, for a discussion of their relationship and an English version of the original poem.
[72] *Revelations* 6:8, *KJV*.

infernal shade, and steers the fool's vulgar tongue to an accidental reckoning: one that for Empson, of course, never comes about.

Empson's foolishness did not consist in being obstinate or dense but simply in getting on with being and loving in the face of darkness and oblivion. It is a theme that animates many of his poems. The Elizabethan fool, with his mercurial disdain for circumstance, embodies, in some ways, both the Buddhist detachment from worldly affairs and the European commitment to living, while his most recognizable distinction is a dark humour which, without his intention, tolls mysterious wisdom. Empson's mathematical poems, his use of mathematical metaphors, seem to work out a balance between these epistemologies, with always a wry step outside them all.

2.2.1 Paradoxes and limits

One idea that we know transfixed the undergraduate Empson was that of contradictory meanings co-inhabiting a verbal unit. Reviewing M. Carta Sturge's *Opposite Things* in 1927, Empson says, "extremely often, in dealing with the world, one arrives at two ideas or ways of dealing with things which both work and are needed, but which entirely contradict one another"; we see it in "the practice of scientists, recent mathematical logic, primitive languages, the doctrine of the Trinity, the corresponding Eastern ideas, and, in fact, out of anything of any importance".[73] This idea, which would evolve into the seventh-type ambiguity, seems to pervade most areas of Empson's interest. We already caught its glimpse when Faustus suppressed the breaths by which he uttered 'not', accepting Satan reluctantly into his soul. Indeed, before embarking on his discourse on the final ambiguity, Empson himself calls on the fallen angel as muse, "to cast upon the reader something of the awe and horror which were felt by Dante arriving finally at the most centrique part of earth, of Satan, of hell".[74]

In one of his poems, "The World's End", faced with the prospect of such a paradox, "blind Satan's voice rattles the whole of Hell" [8].[75] The Devil here is responding in what seems a reasonable way to the condition of the universe as declared by modern mathematical science:

> Alas, how hope for freedom, no bars bind;
> Space is like earth, rounded, a padded cell;
> Plumb the stars' depth, your lead bumps you behind; [4–7]

[73] Empson, *Granta*, p. 46. [74] Empson, *Seven Types*, p. 196. [75] Empson, *Poems*, p. 13.

Because "the curvature [of space] actually leads to a complete bending round and closing up of space",[76] the 'hope for freedom', to escape mundane reality, becomes ultimately futile. Even Milton's Hell held in it the possibility of escape, of unholy rebellion, but the hell fashioned by numbers needs no bars to bind its subjects: here they are immured in circularity, paradox.

Einstein remarks that "the great charm resulting from this consideration lies in the fact that *the universe of these beings is finite and yet has no limits*".[77] The poem that begins with a cavalier entreaty to a lover to fly away with him to the world's end quickly darkens when the structure of that world is disclosed as Einstein's. Empson says "the finite but unbounded universe, popularized by Eddington [...] makes flight seem useless for the lovers".[78] The curtain of futility fell thus on Empson's poetic world. Take the couplet in the next stanza,

> Each tangent plain touches one top of earth,
> Each point in one direction ends the world. [15–16]

There is a sense of frantically scurrying after some end made pointless by relativity—each point portending an end, a vertex in a spherical world, as if Percival were finding the grail at every step, voiding his quest in its perpetual completion. But despite the disturbing premise, for the poem finds a kind of malice, a continual death, at the heart of modern physics's universe, the conjunction in the first line, "Fly with me to all's *and* the world's end" [emphasis mine] suggests a Falstaffian humour in the face of tragedy. All's (everyone's) end, death, is temporal, whereas the physical world's (supposed) end is spatial. That all's and the world's end, though different in kind, cohabit one elusive, vanishing point, suggests already the doom of this flight of lovers in Einstein's space-time, where attaining a spatial end is as futile as, say, completing one reincarnate life in the Hindu samsara. But the line also evokes Shakespeare's "all's well that ends well", a consoling cliché conveying a rather jaunty mood that's amplified by the ribaldry of your plumbing lead bumping you from behind—a tone thus cacophonous in the dim situation, setting the expectation of impudence towards the tragic subject-matter of Empson's mathematical poems.

[76] Arthur S. Eddington, *The Expanding Universe* (London: Harmondsworth, 1940), p. 39. This text was written after the writing of the poem, but has a terse phrasing of the longer exposition on the curvature of space-time in *The Nature of the Physical World* (1928), the publication of which was the immediate of occasion for the poem being analysed (Haffenden, *Poems,* p. 162).

[77] Albert Einstein, *Relativity*, trans. R.W. Lawson (London: Methuen [1916] 1920), s. 31.

[78] Empson, *Poems*, p. 162.

Empson often finds a connection between mathematical paradox and the delusion at the heart of any solemn search, chief among which is the one for knowledge. From the poem, "Dissatisfaction with Metaphysics":[79]

> Adam and Eve breed still their dotted line,
> Repeated incest, a plain series.
> Their trick is all philosophers' disease. [8–10]

The 'dotted line' has a subdued sense of a progression, as in an ellipsis, but more dominantly denotes 'more of the same', as in, 'and so on'. The human lineage—modified by the animal suggestion of 'breed'—is thus accused of a plebeian sameness. If we assume the Malthusian fiction that human beings breed in geometric progression, from 2 to 4 to 8 to 16, and so on, there is an impression in their multiplication of great growth. But the series is plain, simple.

The reader might recall being presented in the classroom with a sequence of numbers with instructions to find the governing rule. There were difficult sequences, such as 1, 2, 6, 33, 49…? And simple ones such as 2, 3, 5, 7, 11…? The assumption that generations multiply in a geometric sequence, 2, 4, 8, 16, 32…—at least we trace our ancestry thus in reverse—gives the human series the appearance of bulking progression. But given the sequence is plain, it shows their proliferation to be trivial, a growth in quantity alone. This sense is supported by line 10, when the poet calls the 'dotted' line of Adam's brood a 'trick'. 'Dotted' can also mean spotty or sordid.[80] These senses compound the poet's scorn for the entire stock, which entertains, through the replaceable members of its plain series, the fantasy or illusion of progress.

Generational proliferation without enhancement is then compared to the philosopher's efforts at knowledge-building. The precise imputation against the philosopher can be understood by recalling that multiple rules can govern a series; that is, one can have different expressions for the same function (a rule that says for every number, what is the corresponding output).[81] In the poem, the fact that many rules can govern the same series, the same exhibition of progress, implies that different metaphysical systems are not

[79] Ibid., p. 17.

[80] There might even be latent in 'dot' a suggestion of the disreputable practice of dowry, as in Mina Loy's phrase, "Virgins without dots" (Mina Loy, 'Virgin Plus Curtains Minus Dots (1915)', in *Modernism: An Anthology*, ed. Lawrence Rainey (Malden: Blackwell Publishing, 2005), p. 417).

[81] See exercises on page 78 of R.S. Aggarwal, *Mathematics for Class 11* (New Delhi: Bharati Bhawan [1990] 2020).

as distinct as they seem, but are rather farragoes of contradiction creating the dialectical illusion of furthering knowledge (as perhaps the male and female of the species couple as opposites issuing an ostensible synthesis), but actually penetrating no deeper nor farther into the nature of truth: all but a plain series.[82]

In this poem, Einstein's theory is introduced in a more resigned tone than the shock of "The World's End." "New safe straight lines are finite though unbounded" [11]. It could be that the poet has come to terms with the paradox at the heart of the new world-picture. Unlike the epicycles of ancient astronomy circling a delusive centre, the paradox at the heart of the new astronomy presents a condition that would put to rest the old toil in the mill for truth (the philosophers' disease). Relativity impresses finitude on all sides by closing the universe upon itself; but by virtue of being circular, the unbounded universe can perhaps set our imagination free from entrenchment in discrete corners of dogma. The philosopher's dis-ease, the guilt and anxiety from a restless repetition of the same, hoping every time for greater value, is eased in the Einsteinian world of bounds with no corners.

The positive spin on this metaphor progresses in later poems. The poet says of an old lady that she is "confident, finds no confines on her sphere" [18–19]. The poems thus seem to gradually reverse the sentiment attached to the paradox, 'finite but unbounded', from imprisonment to freedom. Jacob Bronowski, Empson's co-editor of the periodical *Experiment* explains the metaphor:

> The heart of the metaphor comes from mathematics: it is the theorem that a surface can be finite in extent and yet have no boundaries, no confines [...] if you or the old lady walk all over it you will never meet any boundary and will seem to be going on to infinity.[83]

The changing application of this metaphor returns us to Empson's attraction with aporia. The connection in "To an Old Lady" between the paradox and freedom also leads to a second-order equation that forms across the poems, "The World's End", "Dissatisfaction with Metaphysics", and "To an Old Lady". If the same paradox at the heart of the mathematical world-picture is charged

[82] The original title of the poem was 'Disillusion with Metaphysics', *Experiment*, 1 (1928), p. 48; this supports our reading of material and spiritual growth in the poem as an illusion.

[83] Jacob Bronowski, 'The Imaginative Mind in Science', in *The Visionary Eye: Essays in the Arts, Literature, and Science*, eds. Jacob Bronowski, Piero E. Ariotti, and Rita Bronowski (Cambridge: MIT Press, 1978), p. 27.

with confining and emancipating, there is a darker conciliation of opposites between imprisonment and liberation: that by being reconciled to the epistemic finitude of our humanity, we are unlike Prometheus, unbound: freed from the illusion of progress towards some ultimate truth. This is the kind of seventh-type ambiguity that sent the young Empson into a Dantean mood.

In a letter to Maxwell-Mahon, Empson calls "Dissatisfaction of Metaphysics" a "quite a jolly little poem"; that "it is trying to laugh off the fright of being caught on the flypaper".[84] Despite the terrible burden of the realization, we are to chuckle at our unfortunate fate, of being flies squirming in resin at the mercy of some twisted creative force. In *Granta*, Empson notes a connection between holiness and mirth: "I believe myself poetry is written with the sort of joke you find in hymns" (1927).[85] He naturally does not mean that poems are trivial. What he means in creating and identifying such seventh-type ambiguities is that poetry is written with a kind of inhuman mirth at the calamity of existence, like Zarathustra howling at eternal recurrence. The poet countenances his discovery with humour, demonstrating a kind of fatalism found in the Elizabethan clown, possessed of the belief "that you ought to accept the actualities of life courageously even if rather unscrupulously, and not try to gloss over its contradictions and the depths that lie under your feet".[86] The recognition might be the Buddha's, the response is all Falstaff.

Towards the end of "The World's End", there is an image that seems to fully figure the epistemological tragedy of the human condition.

> Apple of knowledge and forgetful mere
> From Tantalus too differential bend.
> The shadow clings. [13–14]

Since Adam fell to the original urge, thirst for knowledge has clung like a shadow to human doing. When Midas asked Silenus what fate was best for man, he replied, "Oh, wretched race of a day, children of chance and misery, why do ye compel me to say to you what it were most expedient for you not to hear? What is best of all is for ever beyond your reach: not to be born, not to *be*, to be *nothing*."[87] We will settle only in knowing all or annulling ourselves in the river of oblivion. In time, both the anagogical and prelapsarian fantasy, the reaching up for the apple and wading down into the mere (lake, from the

[84] Haffenden, *Poems*, p. 170. [85] Empson, *Granta*, p. 56.
[86] Empson, *Complex Words*, p. 124.
[87] The story is related in Friedrich Nietszche, *The Birth of Tragedy: Or Hellenism and Pessimism*, trans. Walter Kauffman (London: George Allen & Unwin, 1910), p. 34.

Italian 'mare'), come agonizingly close to satisfaction, like the differentials approaching from either side the slope of the curve at a point. The dense word 'differential', containing an aural evocation of 'defer', signifies the deferment of knowledge across time and in the present. It expresses the former through the conceit of the relative universe in the preceding stanzas. The condition of science is to destroy in its imperial march its own world-pictures to erect new ones—Einstein for Newton, Copernicus for Ptolemy, and so on. But the apple bows 'deferentially' towards the outstretched hands of the seeker, tantalizing him into believing after each discovery that the final answer is not far, bidding him to seek further—though satisfaction is always deferred. It expresses the latter through the metaphor of differentiation, which is used in physics to discover instantaneous velocity. Calculus yields this answer, but by abandoning the very identity of an exact moment in time (and the exact slope of a curve with which time is reckoned). That is, it approaches an instant through smaller and smaller intervals of time; or the slope of the curve by a sequence of adjoining, shrinking, secants: converging to an exact limit but never attaining full equality with it—in short, by ingenious approximation. The illusion at the heart of the quest for knowledge is thus exposed in the practice of assuming the calculation of instantaneous velocity to reveal the instant itself. Knowledge of the most basic unit of existence—the instant—, seemingly grasped by Newton or Leibniz, remains yet out of reach for metaphysics.

We began this book discussing the role of propaganda, in the rise of the mathematical world-picture from the seventeenth century on, in confounding the gap between representation and phenomena. The poetry of Empson repeatedly brings that gap to the fore: in "The World's End" the main aspect of the metaphor is the bending too far out of reach, but that image of distance figures, ultimately, the eternal gap between mind and world: the very stigma of the Fall.

2.2.2 Mathematical fictions

The slippage in mathematical physics from representation to reality brings us to another type of mathematical image in Empson's works, mathematical fictions or fictions of mathematics. Fiction here does not mean untruth; it refers to elements of a story that help grasp what challenges understanding. Such fictions are everywhere. Mill, for instance, notes how "the elements of algebra [...] are as full of fictions as English law".[88] Empson similarly connects

[88] J.S. Mill, *Utilitarianism* (Kitchener: Batoche Books [1861], 2001), pp. 5–6.

mathematics and the law through the legal fiction that "ownership of land extends above and below ground";[89] Empson brings out the absurdity of rationalized law in a poem, "Legal Fiction",[90] by picturing land property as an infinitely extending conic section:

> Your rights reach down where all owners meet, in Hell's
> Pointed exclusive conclave, at earth's centre
> (Your spun farm's root still on that axis dwells);
> And up, through galaxies, a growing sector. [9–12]

The poem expresses an old English socialism, perhaps calling up John Ball's mordant dig at the gentry: "when Adam delved and Eve span, who then was the gentleman?" The bourgeois landowner is shown eyeing an endless tract of space above his property, expanding, presumably, with the reach of our measuring instruments—the scientific tenor of 'galaxies' forcing such speculation. See Figure 2.1 below.

The farm imagined as spinning along the rotation axis of the Earth seems to subordinate daily life to the scientific description of our circumstance. But it is the accommodation of a social sense of property by the law to a scientifically accurate definition of boundary—a mathematical fiction spun around the farm by the law—that forces this caricature.[91] The fiction leads, ultimately,

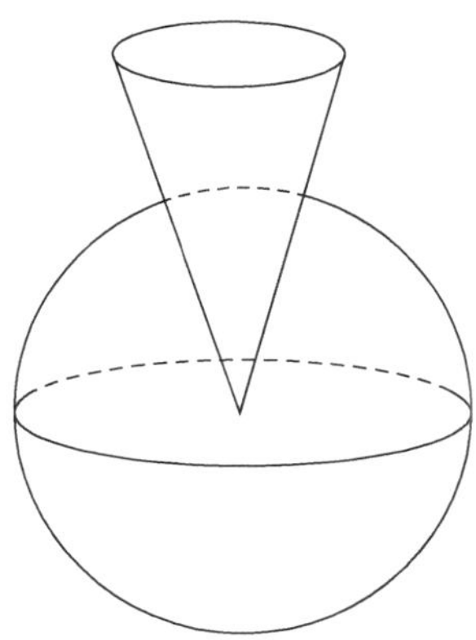

Figure 2.1 Conic-shaped property extending, past the circumference of the Earth, infinitely into space.

[89] From Empson's notes to the poem (Empson, *Poems*, p. 229). [90] Ibid., p. 37.

[91] Law in the Enlightenment is known for having secured and standardized property rights more than, and at the expense of, any other rights, often resorting to the death penalty or transportation to the penal colonies for minor offences against property, or a socialist might say, against the propertied

upon death, to this: "your dark central cone / wavers, a candle's shadow, at the end" [15–16]: when his body sinks to the roots of his plants, the interred gentleman discovers that the cone vanishes to a point where all landowners have collective "short stakes" [1]. Through this idea of a potentially endless domain of private property, the poem seems to ridicule the greed that would demand such a law by its ironic end in a collectivized farm of the dead.

The connection between legal and mathematical fictions plays into the larger theme of Empson's poems, of the absurdity of perfect solutions whose lure often entreats overreach. One of the trends in literary criticism that irritated Empson was the resort to the unconscious, to the probing into unknowable regions of mind, to interpret works of literature. He thought references to the unconscious were low-hanging-fruits of interpretation because unverifiable, and so, often sought as short-cuts to make far-fetched claims about meaning. In *Seven Types*, Empson argues that to understand the puns in Satan's speech to the fallen angels, we must put ourselves in the minds of the Devil's vanquished audience. We must do this, he says, by focusing on the "conscious part of the effect" rather than the characters' unconscious feelings; "for one must continually feel doubtful about antitheses involving the idea of 'unconsciousness' which, like the infinities of mathematics, maybe a convenient fiction or a product of definition".[92] By a 'product of definition', he means that the description of a thing may be wound a certain way to hang together by the assistance of a piece of fiction.

This attitude that can be found in most of Empson's philosophical poems seems an extreme version of a historically English, even nominalist, suspicion of abstractions.[93] This suspicion is not to be confused with anti-intellectualism or the "pride fed by the newspapers in the common-sense 'intuition' relied on by all true Englishmen".[94] It is more like the English scientist who distrusts abstract mathematical theories—favoured by the Continental—in favour of simple models and provisional analogies.[95] It derives from the long shadow of the nominalists over English philosophy, and more recently from Richards, for whom "abstractions, universals, concepts" were often "verbiage or word magic".[96] This suspicion of abstract and well-rounded theories accords with

(see Douglas Hay, Peter Linebaugh, John G. Rule, E.P. Thompson, Cal Winslaw, *Albion's Fatal Tree: Crime and Society in Eighteenth-Century England* (New York: Pantheon House, 1975), pp. 17–19.

92 Empson, *Seven Types*, p. 103.

93 I do not treat English empiricism as distinct from or disconnected to nominalism: I have preferred the latter term to describe Empson because his scepticism towards abstraction does not come with a theory of psychology that appends empiricist metaphysics.

94 Empson, *Granta*, p. 45.

95 Hesse, *Models*, pp. 1–3.

96 John Paul Russo, 'I. A. Richards in Retrospect', *Critical Inquiry*, 8.4 (1982), 743–60, p. 744.

the traces of Buddhism in Empson's work. Gautama too was known for having rejected the abstract questions of his day: about the world's beginning, the soul's relation to body, and the rewards of heaven: the Buddha dismissed these as—and one can see Empson nodding—"the jungle, the desert, the puppet-show, the writhing, the entanglement, of speculation".[97] Whether it is the unattainable "Doctrinal Point" or a "Dissatisfaction with Metaphysics", the poems, 'weary of the knowledge of the mind' and intent to set free the energy of life, seem an intense meditation on the Fire Sermon—so central was this scepticism to Empson that the Fire Sermon became the chosen text to be read at his funeral.[98]

The poem "High Dive", says Christopher Ricks, "is Empson's elaboration of the need to act".[99] In the poem, the poet stands atop a diving board and, forgetting for a moment his bodily predicament, pauses to construct a mathematical model of the swimming pool. In his notes, Empson says the poem is about being forced from "a godlike state of contemplation [...] into action which cannot wholly foresee its consequences".[100]

> Holding it then, I Sanctus brood thereover,
> Inform *in posse* the tank's triple infinite
> (So handy for co-ordinates), chauffeur
> The girdered sky, and need not dive in it [5–8]

The slide in verbal mood from 'brood' to 'inform' to 'chauffeur' pulls out the rug from the mathematician's feet. The mathematician who fancied himself a god atop the springboard or a prophet tiding divine truths—the trinity of the infinite—is shown up as merely the escort of the main act: oblique to actual purpose. The girdered sky suggests the cartesian grid through which the mathematician perceives the space before him: the grided sky, if you will. The mathematician is invested in the actual space before him as much, say, as the chauffeur is in roads of the town: though knowing them better than anyone else, frequents them for no purpose of his own. And given the mannered past invoked through the class relations implicit in the word, there might even be a jesting sense of his world being girdled or repressed.

Veronica Forrest-Thomson reads lines 6–7 as follows: "mathematically the expression is 'triple' because it describes the water's movement in terms of the

[97] Buddha, *Dialogues*, p. 186. [98] Arrowsmith, *Buddha*, p. lii.

[99] Christopher Ricks, 'Empson's Poetry', in ed., Roma Gill, *William Empson: The Man and His Works* (London: Routledge, 1974), p. 186.

[100] From Empson's notes to the poem (Empson, *Poems*, p. 188).

three axes of the system of Cartesian co-ordinates, just as it is 'infinite' because a potential function takes no account of boundaries".[101] The mind can achieve a kind of total comprehension—it can understand and encompass—things with the remote and disinterested genius of Jeeves; and thus, the body 'need not dive in it'. But the quasi-divine work of the brain also finds its nemesis in action: simply from "the splash and eddy made by the diver",[102] the mathematical representation of the pool becomes confounded.

While "High Dive" sets mathematical abstraction and bodily activity in conflict, "Letter I" makes an anthropological point about being in the world, or space, and the disorienting effects that modern mathematical description may have on belonging.

> Only, have we space, common-sense in common,
> A tribe whose life-blood is our sacrament,
> Physics or metaphysics for your showman,
> For my physician in this banishment?
> Too non-Euclidean predicament. [15–19]

Non-Euclidean geometry can be read here as a thing altogether alien to embodied experience of space. Something non-Euclidean is against common sense, obverse to line 15, where space becomes synonymous with common-sense. Although densely packed, the animating feeling of the poem would have been familiar to readers at the time. For those who took the mathematical world-picture seriously, the loss of the Euclidean foundation meant the dominant world-picture was no longer grounded in common sense. The old space, described in the first stanza, as "that net-work without fish" [3], could be grasped by Cartesian geometry, based on Euclidean rules—Empson notes, "the network without fish is empty space which you could measure, lay an imaginary net of co-ordinates over, opposed in verse 3 to the condition when two stars are not connected by space at all".[103] 'Measureability' in the pre-Einsteinian world-picture entailed familiarity of access, and the ability to agree on common facts about phenomena, which in turn allowed communication. Of course, recalling "High Dive", we can say that the ability to measure

[101] Veronica Forrest-Thomson, *Poetic Artifice: A Theory of Twentieth-Century Poetry* (New York: St. Martin's Press, 1978), p. 233.
[102] Empson, *Poems*, p. 188.
[103] Empson, *Poems*, p. 211. The image in the first stanza, that "the system of stars is floating in an ocean", was available to Empson from Arthur S. Eddington, *Stars and Atoms* (Oxford: Oxford University Press, 1927), p. 67 [as has been noted by Haffenden].

something does not make it familiar to embodied experience. So, on a spectrum of alienation, "Letter I" takes the poet even further than the very basic grip of familiarity that he, perched on the diving board, had enjoyed.

Our ability to do physics and metaphysics is a reminder that the world is—a structural realist might have said, at least some aspects of the world are—at base legible. F. M. Cornford says "primitive religion assumed a fusion of Nature (external nature as revealed through the senses) and the supernatural, so allaying our troubling modern split between physics and metaphysics: 'The Nature' of which the first philosophers tell us with confident dogmatism is from the first a metaphysical entity".[104] The poem, pointing to a time before this 'troubling modern split', lumps physics and metaphysics together to suggest that our attempts to secure knowledge about Nature were but consolations to our desperately curious and compulsively social species. Tribes have clung together as political societies amid the wayward cruelties of Nature by extrapolating from the mutual understanding between members of different tribes a belief about the common rationality or legibility of the world: and so, the first act of justice, *dike*, in the Greek city states was to agree upon the tribal laws (premised on their understanding of natural law) that the polis, swollen with new tribes, would inherit.[105] To understand one another, we must share experience of the world; this requires the world to accord with reason and so, be bound by certain regularities. Our striving to communicate and do geometry—to measure earth—have thus been mysteriously connected. This vital connection seems severed after relativity which is interpreted or even dictated to a public in non-Euclidean terms. If the tribal source of physics, where mutual understanding depends on perception of regularities, were accepted, a dominant non-Euclidean world-picture that does not correspond to common-sense perception of phenomena would break down communication and social bonds.

The idea that the new mathematics creates a great gulf between the intellectual and political abilities of man is a common theme in the poems. As Willis noted, "Empson's analogies from mathematics are usually post Newtonian and are used to explore modern man's predicament in a non-Euclidean universe."[106] As we have discussed, philosophers of science and

[104] F.M. Cornford, *From Religion to Philosophy: A Study in the Origins of Western Speculation* (New York: Longmans, Green & Co., 1912), p. 123; Cornford is mentioned as a direct inspiration for this poem in his notes (Empson, *Poems*, p. 211).

[105] Michael Oakeshott, *Lectures in the History of Political Thought*, eds. Terry Nardin and Luke O'Sullivan (Exeter: Imprint Academic, 2006), p. 98.

[106] J.H. Willis, *William Empson* (New York: Columbia University Press, 1969), p. 25; this idea is also explored in more detail in Katy Price, 'Flame Far Too Hot: William Empson's Non-Euclidean Predicament', *Interdisciplinary Science Reviews*, 30.4 (2005), 312–22.

mathematics in the early twentieth century critiqued the inordinately mathematical language of the new theories in physics, which described a world at odds with the structures of experience. Empson's poems avow such critiques; but they use these recognizable views of modern science to make deeper claims. The paradox of a finite and unbounded universe would put a stopper on the grand speculations of philosophy, but its non-Euclidean foundation could threaten human bonds.

"Dissatisfaction with Metaphysics" visualizes a mind, to cope with these uncertainties, projecting a hollow, fictional world onto reality.

> Two mirrors with Infinity to dine
> Drink him below the table when they please. [6–7]

The two mirrors facing each other will reflect one another infinitely many times. But these reflections begin not from substance but reflection itself. There is neither origin nor end; each faint replica leads to a smaller image of disguised emptiness. This is the "self-conscious mind", according to Empson's notes.[107] As consciousness reflects and then further reflects upon its own reflections, it wanders from awareness of immediate circumstance. The perceiver of the present, who actually *is*, cannot be known without setting off this infinite regress of reflection upon reflection (symbolized by the two mirrors). The perceiver is simply part of an event; no amount of reflexive thought can disclose being as is when caught in the swell of activity, having forgotten itself, 'when they please'—just as no degree of differential convergence can reveal the actual instant. 'Infinity', personified with a capital 'I', has gone from mere fiction to a set of reflections, a concrete fiction (like Cantor's infinities of varying size) that looms over the margins of speculative reflection, threatening to replace life with thought. But Infinity is drunk below the table, put in its place, out of sight, when deeper urges are allowed sway. The injection of activity in the verb 'drink' followed by an ambient suggestion of a drinking contest at a raucous pub creates an atmosphere both exciting and obliviating. The fool who acts prevails.

Empson seems to have been working on the idea of infinite deferment in his poems from his early undergraduate days. He says in a *Granta* review (1927): "Any simple relation which may be proposed between subject and object leads at once to an infinite regression; [...] The only analogy so far

[107] Empson, *Poems*, p. 170.

proposed is that of a mathematical limit; it is mysterious".[103] This chimes with the debates at Cambridge about the role of unobservables in physical theory: are theories that build on prior theories slowly becoming theories only of theories, like reflections upon reflections? The poem also evokes the Buddhist belief in the emptiness, Śūnyatā, of phenomena, Rūpa.[109] A consequence of the belief in the illusory nature of form ("also one of his [The Buddha's] chief doctrines") is "the denial of the existence of *atta*, the self, as a permanent unit".[110] Thus, the Buddhist and the mathematical world-pictures met and coiled for Empson at their roots and issue here in the image of reflecting mirrors.[111] But apart from the fool and the Buddha, the line about drinking with Infinity weaves in the 'love of activity for its own sake' as well, by how the suggestion of merriment seems to undermine self-reflection. In a letter to Richards (1933), a few years after writing these poems, Empson says, "Buddhism is relevant to what I am trying to think about, in connection with the Value business."[112] Empson manages to find a complex balance between these ideas through one of the central themes in the *Poems*: that "most of the puzzles of philosophy [...] can be stated as infinite regress".[113]

2.2.3 The mirror image

The two mirrors and Infinity bring us to the final type of mathematical metaphor in Empson's works, the mirror image. The device, broadly speaking, works by showing two things as mirror images of one another but ultimately, in a mysterious way, as different. These metaphors tend to forward the argument that when one thing is shown to be another in a satisfying way, we must always be suspicious: it might well be the imposition of a theorist or a product of propaganda.

In the *Structure of Complex Words*, Empson uses what he calls 'equations' to depict how a complex word communicates its multiple senses; to describe

[108] Empson, *Granta*, p. 33.

[109] These concepts are discussed in the Pali Canon, which we know Empson was reading well before beginning work on his text on Buddhist art (Empson, *Buddha*, p. 9.) and from the fact that the Fire Sermon is in this canon.

[110] Empson, *Buddha*, p. 14.

[111] The use of the mirror as an analogy to Buddhist self-abnegation is made more directly in a poem we discussed in Chapter 1, Yeats's "The Statues" [20–4], in *The Collected Poems of W.E. Yeats* (London: St. Martin's Press, 1958), p. 322.

[112] William Empson, *Selected Letters of William Empson*, ed. John Haffenden (Oxford: Oxford University Press, 2006), p. 61.

[113] From unpublished essay (1935) rendered in Haffenden, *Poems*, p. 174.

"how the two meanings are imposed and which order they are given".[114] Of interest to us here is the Type IV equation which proposes that if 'A is like B', then 'B is like A'. The equivalence can be abbreviated like so: 'A⇔B'. Since the meanings of words in language depend so much on the order—imposed by usage—in which they occur to the reader, this type of equation, where two meanings are simultaneously present, comes about only when enforced by authority. In other words, it is an artificial strain on the natural way of meaning-making. An example Empson gives is Hooker's use of the word 'law' to mean 'both human and divine law'. In other words, some human laws have all the force of divine edicts, and vice versa—an equipotence registered somehow, by the formula 'human law⇔divine law'.[115]

Empson uses an analogy from plane projective geometry to illustrate the relationship between the terms in equations of the fourth-type: "cases can easily be found in which 'B' is like 'A' as well as 'A' like 'B' [...] e.g. a quadrangle and a quadrilateral".[116] In plane projective geometry, a complete quadrangle in a plane is formed by six lines drawn through four points in such a way that every pair of points is connected by a line and no trio of points lies on the same line. A quadrilateral, on the other hand, has four lines on a plane that pass through six points with no trio of lines passing through the same point:[117] see Figure 2.2.

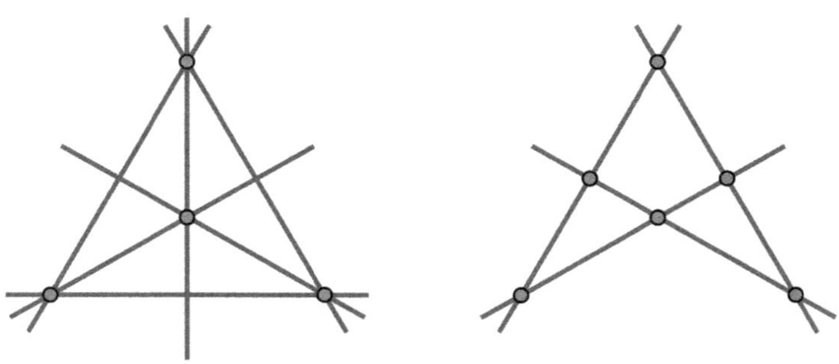

Figure 2.2 Quadrangle (left) and quadrilateral (right).
Source: Image distributed under a CC-BY 2.0 licence, https://commons.wikimedia.org/wiki/File:Complete-quads.svg

[114] Empson, *Complex Words*, p. 46; The specific number of equations is as irrelevant as Empson's ironic claim that there are 'seven' types of ambiguities. As Michael Wood puts it, "if there are seven types, there could be more, and probably are" (Michael Wood, *Literature and the Taste of Knowledge* (Cambridge: Cambridge University Press, 2005), p. 96).

[115] Empson, *Complex Words*, p. 52.

[116] Ibid., p. 51.

[117] Eric A. Lord, *Symmetry and Pattern in Projective Geometry* (London: Springer, 2013), p. 18.

Any proposition in plane projective geometry is as valid as its dual statement in which the terms 'line' and 'point' are interchanged, making them mirror images of one another. A proposition about complete quadrangles, for instance, can be proven by demonstrating the dual statement about quadrilaterals, and vice versa.[118] It seems to be Empson's contention that, within a poem, semantic equivalences of the fourth type allow cognately dual transcriptions of statement.

The extension of projective geometry to characterize the workings of metaphor, how a metaphor uses one thing to talk of another, however, doesn't seem to work. A contemporary critic Scott Buchanan uses projective geometry to diagram the structure of a metaphor used by S.J. Brown. Brown begins with the saying, "you should try to root out your faults one by one".[119] According to Brown, the metaphor in the statement above can be visualized thus: 'a:b = x:y', such that 'faults: soul = weeds: garden'.[120] Empson first points out that Brown seems to miss the entire point of the homily. It is not to suggest that faults can be like weeds in a garden, but to provide a technique for ridding the soul of faults: namely, to do it one-by-one. Rooting—the only verb used in the metaphor and missing in Brown's scheme—is the process by which weeds are removed one-by-one as opposed to digging, which is more indiscriminate.

Empson says, Brown's proportionality scheme "has been put very dramatically by Scott Buchanan in [...] *Symbolic Distance*, which uses the analogy of projective geometry all the way through". Buchanan visualizes the metaphor as follows:

> In Figure 2.3 let, 'O' be a point of projection with rays running out from it to cut a transversal (T) in A, B, C, and D. The intervals thus cut off on the transversal, AB, BC, CD, will have certain relations to each other which can be expressed in ratios, and these ratios may be combined to make more complex ratios of ratios. Draw another transversal (T2) cutting all the rays at some given angle with the former one. There will again be intervals on this transversal cut off by the rays, and the relations between these intervals may be expressed in ratios and ratios of ratios. When this has been done it will be found that the set of ratios of ratios on the first transversal are equal to the set of ratios of ratios on the second transversal.[121]

[118] Lord, *Symmetry*, p. 19. [119] Empson, *Complex Words*, p. 336. [120] Ibid.
[121] Scott Buchanan, *Symbolic Distance in Relation to Analogy and Fiction* (London: K. Paul, Trench, Trubner & Co, 1932), pp. 57–8.

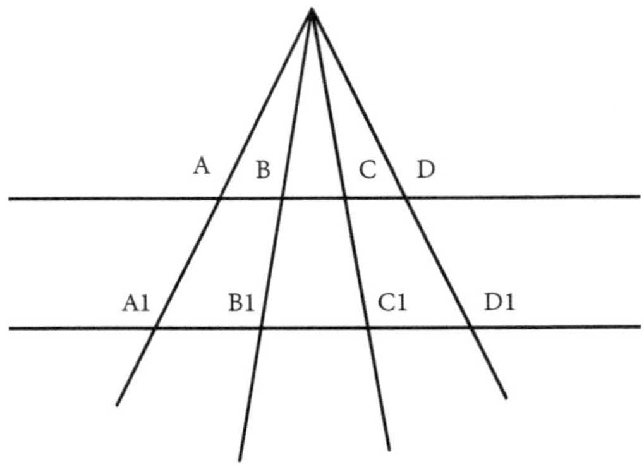

Figure 2.3 Buchanan's visual analogy for metaphors.
Source: Scott Buchanan, *Symbolic Distance in Relation to Analogy and Fiction* (London: Kegan Paul, Trench, Trubner & Co., 1932), pp. 57–8.

This translates to 'AB: BC = A1B1: B1C1,' which is an approximation for 'faults: soul = weeds: garden'. The problem with Buchanan's visual analogy is that the dominant sense becomes the difference in *size* between AB and A1B1, making the process of meaning transfer in a metaphor necessarily visual: the trendy imagism of the time thereby drawing Buchanan into fuzzy regions of psychology. Buchanan calls the transfer between AB and A1B1 "expansion" and the reverse, A1B1 to AB "condensation",[122] and resorts to Freudian and Jungian fictions to justify the assumptions. The process of transfer, he says, begins with "nonverbal archetypes", which then "demand verbal expansion". For instance, the physical relationship between 'mother and son' becomes analogous to the emotional relationship, which is then symbolized as 'AB: BC' along 'T'. This undergoes a projective transformation, or "expansion" at the level of 'family and son' at T2.[123] Empson, profoundly irritated by such psycho-speak, which he felt was a grand haul of empty words, retorts in his characteristic style: the equation formed by the representation (AB/A1B1) is the equivalent of "naughts divided by naughts" (0/0).

> I think at this point we must become irritated with the claims of the subconscious mind no less than with those of the Logos. We had better stick to what the fool of a conscious mind is doing, if we can be sure that it is so distinct from our unconscious wisdom.[124]

[122] Ibid., p. 59. [123] Buchanan, *Symbolic*, p. 64. [124] Empson, *Complex Words*, p. 341.

The fool groping in the present is a fitter topic of study, let the profound fictions of the unconscious remain as unapproachable as infinity. And so, in one go, the two main trends of literary criticism of the time that irked Empson, psychologism and imagism, are tackled by attacking Buchanan's arcane scheme.

In but one area of thought do both the thing and its mirror image become real, and that is the seventh-type ambiguity. We will recall that the seventh type occurs when a phrase or word means opposing things at once. The equation Empson invents to describe the ambiguity is: If a = a1, the meaning is p; if a = a2, the meaning is –p. In the equation, 'a' is the verbal unit as given in the poem, a1 and a2 are possible senses of a, and p and –p are the ultimate meanings of the phrase.[125] We will see Empson himself exploiting this mode of ambiguity in our discussion of "Letter V" but the young Empson hints that he has found one in George Herbert's "The Sacrifice".[126]

> Lo here I hang, charged with a world of sin
> The greater world of the two; for that came in
> By words but this by sorrow I must win:
> Was ever grief like mine?

These lines are uttered by the Christ on the cross. He could be burdened by humanity's collective guilt while nailed to the cross or be the one ultimately responsible for their sins. This popular dichotomy of seventeenth-century theodicy, exemplified, for instance, in Tintoretto's depiction of the crucifixion, is resolved in the poem at the moment upon the cross when the Christ becomes at once man and God: He who induced sin into this world dies for it. So, if 'charged' = laden, then p; and if 'charged' = accused; then -p. Both equations contain the same word, but we are left with opposite senses, such as benevolent and wrathful, scapegoat and tragic hero.[127]

For Empson, the idea of opposite things being true at once, rather than an achievement of philosophical questing, is an emblem of the unattainability of truth. Janus thus becomes the god of English poetry: he glints through the 'crux' upon which Herbert's Christ hangs, signifying both burden and salvation. Paradox is just so the crux of the problem and crux of matter in Empson's deepest musings on the nature of things: of which his poem "Doctrinal Point", our next subject, is the climactic episode.

[125] Empson, *Seven Types*, pp. 196–7.

[126] George Herbert, *The English Poems of George Herbert*, ed. Jacula Prudentum (London: Rivingtons, 1871).

[127] Empson discusses this poem in Empson, *Seven Types*, p. 233.

2.3 The God Approached

We have learnt that among his many idiosyncrasies, Empson was mainly committed to ignorance, and apt to flee from a settling doctrine, deeming any pretension to ultimate answers theological, even tyrannical. He uses mathematical images in the main to explore epistemological limits, trading in the effects of unresolved paradoxes, the mode of cogitation that seemed most to excite the young reviewer of *Granta* and soon haunt the author of the *Seven Types.*[128]

The poet Kathleen Raine recounts from her undergraduate days at Cambridge,

> the impression he [Empson] made upon me—as upon all of us—of contained mental energy [...] This impression of perpetual self-consuming mental intensity produced a kind of shock; through no intention or will to impress, for William was simply himself at all times [...] mild, impersonal, indifferent to the impression to the point of absent-mindedness. Nevertheless his presence spellbound us all. His shapely head, his fine features, his eyes, full lustrous poet's eyes but short-sighted behind glasses and nervously evading a direct look [...] was the head, in any gathering, that seemed the focus of all eyes.[129]

Favoured with enough beauty and misfortune to know the plights of an indifferent muse and his supplicating bard, the young Empson, in the words of I.A. Richards, was able to take "the sonnet-cycle as a conjuror takes his hat"[130]—but he left as fugitive a record of himself in verse as in the memories of his brilliant contemporaries. Of the torments of youth, the miracle of visual delight, the remoter charms of thought, the disappointment assured in all pursuit, I believe "Doctrinal Point" to be his highest expression. Sadly little has been said of its virtuoso performance.[131] A thoroughgoing close reading may have seemed unnecessary to most critics, for the poem appears deceptively easy to summarize, as saying doctrinaire or dogmatic views, as science seems nowadays to hold, are wrongheaded and corrupting. I hope in these

[128] See Section 2.1.

[129] Kathleen Raine, *The Land Unknown* (London: Hamish Hamilton, 1975), pp. 44–5.

[130] I.A. Richards, 'William Empson', *Furioso*, 1/3 (1940), supplement following p. 44.

[131] See Gardner and Gardner, *God Approached*, p. 147; Benjamin Kohlmann, *Committed Styles* (Oxford: Oxford University Press, 2014), p. 71; Kitt Price, 'Empson's Einstein: Science and Modern Reading', in *The Cambridge Companion to Literature and Science*, ed. Steven Meyer (Cambridge: Cambridge University Press, 2018), pp. 97–117.

pages to show why Olson's dictum that a poem "means exactly what it says" finds a votive instance in "Doctrinal Point".[132] Its implications lie beyond paraphrase (Empson's own attempts included), for it is in the act of unfolding that the poem brings its bleak vision to existence.

The New Critical flavour of these suggestions is intended—for much opinion has gathered around Christopher Norris's view that Empson's abiding critical achievement lies in having dissolved the boundaries between poetry and prose. Norris accepts the complaint lodged by the New Critics themselves that the method of the *Seven Types* applies equally to "a piece of infinitely qualified prose"[133]—Norris, of course, presents this as Empson's distinction. He argues that, to Empson, "the language of poetry is continuous with—directly qualified by—the expository language of prose".[134] This view is held to the present, more or less unchanged: Michael Wood, for instance, says "Norris puts this very well when he says that 'Empson's books all seek, in different ways, to make terms between poetry and the normal conditions of language and common-sense discourse', and that ambiguity, for example, 'belongs to a normal, not uniquely poetic order of thought and language'".[135] I do not assume the burden of disputing the settled opinion on 'Empson's books'; but in what follows, we proceed under the assumption that the contemporary view is inadequate to understanding his poetry. When his poems were published in the 1930s, reviewers regarded them as "crosswords", "mosaics",[136] and "riddles"[137]—certainly not confused with prose. The notions of solving the riddle or rearranging the mosaic better approximate the critical labour demanded by his poems.

"Doctrinal Point"

1. The god approached dissolves into the air.
2. Magnolias, for instance, when in bud,
Are right in doing anything they can think of;
Free by predestination in the blood,

[132] From 'Projective Verse' in Charles Olson, *Collected Prose*, eds, Donald Allen and Benjamin Friedlander (Berkeley: University of California Press, 1997), p. 240.

[133] John Crowe Ransom, 'Mr Empson's Muddles', *The Southern Review*, 4 (1938/9), pp. 32–9; p. 33; Frank Kermode, *English Pastoral Poetry from the Beginnings to Marvell* (London: George G. Harrap, 1952).

[134] Christopher Norris, *William Empson and the Philosophy of Literary Criticism* (London: Athlone, 1978), p. 5.

[135] Michael Wood, *On Empson* (Princeton: Princeton University Press, 2017), p. 15.

[136] Louis MacNeice, 'Mr Empson as Poet (review of poems)', *New Verse*, 16 (1935), 17–18, p. 17.

[137] I.A. Richards, 'Empson's "Poems"', *Cambridge Review*, 57.1399 (1936), p. 253.

Saved by their own sap, shed for themselves,
Their texture can impose their architecture;
Their sapient matter is always already informed.

8. Whether they burgeon, massed wax flames, or flare
Plump spaced-out saints, in their gross prime, at prayer,
Or leave the sooted branches bare
To sag at tip from a sole blossom there
They know no act that will not make them fair.

13. Professor Eddington with the same insolence
Called all physics one tautology;
If you describe things with the right tensors
All law becomes the fact that they can be described with them;
This is the Assumption of the description.
The duality of choice becomes the singularity of existence;
The effort of virtue the unconsciousness of foreknowledge.

20. That over-all that Solomon should wear
Gives these no cope who cannot know of care.
They have no gap to spare that they should share
The rare calyx we stare at in despair.
They have no other that they should compare.
Their arch of promise the wide Heaviside layer
They rise above a vault into the air.[138]

Empson described the poem as being about "Deism [...] If you arrive at the point of grace, you no longer require a personal god."[139] What precisely is meant by 'grace' will emerge from the poem's conceit, but the hint of aporia in approaching a god is enough to begin with. Loosely put, the poem opens with a paradox, asserting with certitude the impossibility of certainty. The line is presented as a thesis preceding an argument. I am going to liken its odd structure, of a single-line declarative paragraph followed by staves of elaborate philosophical argument, to medieval disputation. The argument itself is more like metaphysical poetry in its absurdity; but Una Nelly has

[138] Empson, *Poems*, p. 39; Haffenden says the poem was written some time before 1935 (*Poems*, p. 277). The poem is published with the express permission of Curtis Brown, the literary estate of William Empson.

[139] Qtd in John Haffenden, *William Empson: Against the Christians* (Oxford: Oxford University Press, 2006), p. 619.

argued that the medieval technique of disputation was a model even for Donne.[140] One might thus profitably look beyond the usual seventeenth century paradigms for Empsonian argufying and imagine St Anselm posing a prompt such as initiates "Doctrinal Point"—"God created the world in Time", say[141]—and setting his students off to debate. A case presented in the medieval university, not too unlike a modern university debate, would customarily begin with a postulate about the world and proceed to draw conclusions favourable to its compact.[142] The second stanza of "Doctrinal Point" likewise prosecutes its argument through a series of postulations about magnolias, starting in line 2 with the forensic register of 'for instance'. Before interpreting the argument and its conclusions, however, we must recast its mysterious thesis.

2.3.1 I am lying

By calling the first line a paradox, we do not mean the 'language of paradox', which was to Brooks the "language appropriate and inevitable to poetry".[143] The paradox Empson sets is logical and more tenacious; after all, it is a "poem whose subject is being puzzled".[144] Paradoxes depend for their contradiction on the definition of constituent terms. I mean to show that the first line, with constituent terms carefully defined, is a semantic analogue to the Liar's Paradox; and why reading it so is integral to 'how the poem means'.[145]

The Liar's Paradox works through self-reference. It is said to have originated in St Paul's epistolary account of Epimenides the Cretan who said all Cretans are liars. Its simplest version is, "I am lying". If I am lying, then I am telling the truth, and vice versa. In logical terms, the sentence says: "there is a

140 Una Nelly, *The Poet Donne: A Study in His Dialectic Method* (Cork: Cork University Press, 1969), pp. 53–5.

141 Alex J. Novikoff, *The Medieval Culture of Disputation: Pedagogy, Practice, and Performance* (Philadelphia: University of Pennsylvania Press, 2013), p. 35.

142 Ibid., p. 141.

143 Cleanth Brooks, *The Well-wrought Urn: Studies in the Structure of Poetry* (New York: Reynal & Hitchcock, 1947), p. 1.

144 Empson, *Poems*, p. 278.

145 The phrase 'how the poem means' is used in Jonathan Culler's distinction between hermeneutics, what it means, and poetics, how it means (Jonathan Culler, 'Hermeneutics and Literature', in *The Cambridge Companion to Hermeneutics*, eds. Michael N. Forster, and Kristin Gjesdal (Cambridge: Cambridge University Press, 2019), pp. 304–9).

proposition I am affirming and which is false", or "I assert p and p is false".[146] But before interpreting our line in these terms, we must note that the terms of a logical paradox must be shorn of connotation; but linguistic ambiguities are essential to poetry's meaning-making process: so, if an attendant sense of 'god' should render line 1 a falsifiable statement, no paradox obtains.[147] For instance, if 'the god' referred to is Boreas or Zephyr, we might picture him dissolving to mist or wind. We must first show, therefore, how taking account of likely ambiguities in the line will lead us to a paradox.

'The god' approximates to 'The truth'. Owing to the curious juxtaposition of an uncapitalized 'god' preceded by definite article, 'the', the term refers neither to the God of monotheism nor a pagan god. Because the immediate referents of the word are not in play, we must infer its meaning by gathering the plausible attributes. We can with reason assign two positive attributes and one negative: being god, it must (1) be final/ultimate; being *the* god, it must (2) be unique/absolute; and we have explained why it (3) is not a deity. It follows from these premises that 'the god' is the doctrinal point realized, the absolute and final verity of existence, by whichever method—physics or metaphysics—it has come to be known.

The word 'dissolve' stands out in an otherwise lofty Elizabethan line ending. There are more poetic ways to express the unapproachability of 'the god': 'vanishes', 'mingles', 'is consumed'. These, indeed, are senses connoted by 'dissolve', as in William Jones's 1769 phrase, "each gay phantom was dissolved in air".[148] But 'dissolve' indelibly suggests to a post-industrial reader the diffusion of "molecules of (a solid or gas) in a liquid so that they are indistinguishable from it".[149] The chemical dissolution of 'the god' is a tad fantastic—it projects no ready mental image or even logical sense. So, we again look to supplementary senses.

We may begin by modifying the subject to 'the *scientific* truth'. We are prompted in stanza 4 to think of the doctrinal point not from a philosophical or theological lens, as was the case in the medieval university, but a scientific one. The Gardners identify in Eddington's discussion of quantum physics a potential subject for line 1, namely, "Our conception of substance is only

[146] Alfred N. Whitehead and Bertrand Russell, *Principia Mathematica, vol. 1* (Cambridge: Cambridge University Press, 1910), p. 62.

[147] Based on the how linguistic statements can be considered logical in Gottlob Frege, 'Ueber Sinn und Bedeutung', *The Philosophical Review*, 57 (1948), p. 209.

[148] William Jones, 'The Palace of Fortune', *The Works of Sir William Jones*, ed. by Lord Teignmouth (Cambridge: Cambridge University Press [1807] 2013), p. 418; this particular analogue has not been identified by critics.

[149] *OED.*

vivid so long as we do not face it. It begins to fade when we analyse it."[150] While Empson was an undergraduate at Cambridge, Heisenberg's work had established for a wider intellectual culture a definite epistemological limit in our grasp of the basic units of material existence. Eddington writes that the "exact position with exact momentum [of a quantum entity] can never be discovered by us" because the very ascertaining of one alters the other:[151] This formulation understates Heisenberg's extremer view—which is more in agreement with god's dissolution—that both quantities do not even exist, each only coming into existence in the act of measurement. Taking facts of the quantum world as the final residue of reality signified by 'the god', mathematical physics seems to face an impudent nature at the smaller limits taunting and escaping its grasp: the god approached dissolving into the air. As in the apple of knowledge bending away from Tantalus, *hyle* resists and eludes form, sets limits to the ambitions of knowledge. But limiting the line to the findings of quantum theory alone is not commensurate with the larger ambitions of stanza 4.

We might also read 'dissolve' as 'dis-solve': a purportedly achieved solution coming undone: that is, 'the truth' reverting to mystery when approached too closely. An approximation of the opening line might thus read, 'The truth approached dis-solves'. Like the Liar's Paradox, the line first affirms a proposition: that truth is confounded by being solved or when approached, reverts to a riddle. But the declaration, by virtue of its own grammar, becomes the very doctrinal point it denies. If the only quality of 'the god' that can be stated is its unapproachability—for nothing else remains true about a dissolving god—then that very fact is its truth. So, in stating its unreachability, the god's nature has been discovered, and so, in one sense, approached and attained: like the Liar's Paradox, therefore, the opening line both denies and affirms itself. This paradoxical nature of divinity is an old problem wrestled with in apophatic theology but has been extended by Empson thus to the divinatory pretensions of mathematics in its supposed capture of *physis*, which becomes clear by stanza 4.

A god whose molecules, indistinguishable from the air, disappear before the approaching mind also hints at a kind of pantheism; perhaps the atheist Empson also expects his readers to have in their minds Spinoza's *Deus sive Natura* (God-or-Nature).[152] In *Pastoral*, Empson, pointing to a similar line in Marlowe, "their souls are soon dissolved in elements", says, "the doubtful

[150] Gardner and Gardner, *God Approached*, p. 184.
[151] Ibid., p. 154
[152] In reference to the idea that God is Nature, or the totality of existence in Benedictus Spinoza, *Ethics*, ed. and trans. Edwin Curley (London: Penguin [1677] 1996).

word *dissolved* allows pantheism to be still present at the end".[153] He would even suggest decades later that Spinoza's God-or-Nature is one extreme of Milton's God .[154] Line 25 of our poem reads: 'Their arch of promise the wide Heaviside layer'. The rainbow, which was to the Israelites the arch of promise, Heaven's guarantee of their covenant, is equated to the Heaviside layer, an atmospheric layer of gas protecting plants from harmful radiation. The grace of god is thus shown to lie, at least for magnolias and scientific man, within the natural order—not in Manna descending from above. The one Substance of the *Ethics,* like 'the god' of line 1, is unattainable to mind; through investigation, we may discover its many properties, but never its true nature—for "substance is necessarily infinite".[155] Both gods are immanent yet apophatic, remaining tantalizingly elusive.

If, as Norris says, we must not read Empson's poetry as distinct from prose, the line's acrobatics would be lost. A prose statement is an instrument to convey meaning (as is implied by Norris); but meaning emerges here from the meaning-making procedure of the line. According to the New Critical standards, which Empson is supposed to have resisted,[156] poetic statements can be distinguished from prose by virtue of self-referentiality and paradox—both of which have proven essential to unravelling our first line.[157] This is not to suggest a reversion to New Critical doctrine for the study of Empson's poems. It is to stress what we have in the previous section, that the attribution of any stance, New Critical or anti-New Critical to Empson is unwise.

2.3.2 The summer's flower

So much for paradox. To make good our following gambit that the paradox is key to the poem's riddle, we must continue with a full reading. The disputation begins, however strangely, with a comparison between the lives of magnolia and man. Alien to us, magnolias' 'texture can impose their architecture'. The reference of 'texture' is material-and-surface or material appearance. And because architecture is what is over and above the texture of material, that is, its non-material aspects, the design, meaning, and suchlike, it can be taken as referring to form and structure. Distinctions in this vein are common to the

[153] Empson, *Pastoral,* p. 73 (this is noted as a source by Haffenden, *Poems,* p. 279).
[154] William Empson, *Milton's God* (London: Chatto &Windus, 1961), p. 143.
[155] Spinoza, *Ethics,* p. 10. [156] Norris, *Philosophy,* pp. 4–5.
[157] For a list of Brooks's "articles of faith", which he implies are held by other New Critics, see Cleanth Brooks, 'The Formalist Critics', *The Kenyon Review,* 13.1 (1951), 72–81, p. 72.

Aristotelian paradigm of medieval disputation.[158] Recast in the binary of matter and form, we might render the line as 'material cause can determine their formal cause'. The magnolia's architecture is contained entirely in its texture, but there is a gap between architecture and texture in man. A human being's form is their character, their moral nature, their deeds—it does not suffice for man to simply subsist as matter, however radiant he maybe. The gap between our texture and architecture, one might venture, stands for the original rupture between mind and matter—at the birth of higher consciousness, say, or exile from Eden, whichever the reader prefers—that forces us to strive with and against nature to discover the moral way—'the effort of virtue'. But flowers seem good whatever they do.

Line 7 is another statement that is logically confusing. Let us consider a distant analogue, Russell's popular sentence that flouts the law of excluded middle: "The present king of France is bald."[159] France, having no king at the time, could not have had a bald king; but if we take the sentence as false, the negation posits a hirsute king, equally untrue. Line 7 reads: 'Their sapient matter is always already informed'. Just as there is no 'present king', there is no 'sapient matter'. Before writing it off as poetic licence, the oxymoronic juxtaposition of 'sapient' and 'matter' and the factual tone are performative enough to enlist further analysis.

Taken as 'wise inanimate-stuff,' 'sapient matter' is like the 'current King of France': sapient entities maybe 'already informed' but matter is not sapient. Empson's interest in Eastern mysticism—where one might imagine matter burdened with soul—provides too simplistic a resolution. The stress of the metre is on 'their': *their* sapient matter, as opposed to our sapience which resides in our architecture, our mind or soul. That is, an equally potent gloss of 'sapient' is the character of man, the *homo sapien* shown hitherto to lack 'foreknowledge' in the poem. The paradox can however be resolved by noticing hidden senses in 'sapient', buried by its technical use. 'Sapient' derives from *sapēre*, a word for 'wise', but, more readily in the vulgate, for 'having a taste or savour',[160] sapid, a sense employed in Renaissance poetry, as in Alexander Hume's "Of every substance sapient, the sapor and the taste".[161] The poem, however, does not rely on knowledge of arcane etymology. The sense of 'savoury' is proleptically transmitted to 'sapient' by the

[158] Novikoff, *Disputation*, pp. 106–7.

[159] Bertrand Russell, 'On Denoting', *Mind*, 14 (1905), 479–93, p. 479.

[160] *OED*; Haffenden has already made the connection to the Italian but draws none of the wider implications presented here.

[161] Alexander Hume, *Of Gods Benefites Bestowed Vpon Man* (published privately, 1599).

word 'sap' in the fifth line of the stanza. There is firstly the visual repetition of sap in sapient; but the image in the fifth line of a bud safely cocooned in sap readily allows a reading of sapient matter as sap-laden matter. Sapient as 'sappy' resolves the seeming contradiction: the sap in the young buds contains biological entelechy—'foreknowledge'—that already tells them how to flourish. Without the perlocutionary effect across lines—a central feature of poetry—'sapient matter' would not have invoked the germ tissue in sap, as genes merely contain *information* whereas sapient entails *wisdom*.[162] But the gesture towards sap, the reproductive fluid of flowers, forces the consideration of a new, more interesting puzzle: why, among all the species, do the cells of the magnolia endow wisdom: and prefiguring the discussion of scientific epistemology in the fourth stanza, when does information become wisdom? As the magnolias 'in bud / Are right in doing anything they can think of', their biological instincts seem unfailingly correct. One explanation is that flowering plants nurture the birds and the bees; here, the word 'sapient' almost achieves the condition of Empson's 'mutual metaphor' (Type IV equation), whereby two senses are equally imposed by context: the magnolia's matter is 'tasty', and by being tasty, it seems to know how to live and let prosper, making it 'wise'.

So magnolias, by dint of their tissue, are enviably sapient; miraculously, their texture is both good and beautiful, as Ruskin might expect from architecture. As to the differences between magnolias and human beings, it rests on how each comes to acquire wisdom. The couplet, 'The duality of choice becomes the singularity of existence; / The effort of virtue the unconsciousness of foreknowledge', suggests a definition of wisdom as accord with nature: so close an identification with the way of the world as to become indistinguishable from it. From stanzas 2 and 4, we gather that man before the age of science required effort to discover the noble choice, whereas magnolias have always been right, whichever way their foreknowledge bade—they enjoy singularity with existence. The first and last lines of the poem, by ending alike, form a helpful equation. Magnolias 'rise above a vault into the air', the very air into which the god of the first line had dissolved, foreclosing man's approach.

[162] 'Germ plasm' would be the historically accurate term [Ernst Meyer, *The Growth of Biological Thought* (Cambridge: Harvard University Press, 1982), p. 700]. However, "modern gene theory nonetheless retained the genetic determinism that 19th century 'germ plasm' theorists relied on to explain the intergenerational transmission of evolutionarily adaptive characteristics" [D.S. Moore, 'Espousing Interactions and Fielding Reactions: Addressing Laypeople's Beliefs about Genetic Determinism', *Philosophical Psychology*, 21 (2008), p. 335]; also see Lily Kay, *Who Wrote the Book of Life* (Stanford: Stanford University Press, 2000).

The pantheistic god, 'dissolved' in Nature, impenetrable to mind, is briefly approachable to the righteous magnolia, the mind of god even knowable, for indistinguishable from its molecules, the 'always already informed' body coded in accordance with the divine. Such special regard for the magnolia means we mustn't stretch its significance to anything beyond 'beautiful flowers in summer bloom', to avoid an added implication of biological determinism that might reduce the second stanza to a crude Darwinian argument about hereditary information fitting all organisms to nature. Empson, in his sympathetic review of the sceptical biologist Needham, said,

> If you are using the scientific method, you must assume that determinism can always be applied to the matter in hand; if however you are not using the scientific method [...], then you must assume that determinism does not apply to the matter in hand. These actions are not conflicting but complementary [...]. [I]f the organism shows purpose in fitting its environment, so does the environment show purpose in being such as can sustain an organism.[163]

The magnolia is chief among creation not for its evolutionary 'fitness' but because Nature itself seems to conspire with it to bring about, however briefly, a mystical union. Anyone who has appreciated flowers in bloom will recognize the sentiment.

By reserving knowledge of, or oneness with, Nature to the magnolia, the second stanza avoids implying that nature has one law, determinism, with which all instinct is in accord. One imagines the zombie worm to enjoy no such favour. In fact, I would submit that the word 'nature' is meticulously avoided in the poem to prevent such extrapolations. In *Complex Words*, Empson compared the word 'nature' to the Sanskrit 'dharma', both of which can be made to carry the same orders of meaning: "nature of an individual" and "the universal law of righteousness".[164] He quotes Aldous Huxley:

> The Sanskrit dharma has two principal meanings. The dharma of an individual is, first of all, his essential nature, the intrinsic law of his being and development. But dharma also signifies the law of righteousness and piety. The implications of this double meaning are clear: a man's duty, how he ought to live, what he ought to believe and what he ought to do about his beliefs—these things are conditioned by his essential nature.[165]

[163] Empson, *Argufying*, p. 528. [164] Empson, *Complex Words*, p. 69. [165] Ibid., p. 69.

Unlike medieval Sanskrit, in English, the order and degree of connection between the senses Mother Nature and an individual's nature has largely been imposed by context; the cosmos and natural instinct are, given man's exile from paradise in religions of the Book, too removed for a direct mutual equation to develop. But just such a doctrine, says Empson, emerges in the poetry of Wordsworth.[166] Whenever Wordsworth used 'nature', he meant both senses equally—to him, 'the nature of an individual' was unfailingly 'the universal law of righteousness'.[167] And so, an agon between Wordsworth and Augustine, between the normative elevation of the state of nature—the belief that man is born ethical and by society corrupted—and the shadow of Original Sin, is at play in the poem once magnolia and human ethics are starkly opposed: it forms the stage for the next question of whether humanity ought be satisfied flawed or covet the floral boon (i.e., by surrendering to the double-edged powers of mathematics).

The poem seems to say: mind cannot be both Nature and nature for its nature is to seek Nature; but following that nature, it will never actually attain Nature, as magnolias briefly do. Only a flower in bloom, with colour lovely to behold, sap sweet to taste, seems right and fair to all: as in Shakespeare's Sonnet 94, "The summer's flower is to the summer sweet / Though to itself it only live and die" [9–10].[168] The unanswerable beauty of flowers had become to the young Empson almost a philosophical puzzle: human nature yearns for everything that flowers flaunt where flaunting itself is untoward in man. He records once "the source of so vivid a pleasure, of so delicate an impression of character, of so universal an appeal, as is found in flowers".[169] In Empson's famous exegesis of Sonnet 94, he explains lines 9–10 as, "sweet to the summer (said of the flower), since the summer is omnipresent and in a way Nature herself, may mean sweet to God".[170] The magnolias in bud may 'flare/ Plump spaced-out saints, in their gross prime, at prayer', but as with the lovely youth of Sonnet 94, "moving others, are themselves" [3] unmoved. Bewitching all who venture in their presence, the tasty, pretty, and wise

[166] Empson, *Complex Words*, p. 70.

[167] T.E. Hulme had in his popular *Speculations* already drawn the connections between the rise of modern science and what he regarded as the final mutation of Christianity into Rousseauian Romanticism (T.E. Hulme, *Speculations: Essays on Humanism and the Philosophy of Art*, ed. Herbert Read (London: Routledge & Kegan Paul, 1924), pp. 60–1); we also know that Empson had read the *Speculations* whilst an undergraduate (Christopher Norris and David B. Wilson, "An Interview with William Empson", in Bevis, *Versions of Empson*, p. 304).

[168] William Shakespeare, *Shakespeare's Sonnets*, ed. Katherine Duncan-Jones (London: Thomas Nelson and Sons Ltd, 1997), p. 299.

[169] From Empson's notebooks (Haffenden, *Poems*, p. 278).

[170] Empson, *Pastoral*, p. 96.

flowers, having attained Heaven's graces, are—unlike the inflamed saints in their sexual prime—entirely "indifferent to temptation".[171]

A possible pun on the word 'gross' in line 9 to mean 144, sometimes articulated as "dozen dozens",[172] might evoke an image of saints sitting in rows of twelve; but we may attach another equally plausible image to the plump spaced-out saints at prayer: of buzzing bees carelessly drifting, until, ensnared by its heady perfumes, dipping prone into the luring flower. Because line 8 is rather clumsily phrased, yet another interpretation might read the saints as the magnolias themselves, taking 'gross' to signify the 'general' quality of youth. 'Gross' can mean "thick, stout" with a related sense "of a shoot or stalk: thick, bulky,"[173] as in Wesley's "burn to ashes [...] the gross stalks, on which the red Coleworts grow".[174] The Gardners read the 'plump spaced-out saints' as "aptly conveying the posture of the opened blooms, seated on the branches like rows of Buddhas", but identify the problem with their own interpretation: that "the pun of 'gross prime' is tendentious", as "it promotes the implication that the magnolias are well-fed and complacent".[175] The absence of a comma ending line 8 in any edition of the poem allows us to read the verb 'flare' transitively, which, together with 'flames', charges the magnolias in bud with inflaming and ensnaring.[176] Geoffrey Hill seems to have read the verbs just so; he repurposes Empson's lines in *Tenebrae*: "At dawn the Mass / burgeons from stone / a Jesse tree / of resurrection // budding with candle / *flames* the gold / and the white wafers of the feast" [emphasis mine].[177] This reading also makes for a more coherent conceit as the comparison of magnolias to saints is not developed further in the poem, while being 'saved by their own sap' in the previous stanza suggests the ecological relationship with insects.

The repeated 'bud', 'burgeon', and 'blossom', stresses the youth of the flower. It is the transient nature of its appeal that justifies Nature's flagrant favouritism to the magnolia. Empson, reading Sonnet 94, describes "the flower in its beauty, vulnerability, tendency to excite thoughts about the shortness of life,

[171] Ibid., p. 89.

[172] (This sense dates back to the fifteenth century); Steven Schwartzman, *The Words of Mathematics: An Etymological Dictionary of Mathematical Terms Used in English* (Mathematical Association of America, 1996), p. 101.

[173] *OED*.

[174] John Wesley, *Primitive Physic or an Easy and Natural Method of Curing Most Diseases*, (privately published, 1773), p. 112.

[175] Gardner and Gardner, *God Approached*, p. 144.

[176] Slippage is used in Derrida's sense of signifiers suggesting others signifiers without resting on a signified. It might be worth noting that Empson's method in *Seven Types* [pp. 50–1] was inaugurated by an awareness of line-end ambiguity in Shakespeare's sonnets.

[177] From "The Pentecost Castle" in Geoffrey Hill, *Tenebrae* (London: André Deutsch, 1978), p. 9.

self-centredness, and power in spite of it to give pleasure"[178]—possessing these qualities, the magnolias 'know no act that will not make them fair'—a line, incidentally, whose elegant iambic cadence, set in a rare rhyming stanza, amidst the rugged tones of disputation, pronounces the hint of Shakespeare. The poem also distinguishes carefully the 'power' of beauty from worldly 'authority'. In his notes for "Doctrinal Point", Empson says, "Man was given authority over all creatures, but this involves much toiling";[179] he cannot hope to exercise dominion over the land and the seas without discriminating between fair and ugly acts. Returning to Sonnet 94, Empson says, the flower's life "under the power of their own impulses" is sweet to God "so long as they are not in power. [...] This may be the condition of the flower and the condition for fullness of life".[180] Wisdom, then, would come with a recognition that this 'fullness of life' is a fantasy; unattainable, at any rate, to the race sired by sinning Adam. This judgement was already passed, we will recall, in the doleful conclusion to "The World's End":

> Apple of knowledge and forgetful mere
> From Tantalus too differential bend.
> The shadow clings. [13–14]

The god of "Doctrinal Point" thus bequeaths either dominion or grace. Like Sonnet 94's flowers that represent "the full life of instinct",[181] the magnolias in bud "rightly do inherit heaven's graces" [5]. 'Saved by their own sap, shed for themselves', the magnolias would have stood indifferent at the crucifixion. We will recall Empson's note that 'if you arrive at the point of grace, you no longer require a personal god'. The magnolia's sapience is its grace, its oneness with Nature; it needs no saviour from sin nor favour from Heaven. When regarding a flower in bloom, it seems as if designed to please; but the flower lives but for itself; and to promote its own aspect in the flowers-in-potentia of its sap. Gazing at the flower, green with envy then, the lapsed race of Adam thus wonders with Spinoza, what joy it must be to know the union of mind and Nature.

Yet there is something already quite grisly about freedom 'by predestination in the blood'. We may recall the dark mystery which preoccupied Calvin, how "by the mercy of God, not their own exertions, [some] are predestinated

[178] Empson, *Pastoral*, p. 91.
[179] Haffenden, *Poems*, p. 278.
[180] Empson, *Pastoral*, p. 94.
[181] Ibid., p. 92.

to salvation" as are "others to destruction".[182] In the second draft of the poem, alluding to Marlowe's phrase, "Christ's blood streams in the firmament", Empson noted, "X's blood in the firmament"[183]—'X' being shorthand for the Christ.[184] That the poem stresses predestination 'in the blood' rather than soul modifies Calvinist doctrine, from the eternal seat among God's elect to a transitory throne in Nature. We are taking 'blood' as the magnolia's 'sap' because it confers on the flower's destiny the same salvific unction that Christ's blood would have, say, on desperate Faustus. The apparent contradiction in being freed by predestination reminds us of the dark equation we identified in Empson's early poems, between freedom and imprisonment. The poem says the magnolia is free to do as it pleases because its actions are a consequence of foreknowledge and hence predestined to agree with reality. The sympathetic reader will have already begun to regard this condition as not entirely desirable.

The poem distinguishes subtly between desire for and envy of the magnolia. At first, stanza 3 appears to say, from the flaming petals to the blackened branch, that desire leads to ruin. And although we are compromised ultimately by magnoliaesque automation of mind, our delight in the flower's beauty is not itself the font of that ruin. The poem imparts this ethical balance by modulating the verbal force in the lines, 'whether they burgeon, massed wax flames / Or leave the sooted branches bare / To sag at tip from a sole blossom there'. The grammar of the sentence indicates that the branches were 'sooted' before the bud blossomed and (in-)flamed. The 'branch' on which the flower grows can be taken as a metaphor for the fallen world surrounding the flower. Its purpose is to support the flower and allow it to flourish. Recalling Hill's 'burgeons', 'budding with candle / flames', the budding magnolia at the end of a scorched branch can also be imaged as a bright flame at the end of a blackened wick. The main tenor of this metaphor is that one cannot be without the other. On the one hand, the flower bud is aflame because ignited by man's envy; for, as Bacon says, "the act of envy" is an

> irradiation of the eye [...] [T]he times when the stroke or percussion of an envious eye doth most hurt, are when the party envied is beheld in glory or triumph; for that sets an edge upon envy: and besides, at such times the spirits of the person envied, do come forth most into the outward parts.[185]

[182] John Calvin, *Institutes of the Christian Religion*, trans. by Henry Beveridge (Grand Rapids: Christian Classics Ethereal Library [1536] 2002), p. 197.

[183] Haffenden, *Poems*, p. 279.

[184] Price, 'Empson's Einstein', p. 106.

[185] Francis Bacon, *The Essays of Francis Bacon* (New York: Charles Scribners' Sons, 1908), p. 35.

On the other hand, we are charred amidst its flame, as in Sonnet 73, "In me thou see'st the glowing of such fire, / That on the ashes of his youth doth lie" [9–10].[186] Either way, the flower only seems fair in a fallen world already torched by sin as a flame needs a blackened wick.[187] The word 'soot' brings to mind black, insoluble spots descending from the air—a fall that sets the stage for the flower's rise 'into the air'. The verb 'leave' thus only works actively on 'bare' and 'sag'. And even then, an infected will was naturally apt to sag; in other words, since the branch was already 'sooted', it is predisposed to this attitude. To imply a causal link between desire and the Fall, the line would have read something like 'leave the branches sooty bare'. But because, as it stands, the flaw is aboriginal to the branch, the culpability of the inflammatory flowers is tempered; the 'magnolias leave branches bare' as, for instance, 'the clown leaves me sad', despite all efforts to delight.

But in stanza 4, the fragile aesthetic economy between beauty and desire is ruined by the deceptive allure of mathematical perfection. A comparison seems urged between the second stanza, where matter becomes form, to the end of the fourth stanza, where science becomes ethics. Wise Solomon is then described as wearing the worker's cloth, the 'over-all'. Empson notes, "Man was given authority over all creatures, but this involves much toiling and spinning, as when in over-alls."[188] The line recalls the Book of Matthew, "And why take ye thought for raiment? Consider the lilies of the field, how they grow; they toil not, neither do they spin: And yet I say unto you, That even Solomon in all his glory was not arrayed like one of these."[189] The most regal cope is thus outmatched by the careless lustre of petals. Through the Biblical echo, the poem expects readers to remember that Solomon's greatness lay in a life of striving with ethical choice, not an effusion of innate glory. Wisdom from suffering is shown to be fitter for human ethics than always already knowing right. To be sweet to the summer, the flower must 'to itself only live and die'. While the automatic wisdom of the magnolia saves them from envy, for 'They have no other that they should compare' [24], it also deprives them from loving that other, for they 'cannot know of care' [21]. 'Care' can be read here as the crux of ethics, and ethics, as the balancing of individual and collective advantage in one's actions: humans do this because they care about the other; what makes them covet their neighbour's good also makes them struggle for the greater good.

[186] Shakespeare, *Sonnets*, p. 257.

[187] Thus, invoking a common apologia for the problem of evil, namely, that beauty is not possible without its opposite.

[188] Empson, *Poems*, p. 279.

[189] Mathew 6:28, *KJV*.

Via a routine of images, following Solomon and the lilies, the argument thus unfolds by recalling once more the Sonnet evoked in stanzas 2–3. The foreboding end of Shakespeare's Sonnet 94, "for sweetest things turn foulest by their deeds / Lilies that fester smell far worse than weeds" [13–14], looms on the singularity to which we are shown to tend. This is truly a case in which, as Christopher Ricks says, "the past poem is central to though not coterminous with the poem's meaning".[190] "Doctrinal Point" relies on the shared possession of a lyric corpus relaying to imagination the correct vectors of meaning that have crystallized through literary memory. This circumstance, undoubtedly a dent in Norris's theory, surprisingly testifies to Eliot's vision of lyric poetry being involved across time in a sublime semantic entanglement.[191]

2.3.3 Tautology and reference

Roughly speaking, stanza 4 envisions a future wherein total surrender to scientific epistemology brings to pass an abnegation of human mind; pretension to knowledge annuls knowing.

The word 'tautology' in line 14 develops the formal implications of line 1. Because it mirrors itself, a tautology is always true. It is thus the logical obverse of paradox or contradiction, which is neither true nor false. Both violate the cardinal rules for a scientific statement: that (1) it must be capable of being true or false,[192] and (2) its veracity must rest on object conforming to statement. This approximates the basic criteria repeated in popular treatises on logic in the 1920s, such as Ramsey's and Wittgenstein's. Having been taught Solid Geometry and the Theory of Equations by him at King's College, it is possible, as Haffenden speculates, that Empson "fell for a while under the influence of Frank Ramsey".[193] Ramsey calls empirical statements "genuine propositions" and distinguishes these from "necessary

[190] Christopher Ricks, *True Friendship: Geoffrey Hill, Anthony Hecht, and Robert Lowell under the Sign of Eliot and Pound* (London: Yale University Press, 2010); Harold Bloom makes a similar point in Harold Bloom, *The Anxiety of Influence: A Theory of Poetry* (New York: Oxford University Press, 1973), p. 70.

[191] This is the central argument in 'Tradition and the Individual Talent', in G. Douglas Atkins, *T.S. Eliot and the Essay: From The Sacred Wood to Four Quartets* (Waco: Baylor University Press, 2010), pp. 45–58.

[192] We will recall Richards's definition, 'A statement may be used for the sake of the reference, true or false, which it causes. This is the scientific use of language' (Richards, *Principles*, p. 250).

[193] Haffenden, *Mandarins*, p. 104.

propositions" which are "mere tautology".[194] Russell summarizes Wittgenstein's "essential theory" as follows: "in order that a certain sentence should assert a certain fact there must, however the language may be constructed, be something in common between the structure of the sentence and the structure of the fact".[195] Among sentences that violate this axiom, Wittgenstein specifically names "the tautology and the contradiction", which "say nothing"[196]—by 'nothing', he means no thing, as they do not conform to any thing in the world. This is because "the tautology has no truth-conditions, for it is unconditionally true; and the contradiction is on no condition true. Tautology and contradiction are without sense."[197]

We have seen how by contradiction, line 1 comes to negate itself. By referring to physics as 'one tautology', Empson urges a conception of science as a formal system willing the necessity of its own truth—and so we might already imagine its truth dissolving into the air. To make good the charge of tautology, physics must be seen as rending phenomena into measurable parts which are then redefined through mathematical interrelations; and in a parody of discovery, its laws can then be seen as contrivances of equation between these mutually agreeable terms. Newton himself sums up his work in the *Principia* as an elaborate redefinition of terms: "Hitherto I have laid down the definitions of such words as are less known, and explained the sense in which I would have them to be understood."[198] If, for instance, Force is defined as Mass*Acceleration, and M and A are defined as quantities which combine to yield F, then the law can be interpreted as saying F = F. In the dystopian future of the poem, validity is just so established internally with little care for the actual world.[199] This is what J. D. Bernal envisioned in a popular essay predicting the effects of modern science on human morality: "The manner appears to us as physical law but it may well be found to be a tautology."[200] Of course, this is only a problem for those that consider science a metaphysical truth that reveals to us the face of god. If F = M*A merely formalizes a regularity observed among quantified variables that are in turn the artefacts of measuring instruments—as an instrumentalist would view the equation; which

[194] Frank P. Ramsey, *The Foundations of Mathematics and Other Logical Essays*, ed. R. B. Braithwaite, intro. G. E. Moore (London: Kegan Paul, Trench, Trübner, 1931), pp. 4–5.

[195] From Introduction to Ludwig Wittgenstein, *Tractatus Logico-Philosophicus* (New York: Humanities [1921] 1951), p. 8.

[196] Ibid., p. 53. [197] Ibid., p. 53. [198] Newton, *Principia*, p. 77.

[199] See Section 1.2.

[200] J.D. Bernal, *The World, the Flesh and the Devil: An Inquiry into the Future of the Three Enemies of the Rational Soul* (London: Cape, 1970), p. 2.

regularity then assists in the fabrication of machines, further philosophical implications of the formula would not need probing.

The 'tautology' of line 14, read with the contradiction of line 1, may also point to the formalist conception of modern mathematics. The story of formalism in early-twentieth-century mathematics is embroiled, in interesting ways, with the Liar's paradox.[201] Amid the rush to firmly establish the foundations of mathematics, Ramsey through atomic propositions, Russell through logicism, Brouwer through intuitionism, and suchlike, it was David Hilbert's enterprise that led the charge.[202] Hilbert's dream was a legendary list of axioms, a finite, complete, and consistent set to ground all theories in mathematics. Formalism, as Hilbert's programme is chiefly known, requires of a mathematical system only "proof of consistency", disregarding "reference to known acts of experience".[203] Alan Weir summarizes Hilbert's argument as follows: "mathematics is not a body of propositions representing an abstract sector of reality, but is much more akin to a game, bringing with it no more commitment to an ontology of objects or properties than ludo or chess".[204] Formalism takes mathematics itself, we might say, as 'one tautology', which is indeed one of the reasons why Ramsey's essay attempts to disprove it.[205] But it was Kurt Gödel who found definite success with his incompleteness theorem. Loosely speaking, Gödel's theorem states, "This sentence is not provable."[206] It showed that "there is an arithmetical sentence G such that neither G nor not-G is provable in arithmetic".[207] And since one heretical statement in arithmetic could never be proven, it follows that no matter what arithmetical systems we use, statements will always emerge that defy proof using those systems. And so, Hilbert's attempt to make all mathematics 'one tautology', so to caricature it, is undermined: the god approached dis-solves again.

Gödel published his proofs in 1931, four years before "Doctrinal Point" appeared in print.[208] If news of Gödel's breakthrough had indeed reached the keen mind of Empson, however into the Orient it had by then wended, there

[201] L. Goddard, '"True" and "Provable"', *Mind* 67. 265 (1958), 13–31, p. 13.

[202] Engelhardt, *Modernism*, p. 11.

[203] David Hilbert and Paul Bernays, 'The Foundations of Mathematics', *Grundlagen der Mathematik*, 1 (1934), trans. by Ian Mueller, p. 3.

[204] Alan Weir and Edward N. Zalta, eds, 'Formalism in the Philosophy of Mathematics', *The Stanford Encyclopaedia of Philosophy* (Stanford: Metaphysics Research Lab, Stanford University, 2015), online.

[205] Eddington calls formalism a tautology in *Physical World*, p. 131.

[206] Goddard, 'True', p. 13. [207] Ibid., p. 13.

[208] Kurt Gödel, 'Über Formal UnentscheidbareSätze der Principia Mathematica und Verwandter Systeme I', *MonatsheftefürMathematik und Physik*, 38 (1931), 173–98.

might be an added wrinkle of meaning in the poem.[209] We cannot establish Empson's intent as regards Gödel in the crude sense of recovering the conscious will of the poet. But I mean to follow Stanley Cavell in matters of authorial intent:[210]

> [Cavell] notices that a character in Fellini's film *La Strada* can be seen as a reference to the Philomel legend, and he asks what we need to know about Fellini in order to say that the reference was intentional [...]. He imagines a conversation with Fellini in which the filmmaker says that although he has never heard of the story before, it captures the feeling he had about his character while filming, that is, that he now accepts it as part of the film he made. Cavell says that he is inclined in these circumstances to treat the reference as intended. [...] An insight belongs to an artist's intention, on this view, when it fits and illuminates his artistic purposes in a way he would recognize and endorse even though he has not already done so.[211]

It is quite apparent in our poem that reference to Gödel's theorem 'fits and illuminates' Empson's 'artistic purposes' in a manner that he would 'recognise and endorse'. Attainment of the god thwarted in the paradox of the first line, through the resonance of the Liar's paradox, echoes Gödel's theorem and its defeat of formalism, the white whale of modern mathematics. So clear an instance of formal completeness undone also chimes with the tautology of line 14. The god approached with tensors will dissolve; but in dissolving to tautology—an illusory hermetic completion—will leave its devotees with a delusive conviction of attainment.

[209] See Porte, 'An Agreement', pp. 10, 32, who says "The kinds of close reading developed by Graves, Empson, and Richards demonstrate the major tenets of logical modernism. They manifest an entanglement with post-Kantian aesthetic discourse (derived from the Cambridge Apostles and their circle) and an anxiety about the terms of truth in literature (especially truth in poetry) relative to the terms of truth laid out by logical positivism. Furthermore, this disquiet culminates in worries about whether the truths of aesthetics and the truths of science and philosophy and those of literature might ever be reconciled in one complete theory of everything." The influence of logical axiomatics on continental poetry in the aftermath of Gödel's theorem has been explored in Loveday Kempthorne, 'Relations between Modern Mathematics and Poetry: Czesław Miłosz; Zbigniew Herbert; Ion Barbu/Dan Barbilian' (PhD Dissertation: Victoria University of Wellington, 2015), p. 172: "Kurt Gödel published his 'incompleteness theorems' in 1931, a year after Barbu published his final collection of poetry."

[210] Stanley Cavell, *Must We Mean What We Say: A Book of Essays* (Cambridge: Cambridge University Press, 1978), ch. 8.

[211] From Dworkin's reading of Cavell's imaginary conversation with Fellini in Ronald Dworkin, *Law's Empire* (Cambridge: Harvard University Press, 1986), pp. 56–7.

Arthur Eddington asks us to "examine the kind of knowledge which is handled by exact science". Taking a question from a physics exam, he considers a sentence beginning with "An elephant slides down a grassy hillside [...]":

> The experienced candidate knows that he need not pay much attention to this; it is only put in to give an impression of realism. He reads on: "The mass of the elephant is two tons." [...] the elephant fades out of the problem and a mass of two tons takes its place. [...] Never mind what the two tons refers to; what is it? How has it actually entered in so definite a way into our experience? Two tons is the reading of a pointer when the elephant was placed on a weighing-machine. Let us pass on. "The slope of the hill is 60°." Now the hillside fades out of the problem and an angle of 60° takes its place. What is 60°? There is no need to struggle with mystical conceptions of direction; 60° is the reading of a plumb-line against the divisions of a protractor. Similarly for the other data of the problem. The softly yielding turf on which the elephant slides is replaced by a coefficient of friction, which though perhaps not directly a pointer reading is of kindred nature.
>
> And so we see that the poetry fades out of the problem, and by the time the serious application of exact science begins we are left with only pointer readings.[212]

In this vein, stanza 4 throws up an image of a world fast overwhelmed by 'the linkage of pointer readings with pointer readings'. It is, however, in the description of the process by which pointer-readings are shown to replace the physical world that the poem leaves contemporary philosophies of science behind and approaches the visionary.

In lines 14–16, science raises on the actual a simulacral reality stitched from abstract mathematical objects, each standing in for objects in the physical world, as the mass and slope do for the sliding elephant and hill. The tautology of modern science, seen as independent from 'Nature', becomes a ghostly form of Spinoza's true Substance: that "which is in itself, and is conceived through itself".[213] Eddington envisages physics in these very solipsistic terms: "in science we study the linkage of pointer readings with pointer readings. The terms link together in endless cycle."[214] The poem describes how 'all physics', through the endless cycle of pointer-readings, becomes one closed system, 'one tautology', a doctrine that *is* and 'is conceived through itself'. We

[212] Eddington, *Physical World*, p. 127.
[213] Spinoza, *Ethics*, p. 6.
[214] Eddington, *Physical World*, p. 260.

arrive here in the following way. If god is dissolved in Nature, it is coextensive with it. Because only the totality of the one Nature is sufficient reason for itself, anything within Nature is contingent and cannot be adequately explained sans reference without. The comprehension of Nature by any organon—that is, a description commensurate with the one thing that actually exists, God-or-Nature—is impossible because the contingent cannot reconcile the necessary: and the god will dissolve before any such approach. Having worked out these implications from the first three stanzas, we reach stanza 4, where physics becomes necessary and a priori—'tautology', as we have seen, does not depend on external confirmation. In other words, because Nature remains beyond explanatory approach, to achieve the doctrinal point, the discourse of modern physics itself becomes, as it were, Substance. Magnolias attain grace by fusion, physics, thus, by fission. Humanity must thereby select which of these necessaries to occupy: Substance, or the apparition Assumed from Substance. But staring in despair at the veiled flower (line 23), we seem to gamble with the cycle of pointer-readings, the parallel universe of numerical certainty. This certainty does not come from predictions about the world drawn from statistical causal models, as in the mathematical sociology of Isaac Asimov's *Foundations*.[215] A physics become tautology achieves certainty by self-reference; it is by folding in on itself that it taunts the succour of the 'calyx'.

The word 'tensors' in line 15 describes the process by which a simulacral universe of mathematics prescinds from existence. Eddington defined a 'tensor' in his usual manner as a "schedule of pointer readings".[216] But what tensors are is less important to the poem's meaning-making procedure than how tensors work, because lines 13–16 are meant to trigger a process, namely, the transubstantiation of reality to simulacrum. Tensors are typically used in Riemannian geometry when dealing with manifolds. Manifolds can be understood as mathematical tools that allow complicated structures to be described in terms of well-understood properties of ordinary space.[217] In flat Euclidean space, tensors can be thought of as numerical values describing the length, angles, and distances between directed points (vectors); but using manifolds, we can describe more complex behaviours by reducing their local behaviour to this simple flat setting.[218] A collection of interactions and

[215] Isaac Asimov, *Prelude to Foundation* (London: Grafton Books, 1988).

[216] Eddington, *Physical World*, p. 257; this is also the only significance that Haffenden places on the usage (Haffenden, *Poems*, p. 280).

[217] Weyl, *Space*, p. 102. [218] Weyl, *Space*, p. 131.

relationships (such as distances) between vectors on an otherwise unwieldy surface can be accurately described using a mathematical rule (a function), which we can now call a tensor. Empson thus often brings out in his mathematical metaphors a sense of simplifying the world through abstraction and approximation. Differential calculus (from "The World's End") attempts to reach the instant in time by secants, straight lines to describe a curve; tensors attempt to comprehend complex movement on manifolds by reducing the problem to locally flat spaces.

Richard Feynman says in the nineteenth century, tensors were reserved for cases when irregularities in the physical world would otherwise not submit to mathematical description; their use promoted the "feeling that our treatment of physics is complete".[219] Tensors were commonly applied in the study of crystals, the most irregular of solids. The OED, for instance, uses Woldemar Voigt's definition of tensors from his book, *The Fundamental Physical Properties of Crystals in an Elementary Presentation.*[220] Their role in characterizing crystals can illuminate our question of 'how tensors work'. Given the poem refers to the use of 'tensors' in Einstein's relative model of the universe, we can presently think of the physical universe itself as a crystal (or a crystal as a manifold).

When an electric charge travels across different directions on a crystalline surface, tensors can be used to calculate the resulting variations in the quantities being measured. To manage these variations across different directions, a quantity α is introduced. For α to perform its task of maintaining proportionality between the salient measurables, a different α is required for every direction on the crystal's surface. α is thus divided into 'α_{xx}, α_{xy}, α_{xz}' (the combinations of x, y, z stand for different directions across multiple dimensions). The crystal is said to have been "described completely" by the set of all components of α: (α_{xx}, α_{xy}, α_{xz}, α_{yx}, α_{yy} ....): and this set is termed a tensor.[221] In this manner, Eddington argues, "by the use of tensors the mathematical physicist precisely describes the nature of his subject-matter [...]; and those accretions of images and conceptions which have no place in physical science are automatically dismissed".[222] Because the poem uses the term 'tensor' to symbolize the path that 'all physics' takes, we can substitute the crystal in our analogy with the cosmos, seen as, in some ways, the greatest irregular object. The tensor would

[219] Richard P. Feynman, *The Feynman Lectures on Physics, vol. 2*, ed. by Robert Leighton and Matthew Sands (Boston: Addison Wesley, 1971), online version.

[220] W. Voigt and G. Chisholm Young, 'Die Fundamentalen Physikalischen Eigenschaften der Krystalle', *Nature*, 58 (1898), p. 99.

[221] Ibid., 31.1.

[222] Eddington, *Physical World*, p. 130.

then amount to a set of quantities that accounts for every thing across every direction in reality. This set of quantities that claims to account for every wrinkle in the manifold world is the bracketed reality of modern physics that the poem posits—a reality from which all that is beyond mensuration is banished. It is the Assumption of the assumption. 'There is', indeed, 'no need to struggle with mystical conceptions of direction'.

We will recall from our introduction Descartes' argument that things are real to the extent that they display mathematical properties. Line 16 likewise implies that laws in the simulacral reality of physics are ratified by their convertibility to the language of tensors, not their purchase on the physical world. Indeed, if the terms of the law were not mathematical, physics could not be 'one tautology'. But other senses of the word 'law' are clamouring for attention here. The phrase 'all law', in the poem, suggests that mathematical law in modern times supplants all others, including civil, public, and even moral laws. And so, the descriptive becomes the normative. 'One tautology', 'tensors' and 'all law' thus proleptically and retrodictively modify one another to trigger the terminal image of the 'Assumption'.

Line 16, prepared by 'tautology' and 'tensor', reverses the traditional view of scientific law as induced from facts about the natural world. In his review of E. A. Burtt's *The Metaphysical Foundations of Modern Physical Science* (1930), Empson says, the "double stress on order and truth, on the *simplest* rule that covers enough *facts*, is the essence of the scientific method".[223] He means that scientific laws are, and always have been, functions of mathematizability. What is unusual in the circumstance of the poem is that scientific laws have turned into universal laws. Civil and moral laws derived from the 'effort of virtue', imprecise yet profound, become overwritten by mathematical determinism: and the simulacrum legislated by pointer-readings, governed by deterministic laws, thus also becomes peopled by unconscious shades. Unburdened by a duality in choice, man is set to process the veil as spectre, always already informed.

Line 16 seems unnecessarily ungainly in length;[224] this may give us an insight into Empson's temper and attitudes to poetry. The source of prolixity is the word 'they', which seems to undermine an otherwise precise crescendo of meaning to the Assumption. If the line had read 'all law becomes the fact that can be described with them', it would have aptly symbolized the reversal

[223] Empson, *Argufying*, p. 532.

[224] This is noted by the Gardners (Gardner and Gardner, *God Approached*, 145) and Price (Price, 'Empson's Einstein', p. 106).

that the stanza in any case labours to convey. Empson once remarked that despite being induced from observation, theories in modern physics had curiously come to be called 'laws', as if first principles of the world: "it is very odd to call your first principles *a priori* when they are latest news".[225] On closer reading, we see that law become *fact* might have better prepared the reading of line 17 as this ontological exchange—the latest news becoming a priori, the descriptive becoming normative. But the image of reversal obtains despite the phrasing, because the word 'tautology' in line 14 has already declared all physics a priori, mirroring lines 18–19, where man himself, fallen from the Fall, inverts to innocence—to foreknowledge of and singularity with the cosmic drama, whereupon his free will is traded in for a prelapsarian contentedness.

Between lines 14 and 18, the poem repeats thrice, in three lines, various forms of the verb, 'describe', creating an incantatory lilt that brings to pass the final Assumption of the description. Without 'they', line 16 would have said, moral law, the Idea of the Good, becomes mere 'fact'. As it stands, the presence of 'they' prevents a potential contraposition between 'law' and 'fact': the lines simply say, "because 'things' *can* be described mathematically, their mathematical descriptions become the ultimate law which not only predicts how things *will* behave but determines how 'things' *should* behave"; but, as we have said, the artificial necessity of mathematical law is already prepared by the formalist implications of 'tautology' in line 14. Empson has opted for the cumbersome and less poignant line. But given Empson's dismissive attitude to visionary poetry, and insistence that his own poems are simply intellectual exercises, it might have seemed to him unseemly to render a mystical vision complete while interdicting doctrinal points.

In this vein, we might also inquire into the role of the final stanza, which seems largely superfluous. The 'over-all' of Solomon adds little to 'the effort of virtue', the absence of 'gap' between mind and Nature in the magnolias is communicated adequately—and indeed better—by the identification of its architecture with its texture, and the parallel between rainbows and the Heaviside layer seems a trivialization of the idea of Grace thitherto developed in the poem. The attempt to bookend the poem with a repetition of 'into the air' may be seen to structurally resemble the dread tautology of physics, and perhaps faintly to deprecate the poem's own efforts to define terms internally, as it does to 'god', 'air', and 'magnolia'. But as regards the conceit, it adds nothing new. In fact, Empson's notes raise a new suspicion.

[225] From an unpublished essay on Eddington, Haffenden, *Poems*, p. 281.

He says, "I meant here to compare the cope of heaven which protects the earth [...], the cope of the priest-king that symbolizes the protection of heaven, the calyx that protects the growing flower [...], the Heaviside layer that keeps off ultra-violet rays [...], and vaults over tombs under the ground".[226] The notes shed no light on the poem's meaning, and do not appear even designed to; they merely highlight a range of antinomies that metaphysical poetry can juxtapose. H.A. Mason said that Empson's poems "contain a surprising number of ideas brought together, but brought together, it seems, only to show that it can be done"[227]—although the observation is unfair, it seems rather reasonable when applied to Empson's notes. In "Doctrinal Point", the notes rattle through a single stanza with a rather affected rhyme-scheme. They focus on the specimen stanza, as if to say, "look how absurd the conceit of this poem is"; and given that the final stanza adds little of importance to the meaning already conveyed, the gamble (one that gallingly pays off) seems to be that the poem will not suffer when recruited as an exemplar of a poetic movement.

The rhetorical risks in the poem, however, merit the final stanza another explanation: that the master of a medieval university would be expected to close disputation with a summary peroration.[228] But even this explanation does not acquit—one expects from a master's summary a synthesis of insights, not repetition. Norris suggests that Empson's "note functions as an extension, almost a genetic stage, of the poem itself".[229] If this were so, Empson might have noted the relations of magnolias to human being, the effect of tautologies on ethics, or even how the reader is to receive tensors. In fact, the 'genetic stage' of the poem, if it had one, lies in the thesis of the first line, which is instinct with the activity of the entire remaining poem. At the level of form, with a structure inimical to content, the paradox shows that even the most ambitious of answers—nowadays, those of mathematical science—are incapable of releasing man from his fallen state (from which the magnolia in bud seems flamboyantly free). And at the level of content, by declaring scientific teleology unattainable, the paradox halts the conversion of fact to law, and the reprobate inference of 'ought' from 'is'. These irregular flashes of illumination, amid regions of prolixity and profanity, emerge from the poem as if permitted by a poet writing with much the same insolence as Professor Eddington or Magnolias in bud.

[226] Haffenden, *Poems*, p. 278.
[227] H.A Mason, 'William Empson's Verse', *Scrutiny*, 4.3 (1935), 302–4, p. 303.
[228] Novikoff, *Disputation*, p. 141. [229] Norris, *Philosophy*, p. 5.

2.4 Conclusion

One often wonders when reading Empson if modernist verse was at times too clever for its own good. Although Auden says to Yeats, you were silly like us, the fool does not set out to write visionary poetry; visionary poetry happens to him: and it did not happen to Empson. The fool was only ever a persona that Empson wore; it was the mask that symbolized his extreme epistemology and checked his unnatural intelligence. The Buddhist and Ricardian strains in Empson combined forces against the alluring comfort of abstract truths; they were opposite poles of experience that found compromise in the Elizabethan fool, who neither cares to know too much nor stops living. "Doctrinal Point" contrasts the two great themes of Empson's poetry—the Ricardian prescription of being and the Buddhist proscription of knowing: perfect being in the magnolia, and perfect knowing, in mathematics. The latter becomes a barren and hollow perfection that we seek while yearning after the flower's state, wherein being and knowing seem miraculously one.

Empson used logical contradiction and mirror images to work out what bits of the Buddhist scepticism towards knowledge and the European spirit of activity to keep and how best to develop them to come to terms with modern man's various unusual and unnatural predicaments. The poems ask, given the paradox at the heart of the mathematical world-picture, which posits a finite but unbounded universe, should we abandon altogether the hope that has lingered since Ionian science,[230] to make all things come together and make sense? In fact, most of Empson's literary engagement with mathematics seems designed to contend with this very question. The processes by which tensors describe irregular substance and infinity stitches a theory to completion are instances where our restless search for final answers makes us inevitably depart from reality. Through paradox, he shows that modern man's attempts, using mathematics, to create a complete and consistent episteme is doomed to failure. And in the differentials of calculus, he finds an arresting image of our eternal consignment to dissatisfaction in knowing.

We can without loss take away this much from this brilliant eccentric: that if one is not born a Shakespearean flower, it is better to be a Shakespearean fool, who simply acts, than trying to know too much, or worse yet, believing one already does.

[230] At least since Anaximander replaced Thales's water-universe with the 'Boundless'.

3
Eros

By acknowledging the victory of the mathematical world-picture in the preceding three centuries, the poets we discuss in this chapter ask what can be done in twentieth-century verse to rescue those intricacies of feeling and sensuous understanding that language outside the empire of number once could readily impart. This was the main project of T.S. Eliot and what Joseph Duncan calls the metaphysical revival.[1] Michael Roberts and Empson, the poets we discuss in this chapter, can be numbered in the metaphysical revival in that they too hoped to re-associate the estranged sensibilities and wits of verse. They are in many ways more direct heirs to John Donne than Eliot insofar as that they incorporate the scientific and mathematical discoveries of their day into the drama of their poetic world, submitting them then to the forces of eros—recall the erect and loosening legs of Donne's compass that would also chart the courses of planets. From his obsessive uses of mathematical metaphors, we can see that Empson is better placed, with a training in mathematics to boot—as is Roberts—to carry on the metaphysical project of contending with the mathematical world-picture in verse. The metaphysical revival discussed in this chapter is thus not merely a continuation of the romantic reaction against the Enlightenment, as Kermode, for instance, alleges of Pound, Eliot, and Hulme,[2] but an entanglement in which mathematics is deeply attractive as a mode of thinking, for its exactitude, descriptive power, and defamiliarization. A serious attempt is made to answer the difficult questions that arise for poets in rational civilization: how can sensuality be preserved in a poetic idiom competing with science for truth-telling authority? What is love in a mathematical world?

There is an element of showmanship in using mathematical language to invoke passion. Both Empson and Roberts conduct these performances in the main to dramatize the struggle of and for eros under the empire of number. Michael Wood, reading Empson's "This Last Pain", argues that it is

[1] See Joseph Duncan, *The Revival of Metaphysical Poetry: The History of a Style, 1800 to the Present* (Minneapolis: University of Minnesota Press, 1959).

[2] See Part II of Frank Kermode, *The Romantic Image* (London: Routledge and Kegan Paul, 1957).

Agon: Poetry's Challenge to the Mathematization of Reality (1920s–1960s). Anirudh Sridhar, Oxford University Press.
 DOI: 10.1093/oso/9780198985457.003.0003

difficult to gauge whether Empson is writing "a love poem" or "addressing himself".[3] Through his mathematical lens, Empson manages to make the phantom addressee of his poems, one who dissolves in the face of the poet's cleverness, seem distant to sensuous apprehension. The cultivated, or seemingly calculated, absence of passionate resolution in his love poems often acts as a mirror to the plight of poets in a world governed by abstract symbols. Roberts and Empson agreed with Eliot's conclusion that the age after the metaphysicals saw a "dissociation of sensibility" into wit and emotion.[4] Ever since the eighteenth century, says Roberts, "the language spoken by ordinary people has become more and more capable of expressing 'material' facts, and less and less capable of doing any other work".[5] Roberts regarded the main vocation of the poet in the twentieth century as resistance to the Royal Society's project of linguistic refinement. Both looked to the seventeenth century, to the pre-Newtonian world, to fashion an idiom and techne that would once again be capable of expressing eros. This chapter thus takes the metaphysical revival of the twentieth century—as manifest in the mathematical poems of Roberts and Empson—as a central player in the poetic agon against the mathematical world-picture.

3.1 Roberts Recombines Pent Emotion with Stranger Matter

As with many twentieth-century metaphysicals, the model for Michael Roberts's poetry is the sensuality of "Donne, Herbert and Vaughan".[6] He is nostalgic for the English of the Renaissance, for "a prose which spoke not only to the eye and brain but to the senses of touch, taste and hearing. For Donne, prose was nearer to music and to painting or to texture of leaf of carpet than to mathematics."[7] He praises Herbert for his metaphors, in which the "feelings not the measurements are asserted to be congruent".[8] But the seventeenth century also signalled to Roberts the beginnings of linguistic decay: the *Leviathan* of Hobbes and the polemics of the Royal Society were instrumental in setting English on a course to pure denotation, to words shrivelling into signs.[9] In the twentieth century, he

[3] Michael Wood, *On Empson* (Princeton: Princeton University Press, 2017), p. 74.
[4] T.S. Eliot 'The Metaphysical Poets', *TLS*, 669–70 (1921), p. 670.
[5] Michael Roberts, 'The Poetry of T.S. Eliot', *London Mercury*, 34 (1936), 38–44, p. 44.
[6] Michael Roberts, *The Modern Mind* (London: Faber & Faber, 1937), p. 88.
[7] Ibid., p. 90; the description seems an extension of T.S. Eliot's phrases, such as "the massive music of Donne", in the well-known Eliot, 'Metaphysical Poets', p. 670.
[8] Roberts, *Mind*, p. 91.
[9] See 'Materialism and Scientific Language' in Roberts, *Mind*, ch. 3, pp. 63–87.

says, "we habitually describe in abstract scientific language experiences which would once have been described in more vivid and more sensuous speech".[10] But the "desire for a plain, unambiguous language, in which words have hard, definite values, like numbers, is not new. The scholastic philosophers, for example, had aimed at it, and Dalgarno and Bishop Wilkins had tried to invent such a language in the seventeenth century."[11] Roberts was confident that there still lay the debris of seventeenth-century sensibility buried beneath modern English. The archaeology of that eros, understood in this section more as *sensuous passion* than passionate love, particularly in the language of poetry, is the chief aim of his poetic and critical works.

When traversing Roberts's artistic landscape, one is reminded, strangely, of da Vinci's. The water and rock that background Leonardo's paintings, washing clean all mire and confusion, run through Roberts's early verses. Walter Pater once said of Mona Lisa that "she is older than the rocks among which she sits"; "that presence so strangely beside the waters" whispered the wisdom and folly of a thousand generations.[12] Faltering at her knowing look, our poet asks, "Why do you lift the veil and spy / The thing behind our commonsense? // How can you bear to flay the mind / of noble lies and brave pretence?" [9–12].[13] The ever sensible critic seems to confess that he would not refuse a mystical vision should it ever oblige; but remaining unfavoured, he asks of the Mona, "what secret anchorage is yours? [...] What quiet wisdom? Why, O why / May I not find its counterpart?" [13–16].

In his eagerness to share the burden beneath her eyelids, the poet confesses his own failure with words. Roberts's poetic visions are indeed as irregular as his landscapes; but he also knew of another world: having studied mathematics at Cambridge, and taught it thereon,[14] he admits, "I am only a second-rate mathematician, but I have penetrated far enough to admire the generality, rigour and economy of thought".[15] His religious beliefs in the imperfectability of man were tested by insight into the realm of perfect ideas.[16] This was a

[10] Ibid., p. 93.

[11] Michael Roberts (unsigned), 'Examination of Logical Positivism', *The Listener*, 17 (February 1937), 238, p. 238; attributed to Michael Roberts by R. Hamilton (T.W. Eason and R. Hamilton, eds, *A Portrait of Michael Roberts* (Chelsea: College of St Mark & St John, 1949), p. 69).

[12] Walter Pater, *The Renaissance* (New York: The Modern Library, 1873), pp. 129–30.

[13] From 'Mona Lisa' (Michael Roberts, *These Our Matins* (London: E. Matthews & Marrot, 1930), p. 13.

[14] Michael Roberts, *Michael Roberts: Selected Poems and Prose,* ed. Frederick Grubb (Manchester: Carcanet Press, 1980), p. 1.

[15] Michael Roberts, 'Credo: A Note on Poetry and Science', *Poetry Review*, 29.3 (May–June 1928), 192–6, p. 193.

[16] See essay on 'Hulme's Speculations' in Michael Roberts, *Critique of Poetry* (London: Jonathan Cape, 1934), pp. 71–85.

spiritual struggle that shadowed the poet's life, manifesting brilliantly throughout his works. For instance, he denounced modern English for its frigid rationalism—words behaving like numbers—but sets himself the task of making poetic language more precise. This seemingly paradoxical attempt is emblematic of the agon of modernist poetry, where precision becomes at once accusation and aspiration.

Roberts's poems develop as arias on mathematical perfection and apotheoses on failure. In the poem "Nicolas Flamel", the titular character invokes the daemon of Socrates to release him from the cave for a glimpse of the Forms: "Come daemon now, uplift the glass / Reflect the dying candle-gleam, / And show the images that pass / Above my black and broken dream" [1–4].[17] The alchemist pursues not gold but truth behind the veil of shadow. Soon after, "these vessels crumble", for "none can hold / Elixir vitae we have known / Or grasp the true alchemic gold / Or touch the substance of the stone" [25–8]. The poet, trading in the transmutation of commonly alloyed words, thus meets the very fate awaiting the alchemist; and in this defeat lies the recognition of the imperfectability of word and world.[18]

Despite frequent disappointments, the spirit seems once to have favoured the poet on the bitter and bright summit of "Kanchenjunga". In the thin air of the Himalayas, "God's thought" finally descends and "racks cloud-confusion" [13].[19] The "blank precipice / grown final. No analysis / can find familiar footing. Here // All reason ends; ends summer; Time / In eternal winter, watches, thrall / To faltering flesh" [14–20]. At the dying end of his ascent, all thought is wiped clean, and the poet, in desperate delirium, grasps first and final things: and this glimpse is granted, we will note, on banishing analysis and surrendering reason to flesh. It is not the kind of poem one expects from a mathematics teacher of a secondary school;[20] the last time a vision so blank in the clouds ended a lyric, an Irish Airman had foreseen his death: but Roberts was no mild-mannered intellectual: in his university days, he would cross borders into France and Italy looking for brawls with fascists.[21] Janet Roberts said of her late husband, "Michael's life [...] was lived in a heroic temper".[22] The philosopher Quinton even calls him an "unusually muscular

[17] Roberts, Matins, p. 23 (first published in *Poetry* (Chicago), May 1930).

[18] See Michael Roberts, 'Humanism and the Perfectibility of Man', in *T.E. Hulme* (Manchester: Carcanet Press [1938] 1982), ch. 2, pp. 39–56.

[19] Michael Roberts, *Collected Poems* (London: Faber and Faber, 1958), p. 64. (Originally published in *New Country*, 1932.)

[20] Roberts, *Selected Poems and Prose*, pp. 1–2. [21] Ibid., p. 1.

[22] Janet Roberts, 'Introductory Memoir' to Roberts, *Collected Poems*, p. 17.

intellectual".[23] In praise of mountaineering, Roberts once said, "Man can preserve his dignity only by showing that he is not afraid of anything, not even death."[24] It seems his frequent disappearance into hostile country was more than pastime. "Kanchenjunga" begins, "Harsh hills be comfort" [1].

The medium of poetry—in which the jagged lines of mountainsides can be invested with rivalling intricacies in meaning—became for Roberts an "answer to the unanswerable questions: the recurrent 'why' to which our science gives no answer. Poetry [...] rebuilds the shining world that is for ever falling, for ever tarnishing. It reveals the world, it does not number it."[25] The irregular mountainside perhaps demonstrated most prominently to the poet the lacunae in the mathematical picture of reality we saw exposed in Eddington's caricature of the 60° hillslope. Unease about society's fixation on rational thought, logical order, and material advance seems to dog his views on almost any contemporary subject. Roberts's review of Empson's *Pastoral*, for instance, begins, "The prestige of physical science, the natural laziness of human minds, and general ignorance of the difficulties implicit in word 'meaning', all combine to form a mistaken and dangerous conception of the scope of language"—although all this is quite tangential to Empson's book, Roberts makes it seem the central message of the author.[26] Although he thought mathematization beautiful, practical, and even vital, he found it offered but a faint grasp of things: in its procedures he recognized a seductive precision that would set the just use of the human faculties off-kilter. Eros, passion, and emotion, untended by the accurate-minded, had for Roberts in modern times ballooned to vague immensities or shrivelled to a wisp. "Among non-scientists", he says, "the commonest attitude of our time is a materialism, a belief that 'matter' is an ultimate reality, a neutral substance whose motions 'cause' all thought and feelings."[27] Roberts wanted to rout this belief and confer on poetry a central role in the renewal of civilization. He hoped to make of poetry the science of experience. He says,

> The physicist and chemist have been very successful within certain limits and to a certain degree of approximation, in forecasting future events in the world of measurement. [...] In practical experience, however, things which

[23] Anthony Quinton, "Introduction" to Michael Roberts, *T.E. Hulme* (Manchester: Carcanet Press, 1982 [1938]), p. i.

[24] Michael Roberts, 'Poetry and the Humour of Mountaineering', *The Alpine Journal*, 52 (1940), 22–33, p. 33.

[25] Qtd. in Roberts, *Collected Poems*, pp. 17–18.

[26] Michael Roberts [review of William Empson, *Some Versions of Pastoral*], *The Criterion*, 15.59 (January 1936), 345–8, p. 345.

[27] Michael Roberts, 'Science and Human Temperament', *The Adelphi*, 10.6 (September 1935), 381–2, p. 381.

> cannot be measured are quite as important as measurable quantities, and the assumption that we can forecast the behaviour of the world of people and of living things by similar laws and concepts is not necessarily valid.[28]

He even attempts to compare the labours and difficulties of the poetic and scientific methods: "scientific method is no more uniform and scarcely more easily described than is poetic method".[29] In the *Critique of Poetry,* he says, "the fullest appreciation of poetry or of any art may require as much training and effort as the appreciation of mathematics, and the effort is justified by the additional delight".[30]

Besides materialism, Roberts tried to counter the modern extreme of what he believed has perdured "the whole history of human thought", namely, "an attempt to split up experience into a geometrical byzantine mosaic, to substitute for a continuous flux of experience an orderly system of discrete particulars".[31] He paraphrases Hulme to say "poetry is a sensuous concrete language that forces the reader to feel the thing described as if it were actually present [...] It is a kind of algebra in which the x's and y's are not changed back into physical things till the end of the process."[32] Roberts sees in the hermeneutic circle of the poem a faithfulness to the whole of experience that lay beyond the part-by-part recovery of the scientific method. The "unparalleled precision" of mathematical description, he says, applies only to "those parts of experience about which the widest agreement can be established, i.e. measurable things".[33] Be that as it may, another mood would at times compel him to exalt the glories of mathematics: of Maxwell's equations, wrought from vague mounds of experimental data, he says "the precision, elegance and generality of his theory caused it to be accepted by mathematical physicists as one of the most beautiful products of the human mind".[34] Mathematics is thus both admired and abjured for its precision: it will prove a formidable antagonist.

The central role of mathematics in the poetics of Roberts is almost forgotten, which is to be expected, given his work in general is largely neglected. Samuel Hynes reduces Roberts's importance to his having "persuaded the public that there was a left-literary movement, a school of

[28] Michael Roberts, 'The Mathematic Way', *The London Mercury*, 32 (June 1935), 178–9, p. 178.
[29] Roberts, 'Mathematic', p. 179. [30] Roberts, *Critique*, p. 58. [31] Ibid., p. 93.
[32] Roberts, *Hulme*, p. 66 (Roberts is discussing Hulme's *Speculations*).
[33] Roberts, *Critique*, p. 84.
[34] Michael Roberts [review of James Clerk Maxwell 1831–1931], *Time and Tide*, 8.5 (January 1932), 126, 128, p. 127.

Auden".[35] Most who have written on Roberts's poetry see him but as a specimen of the thirties leftist milieu surrounding Auden.[36] Some still regard him as an important anthologist and talented critic of poetry;[37] and others interested in mountaineering notice that aspect of his poetry.[38] The stereotypes of the "editor" and "thirties poet"[39] are disputed in Whitworth's thesis; in the flourishing literature and science field, his is in fact the only serious treatment of Roberts's poems, which is itself not published. The present work will attempt to demonstrate that Roberts's poems should be afforded a central place in the study of modernism and mathematics, not only because their lines are flush with mathematical metaphor, but because they attempt to fuse the rational and emotional world in a manner rarely repeated since Donne. Eros is saturated in the semantic fields of mathematics, landscape, music, and architecture with almost neurotic consistency and precision, whereof it is our present task to anatomize.

We will take Roberts's early poems, from "Mona Lisa" (1927) to "Earth, Impact" (1932),[40] as one lyric corpus. Meaning develops in this phase of poetry by a gradual modification of language across the poems. He writes enviously of the novelist who, unlike the poet, did not depend on "words as they exist in his reader's mind" and could instead build "up those symbols he intends to use: he takes words coloured in the reader's mind by individual experience; he enriches them by use in context after context until they possess for the reader the exact significance the writer wishes".[41] Roberts is not abashed to attempt the feat in verse across his own early collections. The dictional mutations in his early poems are so similar and deliberate that words acquire almost the quality of a private language, as if the poet is trying to undo the damage of three hundred years in the span of a decade. Once these

[35] Samuel Hynes, 'Michael Roberts' Tragic View', *Contemporary Literature* (University of Wisconsin), XII.4 (1971), p. 437.

[36] Kathleen Raine, 'Michael Roberts and the Hero Myth', in *Penguin New Writing*, 39 (1950), 84–98; Valentine Cunningham, *The British Writers of the Thirties* (Oxford: Oxford University Press, 1988), p. 165.

[37] Jason Harding, *The Criterion: Cultural Politics and Periodical Networks in Inter-war Britain* (Oxford: Oxford University Press, 2002), pp. 159–63.

[38] Penny Bradshaw, '"Living at Our Full Compass": Michael Roberts and The Poetry of Mountaineering', *The Alpine Journal*, 116 (2012), 229–37, p. 230; Cunningham, *Thirties*, p. 165; See also Abbie Garrington, *High Modernism: A Literary History of Mountaineering, 1890–1945* (forthcoming).

[39] Michael Whitworth, 'Physics and the Literary Community: 1905–1939' (D.Phil. thesis: Oxford University, 1994), p. 248.

[40] These dates refer to when the poems were composed (Michael Roberts, 'List of Verses', copy in possession of Michael Whitworth; the original is located in National Library of Scotland, acc. 13145/53).

[41] Michael Roberts, 'Broceliande, or the Future of the Past', *Poetry Review*, 20.1 (Jan.–Feb. 1929), 43–50, p. 44.

words with their precisely twisted connotations are bombarded on the reader over the course of many poems, the seeming knots in diction begin slowly to dissolve, or rather resolve into an internal semantic web. Roberts argues that

> Language [...] shows an amazing intricacy which is useless to the traveller until he identifies a few landmarks. Definitions, like compass bearings, are useless until we determine a number of fixed points, our fundamental terms.[42]

Travellers in his poetic realm are likewise expected to find their bearings without key or compass; one uses words as landmarks and begins gradually to be oriented until the terrain becomes familiar. I do not think Roberts wanted his poems to be accessible only to the devout reader of his collections. The poet seems simply to be fascinated with a peculiar outcropping of words which he, almost as a lapidarist, took for his quarry.

To manage this method of reading, I will divide the relevant words of the early poems into clusters. Each will contain a set of words that are semantically and symbolically related. We will see that the repeated use of related words across contexts sharpens concepts and carves for each a unique semantic niche. Although this book is about poetry's agon with mathematics, in order to understand the role of mathematics in Roberts's poems, we must pass through a series of more fundamental clusters where the primeval antinomies of his poetic world are being worked out.

3.1.1 Sureness

The first cluster pertains to the state of sureness. Words in this set are 'perfection', 'precision', 'fact', and 'cause'. In the poem "Midnight", the poet says, "I held my hands to heaven / To hold perfection there" [13–14].[43] 'Perfection' here is an abstraction demanding the artificial stoppage of time: "But through my fingers streaming / Went time" [15–16]. 'Perfection' is sought to cope with the flux of things. In "Nicolas Flamel", the poet asks, "Where is the soul's perfection? Where / Shines the clear lamp of Lully's mind?", and proceeds in the imperative, "Show me the water, earth and air, / And fire's quintessence unconfined!" [21–4]. Distraught with the temporal world, the alchemist

[42] Roberts, *Critique*, p. 21.
[43] Roberts, *Collected Poems*, p. 43 (first published in Fellowship, 1929).

scours the shifting elements for their essence, *ousia*, the residue of the eternal. He finally realizes, we have seen, that 'none can touch the substance', the supposed concrete beneath the flux.

Roberts, a determined nominalist, says, "the collective noun has become an abstraction and its field has changed in the process. Critics, and even poets, set out to define the abstraction, forgetting that it may not be a single positive quality at all"[44]—we may regard the 'substance' of "Nicolas Flamel" as symbolizing these many abstractions. In "Lovely Immortal", 'perfection' is even personified, "darling perfection, whom I scorn, / Lie sleeping, sleeping, softly now!"[45] The poet thus banishes her to the dream-world, though at the end, he says, "You'll turn and leave a wreath to lie / Where I have made my bivouac / Under a black and broken sky" [18–20]. Though the stuff of dreams (we will note the pun on 'lie'), the image of the wreath suggests that the pursuit of heavenly perfection is what delights an otherwise gloomy prospect. Roberts finds this subtle stance in Hulme, whom he paraphrases as saying, "Man is fundamentally and inherently imperfect: he can apprehend perfection, but can never attain it. Man's apprehension of perfection gives him a right and necessary aim in life; but the aim is unattainable."[46]

This ambivalence persists throughout this cluster. Let us take the word 'precision'. In "Johann Sebastian", the poet chides "some fool precisely playing Bach" [15],[47] but in "Note on θ, Φ and Ψ", says "while I, maybe, precisely seize / the elusive photon's properties" [16–17].[48] Whilst the poet seizes a fact precisely, his lover will "pause a learned hour [...] neatly to annotate" [13, 15]. As opposed to the classicist's recondite and laboured annotations, it is clear the poet sees his precise seizing as a more spirited act. "Precision is necessary clarity", he says; "precision is a quality which will characterise the work of a good writer whatever his attitude may be".[49] For Roberts, it is precision that should distinguish the modernist from the romantic: "In those Romantic poets who exalt emotion above intellect, the words 'vague', 'immense', 'infinite' are used as commonly and with as much enthusiasm as intellectual poets use the words 'precise' and 'strict'".[50] Precision is indispensable when vying for authority with science, but at the same time, Roberts holds that the kinds of precision found in formal language must be kept at

[44] Roberts, *Critique*, p. 63.
[45] Roberts, *Matins*, p. 16 (first published in Adelphi, 1929).
[46] Roberts, *Hulme*, pp. 42–3.
[47] Roberts, *Collected Poems*, p. 53 (first published in *There Our Matins*, 1930).
[48] Ibid., p. 56 (first published in *New Statesman*, 1935; written in January 1931 ('List of Verses')).
[49] Roberts, *Critique*, p. 86.
[50] Ibid., p. 93.

bay for poetry to tell its own truth—mathematical precision is not the main quality one looks for in an organist playing Bach. This view is developed in his use of the word 'fact'.

In "Note on θ, Φ and Ψ", the poet belittles science, calling it "one small world of fact" [9]. This is a rhetorical move typical of Roberts's strategy in the agon. He laments in "Credo" (1928) the reduction in the field of truth to mere fact,[51] and continues to develop this sense of the word until *Recovery of the West* (1941), where he says, "[I]f we refuse to make use of any experience save that of the five senses [...] and ascribe no meaning to statements about other kinds of experience, it is obvious that in so far as 'statements of value' are not 'scientific' they are not 'statements of fact'".[52] The word 'fact' is here used to signify 'something taken to be present', the domain of what is, over which the mathematical and poetic world-pictures are in competition. Roberts suggests through his rhyme schemes that activity is not possible without mind separating fact from fiction. In "Perspective", there is an elegant stanza that reminds us of Blake's aphorism on Energy and Reason: "But mind in action is machine / Quarrying stern and stubborn fact; / In rhythmic pulse and discipline / It shapes the firm and final act" [13–16].[53] In order to act, mind must first figure things out about the world; but lines 15–16 sell a rather difficult pairing, that disciplined rational thought influences action through its rhythm and pulse. He repeats the rhyme between 'act' and 'fact' in "Earth, Impact": "volcanic act; / All contact, observation, cuts / Clear shell to sharp projecting fact" [2–4].[54] Just as in "Note on θ, Φ and Ψ", where the 'world of fact' is a "dark and igneous rock" [11], the scientific mind here is like a volcano, casting and recasting its surroundings. In these lines, Roberts achieves a succinct, aspirational, description of science with the words 'contact, observation', which capture the experimental preliminaries of the Baconian method. 'Fact' here is much richer than in "Note on θ, Φ and Ψ". Like rocks jutting from the mountainside, 'facts' are stolid assertions of reality. Unlike the Greek *physis* who "loves to conceal herself",[55] or Earth, her avatar in German metaphysics, who goes invisible to project World,[56] matter in Roberts is not

[51] Roberts, 'Credo', pp. 94–5.

[52] Michael Roberts, *Recovery of the West* (London: Faber & Faber, 1941), p. 129.

[53] Roberts, *Collected Poems*, p. 47 (first published in *Poetry Review*, January–February 1930).

[54] Ibid., p. 62 (written in February–April 1932 ('List of Verses') and later published in *Collected Poems*, 1936).

[55] From Heraclitus's fragment 123 (Heraclitus, *The Fragments of the Work of Heraclitus of Ephesus on Nature*, trans. and intro. G.T.W. Patrick (Baltimore: N. Murray, 1889).

[56] See Martin Heidegger, 'Origin of the Work of Art' in *Off the Beaten Track*, trans. & ed. Julian Young and Kenneth Haynes (Cambridge: Cambridge University Press, 2002), pp. 1–54.

entirely coy:[57] he sees the entire scientific enterprise, its success and importance, as owing to the declarative presence of what physically is. Staring at the colossal mountain, he says, "Stand up, unconquerable fact" [2].[58] Perhaps what is least convincing about his equations is the idea that the intellection behind science is like a volcano; we might be convinced of this metaphor in special cases, Newton, say, or Maxwell, but not as regards the average white-coat. But 'fact', we have seen, extends beyond science proper to any careful observation of reality that permits activity.

Roberts's poetic uses of 'fact' can be seen as a directed act of lexical re-appropriation. Logical positivism tried to reduce the legitimate use of language to statements of fact:

> Its treatment of 'meaning' is severe: it refuses at present to pay any heed to any elements of tone, association, or suggestion in language which cannot be reduced to simple equivalents. The historical aspect of words and concepts must therefore be ignored. Language is regarded as a set of counters, rather than a tree.[59]

And so, Roberts responds by enlivening the word 'fact' itself. While fact is thus rescued from the positivist, the restrictive and deadening qualities of mathematical description begin to be concentrated on the word 'cause'. The elusive "feint, the sudden pause, / the cobweb touch of terror" is "cause / Unknown" [3–5]. The lines articulate a central principle in the poems: spontaneous sensation defeats the causal scheme.[60] Beyond the linear world of "stubborn fact" [14], "we have impulse uncontrolled / And joy not marked on causal map" [7–8].[61] Even the physicist is forced to re-consider causation, in "Rocks Are Immutable", when dealing with the spontaneity of electrons: "Gripped in what causal schemes / Endures each casual atom what / When one perchanced electron screams / A stuttering signal, and is not?" [5–8].[62] Roberts notes how for the quantum mechanist, it is "impossible even in principle to determine the position of an ultimate particle without disturbing its velocity by an indeterminate amount, for every observation is an activity"—upon reaching the fundamental units of matter, Roberts suggests, we must

57 The "fact" must be understood as an outcrop of a much larger body of truth that remains concealed, which science leaves unrepresented.

58 From 'Kanchenjunga'.

59 Roberts, 'Positivism', p. 238.

60 From 'Prelude: It is the First Susurrus' in Roberts, *Collected Poems*, p. 59 (first published in Adelphi, June 1932).

61 From Roberts, 'Perspective', in *Recovery of the West*.

62 Roberts, *Collected Poems*, p. 49 (written in August 1929 ('List of Verses').

"abandon 'causes' altogether".[63] There could even be a sensory effect intended by juxtaposing 'causal' and 'casual' so closely: to rehearse visually—in the forward jump of the 'u'—the subatomic movement of the particle readying itself to wriggle free. The poem uses the aporia of mathematical physics to taunt the measurement-minded: a 'signal' is the one quality an electron offers for measurement, for it cannot be weighed on a scale. At the loosening grip of the causal mesh, it thus spits out a final number as adieu to the faithful realist.

3.1.2 Hardness

The world of facts is related to a second cluster pertaining to hardness. Hardness is to be understood in the geological sense of a becoming and a state, encompassing the words 'rock', 'compact', 'hard', 'jagged', 'frigid', 'stone', 'embalm', and 'congeal'.

In "Rocks Are Immutable", rocks become the symbol of our Democritean foundation, of 'atoms gripped in causal schemes'. The poet seems to imply that the very idea of atoms, as basic units of matter, as hard and unbreachable, derives from the common experience of rock as our secure base, our *terra firma*. "Rocks are immutable and hold / No scope for self-aggrandizement, / Each well-worn pebble churned and rolled / Suffers a like predicament" [1–4].[64] A rock is the object nearest in daily experience to the physicist's substance; both assert a sense of the hypostatically fundamental. Given this denotative restriction, 'self-aggrandizement' can be taken to imply potential for changing signification, which rocks and the physicist's substance do not possess. The words 'stone' or 'marble', as we will see, held for Roberts a latent potential for expressiveness, as raw material for sculpture or monument. But rocks signify for Roberts what Whitworth calls an "inescapable thusness".[65] They were stuck, in a predicament, in the rocky state; broken to smaller pieces, they would still be rocks. They exemplify Aristotle's first predicament of substance: that which is not predicated by any quality, what simply is.[66]

This downward tending earthiness and refusal to animation makes them immutable like 'fact', implied here in the negative sense of dogma. They can be destroyed by mining, as in, "Rocks and boulders are abolished / Under

[63] Roberts, 'Temperament', p. 381.
[64] From Roberts, 'Rocks Are Immutable', *Collected Poems*, p. 49.
[65] Whitworth, 'Community', p. 272.
[66] Aristotle, *Categories and De Interpretatione*, trans. and ed. J.L. Ackrill (Oxford: Clarendon Press, 1963), p. 2a12.

engines brightly polished" [13–14], but never moulded. An engine breaking off stubborn boulders is akin to, we will recall, 'mind in action is machine / Quarrying stern and stubborn fact'—the 'irreducible and stubborn facts' of William James, almost like rigid Kuhnian paradigms, need fresh minds and new instruments to break down. Although the final form of fact is hard, it must be remembered that even rocks are results of metamorphic and sedimentary processes. This subtlety of rock-history becomes tied explicitly to the scientific process, as seen in the phrase, "one small world of fact [...] compact / Within the dark and igneous rock / Of *Comptes Rendus* or *Proc. Roy. Soc*" [9–12].[67] When the rock identified is volcanic, it is charged to hold the memory—as in lava cooled—of intense activity. The vigour of scientific discovery goes stiff in the pages of scientific journals, '*Comptes Rendus* or *Proc. Roy. Soc*'.[68] One might imagine here the mystic vim of Newton's musings on the occult forces beaten to the dogma of gravity by the likes of Samuel Clarke of the Royal Society.[69] This approximates Roberts's views on the nature of good science: Janet Roberts says, "in his view of knowledge as a constantly changing pattern, I think Michael owed much to his training in science; certainly his sympathies were with the empirical thinkers like Roger Bacon rather than with the system-builders like Aquinas whose tightly-woven doctrines left no loophole for admission of new fact".[70] Roberts seems to expect from a good reader an internalization of his associations. For instance, Anthony Quinton, having absorbed the connections, summarizes Roberts's position as follows: a good critic "should seek a 'dry hardness', strive for accuracy and precision in the rendering of a freshly perceived world".[71] Whitworth says,

> The life of John Tyndall, Victorian scientist and mountaineer, was characterised by 'courage, tenacity and intellectual vigour': the first two terms could apply equally to his science or his climbing. Science and climbers cling tenaciously to the facts; they struggle towards an 'objective' which, once attained, ceases to be important: 'The British regard every theory as a stepping-stone to a new experiment'.[72]

[67] Roberts, *Collected Poems*, p. 56.

[68] From Roberts, 'Note on θ, Φ and Ψ', *Collected Poems*, p. 56.

[69] In reference to the less subtle representations of Newton's ideas by Samuel Clarke in his correspondence with Leibniz; see Ezio Vailati, *Leibniz and Clarke: A Study of Their Correspondence* (Oxford: Oxford University Press, 1997), pp. 17, 30.

[70] Janet Roberts, 'Introduction' to Roberts, *Collected Poems*, p. 30.

[71] Quinton, 'Introduction' to Roberts, *Hulme*, p. iii.

[72] Whitworth, 'Community', p. 272.

We may thus summarize three senses essentially related in Roberts's uses of rocks: rigidity/immutability, hardened activity, and sureness (the first cluster). The reference in the passage above is to the third. While the first two derive from rock's common semantic field, the third is a stretch. But that it should be present dominantly in his every use of the word creates across the poems what Empson calls a Type III equation. Empson uses as example Shakespeare's 'fool':

> It seems to me that many of Shakespeare's uses of the word *fool* are an example of it; he takes the symbolism of the clown so far (as I find myself reading him) that in effect he treats the word as meaning 'clown' and nothing else; when he uses the word about ordinary people they are not called foolish but described metaphorically as clowns.[73]

Similarly, in Roberts, every mention of 'rock' comes with the relatively private implications associated with 'fact'.

"Earth, Impact" completes the triangulation of fact-act-rock. The kind of hardness that can crumble can also lend grip to the climber: "Sudden the rock-fall" and the "wounded fingers touch / Granite, and clutch the crumbling fell" [5, 7–8]. "Strip Bare Pretence" similarly says: "Now munch hard fact: it shall not grow / Till you have worn red jagged flesh" [13–14].[74] The impression of hardness on soma, the friction with which the climber, with gashed fingers, is able to propel, is, like the experimental stage of science, a condition necessary for the learning of fact, which in the empirical sciences have a rather a short shelf-life—like the crumbling fell—before being eroded or sedimented by further factoids. Roberts calls for similar equations among his contemporaries; in modern poetry, he says, "what is needed is honest, sincere, and *accurate* sentiment – the sort of sentiment you are prepared to *act* on" [emphases mine].[75] Fact as protruding, involving itself in our activities, resists the semantic appropriation of 'knowledge', after the Renaissance, by the theoretical sciences: "the 'mechanical' world of Newton has become the basis of the implicit metaphor in all philosophic writing: we talk of [...] the field of knowledge as if it were Newtonian space".[76] Fact is not merely a node or terminus in the mathematical world-picture but rather what body

[73] William, Empson, *Structure of Complex Words* (London: Chatto & Windus, 1951), p. 50.

[74] Roberts, *Matins*, p. 24.

[75] Michael Roberts, 'In Praise of Modern Poetry', *The Listener*, 29 (February 1938), 375–6, p. 375.

[76] Michael Roberts, 'Thought in the Seventeenth Century', *The Philosopher*, 12.4 (October 1934), 170–2, p. 172.

uses in contending with the world: the term is thus restored to what it was in the age of Bacon before the fellowship of Newton subsumes it within the grand metaphysic. The fact-hardness nexus and its extension to 'action' is a central part of Roberts's larger programme to undermine enlightenment entrenchments and involve the senses in all the faculties of thought.

'Stone', on the other hand, comes with hardly any of the connotations of rock; there is perhaps a vague similarity to volcanic rock, if carved stone is seen as bespeaking the creative labour of masons—but this connection is never made explicit. Unlike 'basalt' or 'igneous', which are stretched into geological metaphors for scientific activity, the everyday use of 'stone' perhaps carries latently a suggestion of carving or moldability. Roberts is particular about the distinction. When a poem of Edwin Muir says, "often hewn steps in the steepest mountain-side", Roberts is irritated by the suggestion that rock is hewn, and sneers, "steps are cut in ice, not hewn in mountain-sides".[77] In "Artificer in Torment", the poet begins, "He would express in quiet stone / The flame that bites beneath the bone, / He would congeal the dying storm / To grey, slow torment, frigid form" [1–4].[78] The word 'quiet' animates 'stone' in a manner that 'hunched' and 'stubborn' had not 'rock'—a quiet person, for instance, can brim with interior drama she thinks prudent to suppress or be like those "Who, moving others, are themselves as stone".[79] The word 'frigid' belongs in this cluster. It is worked upon to signify bitterness, alienation, repose, and other senses remote from his use of rock. Because the quality that makes stone distinct from rock is the potential for animation in art, its true substance lies not in matter but form, which also explains why the alchemist can never ' *touch* the substance of the stone'. The phrase 'to congeal the dying storm' is odd and doesn't readily throw up any image, particularly as it is the stone that is purportedly congealed. Roberts is at times careless about the ramifications of metaphors and, overlooking the poem at hand, gets ensnared in his own elaborate semantic web. He assumes the direct sense of congealed storm will be taken as inner torment expressed in stone. The only sense in which a storm can be imagined as hardening is—with the assistance of 'frigid'—as freezing. But neither freezing nor congealing appertain to stone, which unlike cooling lava, is already hard. And with the reference to 'cathedrals' in the next stanza, the imagination is not even suffered to posit schist, a stone some Asian civilizations found kept soft enough for carving

[77] Michael Roberts, 'Poetry and Mountains', *The Spectator* (March 1934), 420, 422, p. 420.

[78] Roberts, *Collected Poems*, p. 44 (first published in *These Our Matins*, 1930).

[79] William Shakespeare, *Shakespeare's Sonnets*, ed. Katherine Duncan-Jones (London: Thomas Nelson and Sons Ltd, 1997), p. 278.

idols before it congealed. A more elegant image of hardening can be found in his use of 'embalm', in the phrase "Time / embalms corruption yet uncast" [13–14].[80] Its beauty derives from the ambiguity—Empson's seventh-type—in the word 'uncast', which can mean cast in stone or cast out, rendering the opposite senses of time preserving all sin or simply those not yet expunged.

3.1.3 Process

The next cluster is of process: of movement, activity, and their relation to the inner life of man. The cluster consists of words that invoke continual change, impelling a quickening of the imagination, with nouns like 'wind', 'chaos', and 'fire', verbs like 'corrode', 'act', and 'stream'. We have already seen 'Time', in "Midnight", 'Through my fingers *streaming*' [emphasis mine]. Time is the creator of movement and change: the time in "Time and Crystal Image" "sweeps, swerves, leaps".[81] Activity, under the auspice of time, is the other great principle of the poems. Challenging Mathew Arnold's famous line, Roberts says, "rhetorically we claim 'to see life steadily, and see it whole,' and sometimes the concealed metaphor of 'life' as something static and unchanging makes us forget that the vision must continually change".[82] The poems, on might say, stage an agon between hardness and fact, on the one hand, and movement and impulse on the other. With the firm grip of granite in "Earth, Impact", the climber realizes, "a world of action waits / Vital the chance, the random rain" [11–12]. The word 'random' distinguishes the active aspect of reality as unavailable to rational inquiry or arrangement in fact. The same shift occurs in "Perspective", from the hard world of Euclid to 'impulse uncontrolled / And joy not marked on causal map'.

In "Artificer in Torment", the poet says, "He would express, could eyes corrode, / In living rock" [13–14]. From 'stone' in the first stanza, he reverts in the last to 'rock', because here he seeks only the sense of solidity—not potential for animation –, to show in the face of unconquerable hardness the artist's triumphal act of 'corroding'. Roberts gambles too much with 'rock' in this stanza. He brings it back to act immutable, to increase the influence of the verb 'corrode'. But in the context of the artificer, 'rock' must perforce refer to sculpture; but the semantic field of 'rock', both in common use and his own

[80] From Roberts, 'Nicolas Flamel', *Collected Poems*, p. 45.
[81] Roberts, *Collected Poems*, p. 48 (first published in *Poetry Review*, January–February 1930).
[82] Roberts, *Critique*, p. 50.

poetic symbolism, does not allow for the association—even the most muscular of poets, Dryden, has his Aeolus carve "a spacious cave of *living stone*" [emphasis mine].[83] Roberts has been cornered in his own baroque patterns, and forced to animate 'rock' in the most literal and crude fashion, with the adjective 'living'. The articulation of vitality seems at times difficult to achieve with Roberts's impersonal register: for instance, "There is no quiet in the earth, / In the green root of history, or the leaf / Bending toward the earth; [...] / Cumulus, cirrus, moving". In this poem, he wants to exhibit the steady growth of life on Earth as instinct with biological entelechy, but it comes off rather flat; he seems to recognize this, for he's resorted again to literal means with the word 'moving'. But this is not always the case. Changing but one word in the action of "Artificer in Torment", "Nicolas Flamel" elevates the phrase to "corrode my living past" (a 'past', we will recall, which had been embalmed) [16]. The temporal mingling of 'living' and 'past' creates an effect considerably more striking than 'living rock'. As with the chemic energy of 'corrode', the natural forces wind, air, water, and fire work in the poems to shiver hard things. In "Artificer in Torment", "wind and weather wreck the blind / and bleak cathedrals of mind" [5–6]. And in "Nicolas Flamel", we will recall, 'earth' is flanked by 'water', 'air', and 'fire': 'show me the water, earth and air / and fire's quintessence unconfined'. It is as if the alchemist so far alters the nature of his situation—and this is what made him so attractive a figure to Roberts—that stasis is hemmed in by movement; as if a dam made of water were to contain concrete.

The contrast between the present cluster and the prior two is more marked with the word 'chaos'. Roberts argues, again resonant with Blake's aphorism, that:

> Individual human perceptions only approximate to the 'fact', the reality which is the highest common factor of experience of us all; by the unconscious selection of the new stimuli which most resemble the old, the old response is evoked and an air of pattern is given to what might well be chaos. The human mind may well be a kind of sorting machine, imposing order willy-nilly on experience and selecting as scientific knowledge those elements in the experience of others which most resemble its own.[84]

[83] John Dryden, *Virgil's Aeneid* (New York: P.F. Collier [1697] 1909), p. 1.

[84] Roberts, Critique, p. 94; Roberts uses the phrase 'sorting machine' from Karl Pearson, *Grammar of Science* (Briston: Thoemmes Antiquarian Books, 1991), p. 106. It is later used by Hulme in *Speculations*, p. 228.

The poem "And I have Turned to Westward" affirms chaos defiantly, in the line, "chaos, shall be my solitaire" [32]:[85] 'solitaire', a type of diamond, the hardest of materials, stands here for solace, support as in the rock grasped by the climber, granted, paradoxically, by chaos glinting through the dim world of inert fact. We will note again that the paradoxical effect can only work if hardness is received readily as signifying rationality and fact. In "Rocks are Immutable", contradicting the titular statement, the poet asks, "Or grant pure chaos, flickering chance" [13]. Solidity—the causal scheme—and chance are combined in the lines, "Then mark the shock of circumstance, / The day from solid darkness hewn" [15–16]. The cluster of solidity is often appended, in this way, with 'darkness'—that 'darkness' is 'hewn' binds it to the hardness cluster. The poems draw on the common attribute of opacity in solid objects and darkness—as in the title of the poem, "Darkness in Stony Places".[86] Order is to darkness what chaos becomes to brightness: the 'day' emerging from 'darkness' is charged by 'the shock of circumstance'. In "Midnight", 'chaos' is repeatedly contrasted with darkness. "The chaos of the mind" [20], which the poet implies is the natural way of being, is "screened from outer darkness" [19], as if the two principles will suffer from contact. In the phrase, "burning / In chaos, proudly bright" [3–4],[87] the Blakean collocation 'burning bright' is split and conjoined by the word 'chaos', on the one side, and the Yeatsian 'pride' of the organic world of becoming in "The Tower", on the other. Light and dark come together in his Miltonic phrase "Shining Dark", in which 'dark' is mapped onto "firm insubstantial Earth" [5] and the 'shining', onto "living pulse" [6].[88]

The two principal forces, movement–activity–passion, and hardness–fact–rationality, are distinguished in Roberts from familiar antecedents like the Apollonian and Dionysian of Nietzsche, by how lightness and darkness are distributed. Apollo is logic, rigidity, the columns of Doric order, but he is above all light; whereas sensual Dionysian chaos tends to unfold in the dark. We know Roberts thought in cognately grand terms. He says in *Critique of Poetry*, "although we find logical order more satisfactory than chaos, and harmony between our wishes more desirable than incompatibility, order and harmony are never final: they need continual revision, for integrity is an activity, not a final state".[89] The strange reversal in Roberts's poems of these ancient forces is curious: it is, I believe, to be taken as indicating how these

[85] Roberts, Matins, p. 26 (first published in *Adelphi*, December 1929).
[86] Roberts, Matins, p. 20 (first published in *Poetry* (Chicago), May 1930).
[87] From Roberts, 'Midnight'. *Collected Poems*, p. 43. [88] Roberts, *Collected Poems*, p. 75.
[89] Roberts, *Critique*, pp. 24–5.

principles came to seem in the modernist period. For instance, when speaking of the past, Roberts consciously suspends his regular associations. In "And Is Familiar Country", "Abelard [...] walks through fields of vision, bright / with premiss and white predicate" [33–6].[90] To Abelard emerging from the Dark Ages, logic appears as bright as to the Athenian breaking from what Yeats calls 'Asiatic vague immensities'. But by the twentieth century, the rational, logical regime of science must have seemed as large, misshapen, uncontrollable, and oppressive as the devilish congeries of Persia had to the Greeks. Chaos seems to Roberts the light of hope in rational society, and fire and water as necessary to hem the stolidly inching Earth.

3.1.4 The static and the dynamic

The elevation of irregularity and disorder in the poems is to be understood as an attempt to bring equilibrium to the rational idiom of a culture ossified under the rule of the mathematical world-picture. It was never in Roberts a reversion to romanticism. He in part accepts Hulme's diagnosis of romanticism when he says, "in so far as romantic poetry contains vivid and accurate sensuous description, it resembles all great poetry, but when it is yearning after the infinite and claiming attention and admiration for its own soulful sentiments, it is specifically romantic".[91] His poems, whether invoking the hard or the fast, aim for 'vivid and accurate sensuous description'. Yet there are moments when the two opposite principles of Roberts's poetry are concentrated in a single phrase. This cluster is a synthesis of the prior two and may be termed the static and the dynamic. It consists of phrases such as 'frozen spring', 'crystal air', and 'burst vulcanite'. "Jura in Islington", in an elegant line of chaste diction, imbues to rocky terrain a certain bucolic vivacity: "And water leaps from rock to rock" [27].[92] Despite the frequent failings of the poem, the ending couplet of "Artificer in Torment"—"He would unlock the frozen springs / Of pain in dumb unliving things"—has the epigrammatic terseness of a final couplet in Shakespeare. Facing the limp Christ of Michelangelo's "Deposition", or Bernini's "Teresa" in ecstatic agony, it is difficult to believe that pain was not somehow native to the stone anchoring the scene. One likewise does not doubt that the water frozen in spring did not

[90] Roberts, *Collected Poems*, p. 67. [91] Roberts, *Hulme*, p. 64.
[92] Roberts, *Collected Poems*, p. 55 (written in October 1930 ('List of Verses').

once flow. 'Frozen spring' captures the Renaissance Neo-Platonism which saw in inanimate stone the entelechy of ordained form.

In "And These Our Matins...", 'matins', or midnight prayers, sound of "stony rills" [7] and "bony tune" [11].[93] The expressions are meant to evoke the "bleak ecstasies" [5] of midnight choral, and do so, appropriately, through sound more than image. The sound of streamlets breaking on stony passes is polyphony comparable to a church choir performing Byrd or Tallis, but the word 'rill' also works phonetically to call up the sinewy, striated columns of cathedrals. 'Bony tune', if pictured, becomes sensational, like Breughel's "Triumph of death"; yet, as a symbol of the static and the dynamic, of life-in-death, the phrase intensifies the aural atmosphere of night, the slow hymn of choral, building to the haunting line, "uneasy murmur breaks to sound" [18].

Roberts says,

> In Shakespeare dynamic imagery predominates over static: it is suited to the emotional presentation of the ethical passion, and we therefore find it in Donne and Hopkins; for that ethical passion is not like contemplation, something timeless and geometrical, it is essentially something existing in time and leading to action. Even in describing things which are at rest, Hopkins gives this air of ceaseless motion.[94]

Unlike the restless images of Hopkins, Roberts deliberately checks his impulse to motion. The dance between the static and the dynamic emerges from contrasting collocations, even across entire stanzas. Take, for instance, the somewhat pleonastic phrase, "rigid vulcanite".[95] Although not the hardest mineral, 'rigid' can hardly be said to add information not already conveyed by 'vulcanite'. But read in stanza, 'rigid' acquires multiple functions: "And Time has burst the wet ravine / To crocus-blue and cherry white / Glass and rigid vulcanite" [18–20]. 'Rigidity' serves first as reassurance along the steep walls—a firm respite amid precarious shards. But the mastery of the phrase unfolds across the lines. In the first line, the sudden contraction of geological Time by 'burst' seems as vertiginous as the resulting 'wet ravine'. The contraction is then held together by the movement from 'burst...to...rigid'. By 'held together', I mean specifically that the sublimity in capitalized geological Time, is injected with eros in the urgency of 'burst', which is then stilled by the

[93] Roberts, *Matins*, p. 36 (first published in *Poetry Review*, May–June 1929).

[94] Roberts, *Critique*, p. 91.

[95] From Roberts, 'Jura in Islington', *Collected Poems*, p. 55.

closing 'rigid': the lines are thus able to convey at once both the devastating and enduring acts of Time.

The brilliancy of white glass along the time-carved gorge is also a sensory effect created by Roberts's image of crystals—as in the titular phrase of the poem, "The Broken Crystal".[96] A subtle distinction between stone and crystal makes the latter not a feature of the hardness cluster. The timelessness of crystals is contrasted with their vulnerability. For instance, the poem "Time and Crystal Image" expresses the desire to preserve popular deeds from oblivion. The aged poet of "Nicholas Flamel" likewise asks in futility to defy the decay of time, "show / Some blossom that defies the sun / show me no more of crystal snow" [5–7]. In "Time and Crystal Image" 'crystal' is a surface upon which heroes challenge the forward, obliviating march of Time, as when great events, future and past, are shown on crystal balls. Contrastingly, in refusing 'crystal snow'—for it does not defy the sun—the poet of "Nicolas Flamel" notices its vulnerability. 'Crystal snow' is precipitation become still while falling: the timelessness of the event itself becomes ultimately brief—like Empson's flower, the crystal's ethereal beauty is bound to its fragility. The timeless and precipitous image of crystals reappears in the poem, "Andromeda": "By crag or sudden fell, / The crystal air is fairest, / Glim-glossy filoselle" [10–12].[97] The images on 'crystal air' arrive through lucent shimmerings of silky fabric, its 'filoselle', which also calls up that fateful thread of the weaving goddesses, spelling the same doom written in the stars.

3.1.5 Matrices, quaternions, and vectors

In the fifth cluster, the two opposing principles of Roberts's poetry, hardness and activity, find corresponding mathematical symbolism.

Roberts repeatedly uses the image of a matrix. Take, for instance, the line in "Earth, Impact", "And firm rock-final matrix" [37]. Matrices look like latticed lines; so, their structures are visually insufficient to impress solidity; *eo ipso*, they lie more in the first than the second cluster. To twenty-first-century eyes, saturated with rows and columns, the 'firm' may seem superfluous. But the poet manages to communicate a subtle ambivalence through the seeming redundancy. Based on our development of the word 'rock', we can venture that rock-final indicates a firm grasp of hard, immutable fact. This means the

[96] Roberts, *Matins*, p. 25. [97] Roberts, *Matins*, p. 38.

matrix cannot contain senses in the third cluster, the 'world of action' that 'waits vital chance'—their structure is too rigid. The other connotations of matrix in the poems are speculated by Whitworth as denoting "womb, [...] the typefounder's matrix [...] and mathematical matrices";[98] despite the ambiguity of reference, he settles on the implication that "the phrase 'rock-final' offers an illusory security"—[99] the poem's message, however, seems less caution than reminder that facts are partial answers, useful but not ends in themselves. The phrase in stanza slips readily from hardness to action: "And firm rock-final matrix, deep / And leafmould answer, winter thorn, / Action". The final stanza thus recalls the fact-act coupling of the first. Roberts says, "Imperfection is life. So long as we can symbolise our feelings and experience in patterned words we have evidence of pattern [...] there is scope for action".[100] Once his poems segregate modes of thought, they begin in this way to show nice ambiguities that reconcile opposing forces.

In "Note on θ, Φ and Ψ", we have discussed the words surrounding 'matrix' but will now see why its presence in the third stanza is pivotal.

> And find for one small world of fact
> Invariant matrices, compact
> Within the dark and igneous rock
> Of *Comptes Rendus* or *Proc. Roy. Soc.*
> [...]
> While I, maybe, precisely seize
> The elusive photon's properties
> In α's and δ's, set in bronze-
> bright vectors, grim quaternions. [9–12, 17–20][101]

Matrices are in mathematics tools for ordering and arranging numbers that allow for complex and many-faceted calculations. As we have seen with the prerequisite of 'fact' for 'act', structure itself in the poems is not accused of torpor. After all, setting values in matrices is but the first step in further calculation, theorization, and discovery. But in this poem, the conversion of discovery to matrices is narrated as an act explicitly of confining and congealing, like hardening lava. Thus, the relatively neutral term matrix requires a distinguishing property, 'invariance'. The *OED* defines invariance in mathematics as follows:

[98] Whitworth, 'Community', p. 280. [99] Ibid., p. 280.
[100] Michael Roberts, 'Symbolism and Romance', *Poetry Review*, 21.1 (1930), 34.
[101] Roberts, *Collected Poems*, p. 56 (first published in *New Statesman*, March 1935; written in January 1931 ('List of Verses').

invariant transformations are those in which the object remains "unchanged by a specific transformation or operation".[102] So the technical definition does not differ significantly from its common sense of remaining the same through some process. The descriptor 'invariant' thus serves to administer a dulling sensation from 'grasping a postulate' to setting down in mathematical representation. Roberts says, "Facts may be represented mathematically but 'working models' are utterly impossible".[103] At the same time, we are urged not to dismiss the romance in *doing* science. Roberts declares that "mathematics and mathematical physics have a beauty as precise and classical as that of Bach or Dürer, but it resides in its *method* not its subject matter" [emphasis mine].[104]

William Hamilton's discovery of quaternions marks, symbolically, an important stage in the mathematical picture's intellectual conquest of the world.[105] Quaternions are algebraically four-dimensional expressions that can be used to describe properties of three-dimensional space and to model higher dimensional rotations.[106] Quaternions can in some ways be seen as capturing three-dimensional space. From being limited to describing lower-dimensional rotations, like that of the clock hand, mathematicians could with quaternions describe more complex movements like that of a globe. Despite its potency as a tool, like dark matrices, quaternions are described as 'grim'. This reinforces the notion that most forms of mathematical representation in some primal sense sap the vitality of intuition and dim the brightness of ideas. Although vectors are often set in matrix form, the word 'vector', described as 'bronze- *bright*', carries in it the sense of "movement and directed magnitude",[107] bringing the word into the third cluster, whose relation to brightness we have already discussed. Setting in bronze has a sense of finality. But bronze, once cast, continues to express hidden energy through the aesthetic response and interpretive allure the statue entreats. Vectors, ambiguous like bronze, have fixed magnitude but also possess direction. They thus come to signify a kind of formalized activity.

The scientific method is shown to begin in seizing the elusive light of an idea. The hint of anagogic enlightenment is tempered by the technical register of 'properties' and 'photon'. Light, that unapproachable bliss in Milton, the

102 OED.

103 Michael Roberts, 'On Mechanical Hallelujahs, or How Not to Do It', *Poetry Review*, 19.6 (December 1928), 433–8, p. 438.

104 Ibid., p. 438.

105 Charles Jasper Joly, *A Manual of Quaternions* (London: Macmillan Co., 1905), p. v.

106 W.H. McCrea, 'On Matrices of Quaternions and the Representation of Eddington's E-Numbers' *Proceedings of the Royal Irish Academy. Section A: Mathematical and Physical Sciences*, 45 (1938), 65–71.

107 OED definition of mathematical vectors.

mysterious "bright effluence of bright essence increate",[108] is surrendered to Newtonian theory by Pope, in his "prismatic glass" that "Its gaudy colours spreads on every place",[109] and finally becomes mere object, in the modern notion of a 'photon'. The elusiveness of what is being seized is first captured in what was considered "Hamilton's admirable notation", the mysterious Greek character "α's and δ's",[110] which are bronze-bright vectors, which in turn populate the grim quaternions. The brightness and grimness of vectors and quaternions show the process of discovery as beginning in the light of cluster three and receding to the darkness of clusters one and two. The stanza expresses a typically English attitude to science which shrinks from the finality of mathematical theory, always utilizing mathematics to order and model phenomena sufficiently to further exploit part-understandings for an active advance, like the climber mapping the daunting bluff to aid his ascent.

"There is no sense in which the content of the science is essential to the poem", says Whitworth: "the speaker might equally well have been studying Mendelian genetics without the poem being substantially different".[111] It is germane to the poem, however, that in the history of mathematics, quaternions replace Euler's mode of representing rotation. Euler's angles were typically denoted by 'θ, Φ, and Ψ', the very symbols selected for the poem's title.[112] Perhaps the poet expects his readers to think of quaternions, the vectors in four-dimensional space, as freeing mathematics from earlier methods of modelling rotation that relied on Euler's angles to now be able to capture more complex movements: but that very enhancement of ability as another step towards, as Empson and Eddington imagined, the slight but complete mathematical capture of reality.

Hamilton himself saw the quaternion as a seminal step in this direction:

> Time is said to have only one dimension, and space to have three dimensions. [...] The mathematical quaternion partakes of both these elements; in technical language it may be said to be "time plus space", or "space plus time".[113]

[108] Book 3 of *Paradise Lost.*

[109] *Essay on Criticism* [313–14], in Alexander Pope, *Alexander Pope: Selected Poetry*, ed. Pat Rogers (Oxford: Oxford University Press, [1749] 1994), p. 8.

[110] Joly, *Manual*, p. 3.

[111] Michael H. Whitworth, 'Strange Synthetic Perfumes: Investigating Scientific Diction in Twentieth-century Poetry', in *Science in Modern Poetry: New Directions*, ed. John Holmes (Liverpool: Liverpool University Press, 2012), p. 93.

[112] Lokenath Debnath, *The Legacy of Leonard Euler: A Tricentennial Tribute* (London: Imperial College Press, 2010), p. 330.

[113] R.P., Graves, *Life of Sir William Rowan Hamilton*, vol. II (Dublin: Hodges, Figgin and Co., 1885).

Quaternions can in Roberts thus be seen as prefiguring the mathematical world-picture's eventual conquest of reality with Einstein's space-time. Himself an amateur poet, Hamilton pens a poem, "The Tetractys", to celebrate his discovery of the quaternion. He sees the feat as placing him in an ancient, illustrious lineage with Pythagoras, whose tetractys, the ten points arranged in four lines like bowling pins, conferred on the Brotherhood the belief of having found the secret of cosmic design.

> Or high Mathesis with its "charm severe
> Of line and number," was our theme; and we
> Sought to behold its unborn progeny,
> And Thrones reserved in Truth's celestial sphere;
> While views before attained became more clear:
> And how the One of Time, of Space the Three,
> Might in the Chain of Symbol girdled be[114]

Perhaps it is the view previously granted by Euler that became clearer with quaternions. This feeling of having attained Truth's celestial sphere with the algebraic potential of quaternions could be what led contemporaries like Lord Kelvin and Oliver Heavyside to, deeming that feeling illusory, suspect the quaternion as an evil.[115] In our poem, however, it is the girdling of the cosmos in mathematical symbol that merits quaternions the appellation 'grim'.

Roberts thus uses mathematical terms primarily in relation to other clusters—I hope the reader will now forgive our long preamble to the discussion of mathematics. To a mathematician, quaternions would seem a bright and fertile new way of dealing with numbers: but in Roberts' poems, the definition of terms in their native discourse is subordinated to their role in his linguistic-semantic web. With his carefully developed symbolism of hardness, lightness, and darkness, Roberts is able to establish cognately fine distinctions between his mathematical metaphors, to develop his complex attitude to different modes of thought.

Just as he modifies mathematical terms with varying scales of darkness to suggest a deadening effect created by mathematization, Roberts uses words like 'formal' and 'hermetic' to express the elite exclusivity that mathematics had acquired in its imperial career, after its humble role as a language of

[114] Graves, *Hamilton*, p. 524.

[115] H. Piaggio, 'The Significance and Development of Hamilton's Quaternions', *Nature*, 152 (1943), 553–5.

common sense in Bacon and Spratt. "In mind's dimension, bring to birth / In formal symbol, Venus, Mars, / and map the orbit, map the earth" [5–8].[116] The idea of formal symbols that are scrutable to few yet used in the conquest of 'the earth' and its 'orbit' is also conveyed through the word 'hermetic'. Like 'formal', 'hermetic' suggests being closed off from daily experience. In "And These Our Matins...", we have the phrase "hermetic matrices" [32]. There is also the curious phrase in "Nicholas Flamel", "hermetic ordinall" [20]. An ordinal is a book setting forth the rules and regulations of a practice. The line "Can we save / One poor hermetic ordinall?" [19–20] refers, perhaps, to Thomas Norton's poem, "The Ordinall of Alchemy", a primer on the "purer ways" of the ancients in matters occult.[117] But the term 'ordinal' likely also has a punned mathematical sense. An ordinal number is the place of a particular term, as in a series, there is the first, second, and third. The mathematical and alchemical senses might combine to point to the hierarchy or first truths of alchemy. For instance, Norton says, "Understood in the light of this [the works of the Ancients], my Ordinal, the truth of the matter, is fully set forth".[118]

In modern mathematics, the term 'ordinal' takes on more complex meanings. Georg Cantor introduced in 1883 the notion that infinities can vary in size.[119] For instance, consider the power set: if we have a set X containing the items {1, 2}, the power set P(X) = {∅, {1}, {2}, {1,2}} would contain all the subsets of X, including the empty set and X itself: and, we will note, be necessarily greater in size. Just so, if we imagine an infinite set like that of all natural numbers, {1, 2, 3, 4...}, its power set must be greater: infinities can thus vary in size and rise in scale, if we keep abstracting power sets from power sets. In a graph, we can treat the X line (with Y = 0) as the number line, and each rising level of Y can be used to represent a rising scale of infinity. Now an ordinal in modern mathematics can refer to a place on this celestial map of infinities.[120] A simple coordinate on the graph can be thought of as an ordinal: which suddenly acquires a greater-than-infinith place because entire infinite rows of coordinates would typically come before it.[121] In this way, a hierarchy of infinities can itself be ordinalled, as it were, according to the rising cardinal,

[116] Roberts, *Collected Poems*, p. 65 (first published in *New Country*, 1932).

[117] Thomas Norton, *The Ordinall of Alchemy, ed. Elias Ashmole Thomas, and Eric John Holmyard* (London: Edward Arnold, 1928).

[118] Norton, *Ordinall*, p. 66.

[119] Georg Cantor, 'Ueber unendliche, lineare Punktmannichfaltigkeiten', *Math. Ann.*, 21 (1883), 545–91, Section 5.

[120] Josè Ferreirós, *Labyrinth of Thought: A History of Set Theory and Its Role in Modern Mathematics* (Basel: Springer Basel AG, 2013), pp. 271–4.

[121] Joseph Warren Dauben, *George Cantor: His Mathematics and Philosophy of the Infinite* (Princeton: Princeton University Press, 1990).

so that one can speak in theory of the fifth or sixth infinity, say, though in actuality, we do not know how many there are in all and in between. If we look at the ordinal as a coordinate in the map of infinities, given it is bracketed, it could be thought of visually as a 'hermetic ordinal', but also hermetic because in identifying the ordinal among the hierarchy of infinities, we are wandering into fugitive regions of the mathematical occult. The one hermetic principle of alchemy is 'poor' because illusory, just as the region of infinities is an artefact of pure reason beyond the realm of intuition in which counting can happen. Such a reading would be in keeping with contemporary philosophical reflections on the transfinite, such as those of Brouwer and Ortega y Gasset, who saw the ordering of infinities as an act of mathematical imperialism that sought to prise the act of counting, fundamental to human life, from the realm of sensuous intuition.[122] Brouwer thought it would be best to avoid thinking of the natural numbers as sets altogether in order to avoid Cantorian mysticism.[123] Unlike the philosopher who asks if theory can save the phenomena, the occultist asks if the theory, the abstract, the ordinal can be saved for its own sake. But, as we have seen, Flamel fast realizes that 'none can [...] grasp the true alchemic gold / Or touch the substance of the stone'.

We have so far mostly seen mathematical concepts in collocation with clusters one and two—this is almost entirely the case with the prolific matrix. But outside the poems, in literary criticism, Roberts seems to find greater potential for matrices. I refer to his critique of Richards's Benthamite theory of counting impulses:

> Arithmetic and elementary mathematics come to be regarded as exceptionally reliable reasoning and therefore an attempt is made to force all reasoning into that mould. Mr. I.A. Richards, for example, speaks of the 'number' of impulses as if they could be counted like potatoes or 'events'. The pure mathematician [...] is not liable to fall into any such error. If the new romantics require a powerful new method of investigating the field of experience which is qualitative, not metrical, they might well start with a study of the properties of matrices: it would at least enable them to avoid that creeping arithmetisation which threatens to fossilise the "emotional" field, as completely as rationalism did two centuries ago.[124]

[122] See Dirk van Dalen, *L.E.J. Brouwer—Topologist, Intuitionist, Philosopher: How Mathematics Is Rooted in Life* (London: Springer, 2012), p. 114; and Lior Rabi (2016) 'Ortega y Gasset on Georg Cantor's Theory of Transfinite Numbers', *Kairos*, 15, 46–70.

[123] Dalen, Brouwer, p. 114.

[124] Michael Roberts, 'The Loneliness of Mathematics', *Adelphi*, 1.6 (March 1931), 510–11, p. 511.

Given Roberts never shows us just how he would use matrices for poetic analysis, I will now attempt the same on one of his own poems. Matrices, we have said, allow a set of terms to operate on another. For instance, two matrices may be added as follows:

$$\begin{bmatrix} 1 & 4 \\ 6 & 2 \end{bmatrix} + \begin{bmatrix} 5 & 3 \\ 8 & 7 \end{bmatrix} = \begin{bmatrix} 1+5 & 3+4 \\ 6+8 & 2+7 \end{bmatrix} = \begin{bmatrix} 6 & 7 \\ 14 & 9 \end{bmatrix}$$

Each term in the first matrix is added to a corresponding term in the second to yield a third. Poems in some ways work as microcosms of Roberts's word-modifications across collections: that is, words are introduced to then be modified by others through their deployment in rhythm and syntax, to then yield more refined senses. This allows us to experimentally replace numbers with words in the matrix form. In Roberts's poem, "Perspective", the squareness of the first and last stanzas in their tetrametric structure has prompted me to select them for the experiment:[125]

The music of no temporal ear
Is Euclid's world invisible,
In calm, pellucid, liquid air
The mental mode is sculptural.
[…]
And music of no temporal ear
Is all that world invisible;
In bright and frigid crystal air
The mental mode is sculptural [1–4, 21–4]

The first and last lines of both stanzas are the same. In lines 2 and 22, the music of the 'temporal ear' sees 'Euclid's world' and 'all that', respectively. 'All that' refers to the accretion of descriptive imagery from the intervening stanzas. The distance between 'Euclid's world', the world of geometry, and 'all that', the world conjured by the poem, can be scouted by observing the differences between the third line of each stanza. Both lines 3 and 23 have a common setting, the 'air', upon which we now place our matrix:

[calm pellucid liquid] + [ ] = [bright frigid crystal]

[bright frigid crystal] – [calm pellucid liquid] = [ ]

[125] As Whitworth notes, the stanzas are also reminiscent of Read's poem, 'Equation', which Roberts admired (Whitworth, 'Community', p. 274).

'Calm' has become 'bright', 'pellucid' has become 'frigid', and 'liquid', 'crystal'. We can now deduce what operations were performed on each word in the first stanza to yield the corresponding terms in the last. The attribute of translucency obtains between 'liquid' and 'crystal', but the refractive property of the translucent medium is intensified in the crystal, such that the vision, musical or sculptural, is in the final stanza not placid but dynamic. While the properties of 'crystals' in the poems are familiar to us from the fourth cluster, 'liquid', the scientific, abstract and general noun for all manifestations of water, modified by 'calm', does not seem to belong to the third cluster of rivers and movement. The transformation from 'calm' to 'bright'—that liquid when crystallized becomes bright—supports this reading, as 'brightness', we have seen, suggests anarchic activity. 'All that', we know from the intervening stanzas, consists of 'impulse unconfined', 'scourged love', and 'rhythmic pulse'—with these added, the 'calm' world of Euclid becomes 'bright'. 'Pellucid' denotes both transparency and lucidity; because 'Euclid's world' is not complicated by circumstance, it is the picture of simple clarity. Transparency cannot be reconciled with darkness, and by extension, 'Euclid's world' is not to be confused with the world of rocks and facts—it is nearer the unearthly music of the spheres, the 'clear lamp of Lully's mind'. This extends to other geometric jargon in the poems. For instance, when in "Note on θ, Φ and Ψ", the poet 'grasps a postulate', he is being as creative as an artist. 'Frigid', on the other hand, as we saw in "Artificer in Torment", comes loaded with bitterness, alienation, and silent repose: "a certain frigid beauty" that Roberts finds in the verses of Pope.[126] Moving vertically from 'pellucid' to 'frigid', and horizontally from 'bright' to 'frigid', we begin to develop an idea of the 'sculptural mode' as abstract precision (moving down) and frozen emotion (moving across). We can see a similar freezing of emotion in Sterne's description in *Tristram Shandy* of "a thin, blue, chill, pellucid chrystal with all its humours so at rest".[127] The poem might also draw from Hulme's descriptions of geometric art, from the haughty and angular effigy of an Egyptian Pharaoh, say; and perhaps also from Yeats's "The Statues", which when 'live lips meet plummet-measured face', captures frozen passion in proportioned stone.

"Geometry has never appealed to me", says Roberts: "it is, in its lower reaches at least, at once less subtle and more restricted than analysis".[128] In "Earth,

[126] Michael Roberts, 'Pope and English Classicism', *Poetry Review*, 21.3 (May–June 1930), 61–70, p. 162.
[127] Lawrence Sterne, *The Life and Opinions of Tristram Shandy, Gentleman* (London: J.M. Dent and Sons [1759] 1917), p. 441.
[128] Roberts, 'Credo', p. 193.

Impact", geometric imagery is used to express this restrictiveness. "Down to the last abstraction, Earth / Fulfils her geodesic curve" [33–4]. A geodesic is the shortest distance between two points on a curved surface; more generally, the path of least resistance, an effortless course. The poem could imply that the Earth behaves in ways commensurable to our condensed 'abstractions'; Earth, in revealing to us her patterns and regularity, by 'fulfilling' the mechanical destiny set out by astronomy, becomes contained in our geometry. But this might be illusory: "Mathematics in general like Euclidean geometry in particular is based on human experience, and the apparent scientific unity of the world may be, as Hulme suggested, due to the fact that man is a kind of sorting machine."[129] The phrase 'geodesic curve', especially its non-Euclidean tenor, expresses this ambivalence in its appeal to sensual flair in the image of buxom Earth fully filling her elegant curves or pirouetting in the soft curve of her orbit. The poet seems to regard the spinning Earth with the same inclination as he might his departing mistress. This reading is prepared by the line, "fertile earth / Fulfils her rhythm" [17–18]. The sensuous potential of geometric imagery is again invoked in "Jura in Islington", which begins, "Sing in my heart, you pointed pines, / And trace your geometric lines" [1–2]. Geometry more directly signifies beauty here; the lines of landscape are the curves of human form.

In *Critique of Poetry,* Roberts says, "because some non-visualisers (such as Pascal) find aesthetic satisfaction in mathematical elegance, he assumes that the most abstract words can become associated with emotions. Geometry may be studied by both mental types, diagrammatically in the one case, and, in the other, analytically".[130] In his poems, Roberts thus finds for mathematics a wide range of significance across all the previous clusters. As with 'fact', Roberts loads mathematical words densely with meaning, as if to answer those who would dare make language mathematical. In his poems, matrices gird and grid the landscape of reality, shapes impassion and control, and curves trace its sensuous outlines.

3.1.6 Eros and logos

As critic, Roberts fought all his life against the separation of reason from passion, and precision from imagination. He lamented that "the languages of poetry and science are still separating; the scientist is making his more

[129] Roberts, 'Loneliness', p. 511. [130] Roberts, Critique, p. 50.

precise, restricted and mathematical; the poet is making his more widely symbolic, more allusive, more capable of organizing emotional states".[131] As poet, he managed to expand semantic fields, recover lost significance, impose unexpected meanings, and ultimately, in his poetic landscape, blur the boundaries between the supposed two cultures.

Roberts's poetry brings to bear the full sensual and intellectual range of words. The moral system of the poems is not entirely rid of the Paterian, Ricardian view of art. Although Roberts rejected Richards's calculations, his poems associate the good with a wider, though curated, range of impulses. For instance, 'hardness', we will recall, can suggest rigidity, sublimity, or even form *in potentia*. By distending clusters and clashing one against the other, the poems seek balance with and fidelity to the manifest modes of experience.

As he traces the alienation of eros from logos in modern times to the seventeenth century, we might view Roberts's project as one of semantic revanchism. He seems in this regard to follow Coleridge:

> Throughout his work, Coleridge takes terms from special sciences, words which have a definite function as counters in an exact calculus, and fogs them, adulterates them, in a desperate effort to give back to them all that richness of meaning which the scientist has carefully taken away in order to state his neat and syllogistic science.[132]

Besides 'adulterating' mathematical terms, Roberts tried also to reunite the passions with the virtues—accuracy and precision—reserved for the mathematical sciences. Fact and passion in his poems are as intervolved as light and shade.

Cunningham says Roberts "forces together his own activities as poet, mathematician, philosopher, critic and climber, and seeks pervasive analogies between writing and climbing".[133] If a creed can at all be found in the poems, it is that mountaineering, mathematics, art, and language are varying manifestations of the same underlying enigma. He said once, "I believe in the rational order of nature, in the economy of thought, and in the economy of symbolism. These are the fundamental principles of metaphysics, mathematical science and aesthetics respectively."[134] Logos and eros are

[131] Michael Roberts, 'Beyond the Golden Bars', *Poetry Review*, 20.2 (March–April 1929), 109–18, p. 117.

[132] Michael Roberts, 'The Two Coleridges', *The London Mercury*, 32 (1935), p. 293.

[133] Cunningham, *Thirties*, p. 165.

[134] Roberts, 'Credo', p. 196.

separate from and incomplete without the other, and so always in dialectic tension in his poems.

The poet wonders in a "Reverie by Candlelight", what it would be if mind and nature were united: "Mark how each word must strain and bend / To ape the vision that is caught: / Would one perfection bring the end, / And swing full circle, all to nought?" [17–20]. To auger the 'one perfection', incarnate this logos, the poet will suppress his "vague unbounded lust". And to express (catch) the final vision, the fusing of mind with nature, the poet must find union between word and world: and so, strain and bend words like a magus. The Cambridge Platonists, about whom Roberts writes in the *Modern Mind*, speculated that man would no longer need language to converse were he an angel.[135] Once mind recovers its prelapsarian oneness with nature, there would be no struggle to fit words to thought: all would know all, and words would become useless. As he—somewhat suicidally, given his vocation as poet—dreams of closing this gap, he like Empson's cavalier lover at "The World's End", sinks to oblivion, and brings 'all to naught': oneness with nature thus annuls knowing.

Knowledge, therefore, can only be of imperfection: and comes from a life of trying to attain perfection. The impious hope of uniting mind with matter by reconciling logos with eros thus remained with the poet to the last. In "And these our Matins", he prays: "And these our matins / rise and ring, / Austere and cold, the stony hills, / Till words can shatter / living rock / And pent emotion recombine / with stranger matter, torn from these / its set Hermitic matrices!" [25–32]. Words shatter rock, the world of fact. Like crystals forming around dust, the polluted words of poetry allow matter to shine through the dark matrices. And so, in a sense Earth, released from hermetic embalmment in the mathematical scheme, once more regains her strange (and ever-estranging) allure. Just so, eros, the pursuit and failure, disclosed to Roberts the truth of substance: its beauty and unattainability to man.

3.2 The Metaphysical Poem

We turn now from the immanent god of the rocks to the heavenly spheres from which a mathematical poet will deign to apply for erotic love. We will read closely a most curious love lyric by Empson, which bids us imagine how

[135] See, for instance, Benjamin Whichcote's aphorism 732: "If it were not for sin, we should converse together as Angels do", in *Moral and Religious Aphorisms* (Norwich: Fr. Burges, 1703).

troubadour praise and persuasion might seem if the powers of the poet were reduced to geometric description.

I have attempted to show in this book that one of Empson's enduring achievements as poet is in demonstrating to what extent the poetic form can be an arena for the contestation of philosophical ideas. John Haffenden says, "Empson believed that the best poetry (in particular his favourite metaphysical poetry) handles philosophical and ontological problems, being interested in far more than the process of mind—the apotheosis of the individual sensorium—and striving beyond egoism to articulate ideas and statements of meaning."[136] This is was what first struck the readers of Empson's literary criticism, which treated poems as full slices of past intellectual worlds, whose lineaments would then be displayed in his close readings. His own poems seem like catchments of the intellectual history of his time: but they are not merely philosophy, for their episteme lies in their techne. To come to terms with this poem, "Letter V", the reader must come armed with contemporary debates in mathematics, science, philosophy, and art, but to see it whole, navigate a whole armada of Renaissance techne: catachresis, endophora, prepositional and pronomial ambiguity, and indeed, the seven types of ambiguity, and more.

Such a haunted and neurotic project throws up unique challenges to literary criticism. Reading, re-reading, and back-reading, all effect subtle alterations to the poet's emotional state, which in turn affect the eventual, albeit fleeting, suggestion of the poem. Close reading today proceeds primarily through paraphrase, which proves inadequate in conveying the mechanics of prolepsis and retrodiction in the poems. Empson's own paraphrastic endnotes often seem comically facile. Roberts once said that to Empson, "meaning is not any of these readings [the series of paraphrases], nor is it their arithmetical sum, it is the result of having all these in mind at one time".[137] We will try in this reading to keep the various vectors of meaning in "Letter V" before the reader at all times by trailing closely the work that gets done by individual words acting upon each other.

From a surface reading of tone, we can see that the poem is trying to demonstrate how a mathematical love poem is absurd. To draw meaning from absurdity without reducing the subject to a farce is a tricky task. To achieve this, Empson finds a convention where it would not seem merely perverse to

[136] Haffenden's introduction in William Empson, *Argufying: Essays on Literature and Culture*, ed. John Haffenden (London: Chatto & Windus, 1987), p. 16.

[137] Roberts, *Critique*, p. 26.

sing geometric praise: a largely forgotten Renaissance technique of periphrasis. The central argument of the lyric thus becomes a contest between geometric periphrasis and poetic metaphor. And so, the agon resumes.

> "Letter V"
>
> 1. Not locus if you will but envelope,
> Paths of light not atoms of good form;
> Such tangent praise, less crashing, not less warm,
> May gain more intimacy for less hope.
>
> 5. Not the enclosed letter, then, the spirited air,
> The detached marble, not the discovered face;
> I can love so for truth, as still for grace,
> Your humility that will not hear or care.
>
> 9. You are a metaphor and they are lies
> Or there true least where their knot chance unfurls;
> You are the grit only of those glanced pearls
> That not for me shall melt back to small eyes.
>
> 13. Wide-grasping glass in which to gaze alone
> Your curve bars even fancy at its gates;
> You are the map only of those divine states
> You, made, nor known, nor knowing in, make known.
>
> 17. Yet if I love you but as Cause unknown
> Cause has at least the Form that has been shown,
> Or love what you imply but to exclude
> That vacuum has your edge, your attitude.
>
> 21. Duality too has its Principal.
> These lines you grant me may invert to points;
> Or paired, poor grazing misses, at your joints,
> Cross you on painless arrows to the wall.[138]

The poem's absurdities work as parodies on the conventions of sensuous love poetry, whereby its subject is eventually rendered in a language opaque to carnal appetite. Although four "Letter" poems precede, the fifth, with 'envelope' and 'enclosed letter', is the first to draw attention to its title. The

[138] William Empson, *The Complete Poems*, ed. John Haffenden (London: Penguin, 2000), p. 60. This poem is published with the express permission of Curtis Brown, the literary estate of William Empson.

poet says he will not address the 'points' but focus on the surrounding envelope; he will respond not to what is inside the 'enclosed letter' but the idea they suggest, the 'spirited air', where he will conduct an experiment to convert matter to form, the actual to the real. It is a rather slanted continuation of the theme in the other "Letter" poems, of the troubles to communication and cohabitation in modern times.[139] The experience of space-time having been warped by relativity, "Letter I" begins the series with a 'non-Euclidean predicament' to human intimacy. It asks, now that the world, the common stage for love, is no longer recognizable to common sense, how are we to communicate, find common ground? By "Letter V", the poet realizes that commitment to the mathematical world-picture brings his lover's very existence eventually to question. Writing from the prison of abstraction, the poet of "Letter V" eventually confesses the futility of his penumbral adoration.

3.2.1 Metaphysical revision

In the tradition of humanist love poetry, the mistress of "Letter V" is unattainable. Although sex is hardly relevant in the poem, we will require it for pronoun shorthand: interpreters have mostly used 'she', but 'he' seems more appropriate, when the arrows of the final line bring to mind those transfixing St Sebastian.[140] Kitt Price identifies the addressee of the "Letter" poems as a certain Desmond Lee, if such biographical particulars should help grasping an object so general.[141] Having 'less hope' that this boy will 'melt for me', the poet addresses instead his ideal Form. The Renaissance mistress unattainable to troubadour persuasion is an attractive yet fraught model for the youth of "Letter V". At first, Petrarch's platonic Laura seems to adequate 'the Form' with 'Cause unknown' [17–18].[142] Leah Whittington records a shift in Italian love poetry upon the arrival of Neo-Platonism in Florence: "the supplicatory mode of addressing the beloved, with its concomitant hope for an erotic response" was replaced by "laudatory speech

[139] Katy Price, 'Monogamy and the Next Step?: Empson and the Future of Love in Einstein's Universe', in Mathew Bevis, ed., *Some Versions of Empson* (Oxford: Oxford University Press, 2007), p. 245.

[140] St Sebastian was a gay icon according to Christopher Ricks's introduction to T.S. Eliot, *Inventions of the March Hare: Poems by T.S Eliot 1909–1917*, ed. Christopher Ricks (London: Faber & Faber, 1996), pp. 267–70. Ricks, however, is adamant to domesticate Empson's love poems in 'Empson's Poetry' in Roma Gill, ed., *William Empson: The Man and His Works* (London: Routledge, 1974), pp. 145–207.

[141] Price, 'Monogamy', p. 244.

[142] Robert Valentine Merill, 'Platonism in Petrarch's "Canzoniere"', *Modern Philology*, 27.2 (1929), 161–74, p. 161.

[...] celebrating the lady's divine virtue".[143] The poet in "Letter V" likewise has 'less hope' for carnal success, and so as a matter of course, moves on to 'the divine states' to which this transient beauty points. However, the 'ladder of love' fast exhausts its explicative potency, as the deeper concern of "Letter V" lies beyond the adoration itself to the psychological effects of loving vacant Form. So, we must look further. After Shakespeare was done toying with the Petrarchan tradition, his "mistress' eyes" being "nothing like the sun", the Metaphysical poets began to recognize in the unapproachability of the lady a limitation to language itself. Sensing this, Donne, according to Frank Doggett, modified the trendy Neo-Platonism of his day by restoring sensuality to the thitherto pious exaltation of virtue.[144] Read against the voluptuary verse of the seventeenth century, "Letter V" seems a conscious reversal of metaphysical aims, as it attempts to impress how distant to rational mind the body seems in its twentieth century situation—the 'grit' of line 11 placing the incarnation of heavenly beauty further from sensuous apprehension than perhaps it has ever been in a love poem.

The Gardners say, "what force the poem has comes not from Empson's intellectual attempts to define the indefinable, but from the reader's sense of an amatory failure".[145] To endure rejection, the poet petrifies his lover into abstract form. The boy is thus outlined through periphrasis, held in place, rather than enfleshed by metaphor. Perisphrasis, it has been argued, was in Renaissance poetry used to express inexpressibility. Dorothy McCoy defines it as "a figure which usually is considered to elaborate, elevate, or decorate".[146] The term has historically signified both euphuistic excess and circumlocution. In 1560, Thomas Wilson illustrated the relation between the two senses; periphrasis, he says, can be a "large description, either to set forth a thing more gorgeously, or else to hide it".[147] From John Hoskins's use of the term in his theories of rhetoric, 'periphrastic' has also come to mean "strange and admirable".[148] The poet of "Letter V" uses periphrases—'envelope', 'paths of light', 'detached marble'—to speak indirectly of, and in some ways to hide the physical thing his lover is—the 'locus', the 'atoms of good form', 'the discovered face'; but in the process, the object of his elevated circumlocutions comes to seem 'strange and admirable'. McCoy says the "periphrastic treat-

143 Leah Whittington, *Renaissance Supplicants: Poetry, Antiquity, Reconciliation* (Oxford: Oxford University Press, 2016), p. 90.

144 Frank A. Doggett, 'Donne's Platonism', *The Sewanee Review*, 42.3 (1934), 274–92, pp. 281–2.

145 Philip Gardner and Averil Gardner, *The God Approached: Commentary on the Poems of William Empson* (London: Chatto & Windus, 1987), p. 147.

146 Dorothy Schuchman McCoy, *Tradition and Convention: A Study of Periphrasis in English Pastoral Poetry from 1557–1715* (The Hague: Mouton & Co, 1965), p. 14.

147 Ibid., p. 175. 148 Ibid., p. 24.

ment of a word or an idea is often unwieldy. Its failure is likely to be more evident and its triteness more annoying than that of other figures."[149] This is evidently so in "Letter V"—but while critics at times cite the failure to dismiss the poem, we will see it as an effect artfully achieved.

The seventeenth-century poet John Cleveland may have been the most direct inspiration for the poem. Cleveland's periphrases were elaborate and often intended for comic effect. Herbert Grierson, in his introduction to *Metaphysical Lyrics and Poems*, regards Cleveland an important figure to the metaphysical movement.[150] Colin Burrow argues that Cleveland's

> 'Hecatomb to his Mistress' is perhaps the point at which metaphysical poetry first shows signs of defining itself as 'metaphysical' [...] In a passage which Johnson cites in his definition of 'metaphysics' in his *Dictionary* of 1755, Cleveland urges poets to set aside hyperbolic praise of their mistresses, since his mistress exceeds any possible extremity of comparison: 'Call her the metaphysics of her sex, And say she tortures wits'.[151]

"Letter V" seems consciously to modify the metaphysical tropes in "Hecatomb" to create meaning as if by distinction.

The poet of "Hecatomb" begins with an arrogant dismissal of his peers, telling them to "Be dumb ye beggars of the rhiming trade" [1].[152] His derision is directed first to poetic metaphor; and he then prays the reader suffer him 100 lines of circumlocution to demonstrate why periphrasis is better fitted than metaphor to capture the real essence of his mistress.

> My text defeats your Art, ties Nature's tongue,
> Scorns all her tinsil'd metaphors of pelf,
> Illustrated by nothing but her self.
> [...]
> So is it with my Poetry and you.
> From your own essence must I first untwine,
> Then twist again each Panegyrick Line. [8–10, 14–16]

[149] McCoy, *Periphrasis*, p. 12.

[150] This was possibly Empson's first encounter with the poet; Herbert Grierson, *Metaphysical Lyrics and Poems of the Seventeenth Century* (Oxford: Oxford University Press, 1921); Haffenden affirms Empson's intimate familiarity with Grierson in John Haffenden, *William Empson: Among the Mandarins* (Oxford: Oxford University Press, 2005), p. 361.

[151] Colin Burrow, ed., *Metaphysical Poetry* (London: Penguin, 2006), p. ii.

[152] John Cleveland, *The Poems of John Cleveland*, eds Brian Morris and Eleanor Withington (Oxford: Oxford University Press, 2013), p. 50.

"Letter V" begins just so, rending its addressee into attributes that are essential and expendable, preparing the material, as it were, for periphrasis. In presenting his poem as the original and her beauty as the artwork, Nature's metaphor, if you will, Cleveland is reversing with arch conceit a well-trodden seventeenth-century tradition, most famously exemplified in Shakespeare's Sonnet 111, of verse backwardly acknowledging its own distortions, confessing in its art a tendency to veer from the notional object of praise. And so, in "Letter V", "you" becomes the "metaphor", the 'Art', while "These lines you grant me" become the truth, the 'panegyrick line' that Cleveland twists and unwinds from his mistress.

> As then a purer substance is defin'd
> But by an heap of Negatives combin'd,
> Ask what a Spirit is, you'l hear them crie,
> It hath no matter, no mortalitie:
> So can I not define how sweet, how fair,
> Onely I say she's not as others are.
> For what perfection we to others grant,
> It is her sole Perfection to want. [31–8]

After an initial bluster, the poet finds that to love pure spirit poses a challenge to the familiar diction of poetry—sensory words like 'sweet' and 'fair', as it happens, do not adhere to ghostly form. Similarly, in "Letter V", the 'paths of light' lack conspicuously the properties of the 'atoms of good form'—however sharp the poet's protractor, the mistress's outlines are never quite buxom enough.

In the manuscript version of "Letter V", the poem had the line, "I have loved the sea-mark of a standing negative".[153] This seems to have been an attempt to develop line 7, 'I can love so for truth, as still for grace'. Taken together, the lines assure the quizzical reader that he *can* love a bare truth, for he has already loved a symbol without instance: a 'sea-mark' is the high tide line and a navigation sign for sailors lost at sea; and a 'standing negative' seems contradictory, a 'negative', non-existent and to sense inconceivable, asserting a standing presence. In other words, the poet not only infers what is absent—the negative—by a mark, the silhouette of the tide line—but he is able then to love it. The unpublished line seems also to also prepare the 'vacuum' in line 20 that 'has your edge'—of which hereafter.

[153] Empson, *Complete Poems*, ed. Haffenden, p. 284.

'Metaphors', in "Hecatomb", are 'pelf' and in "Letter V", 'lies', and so are replaced with periphrases—the writing around a 'standing negative' to disclose, however tangentially, 'a purer substance [...] defin'd / But by an heap of negatives combin'd'. Defining through negatives—or showing by concealing—we will recall, is how periphrasis was generally used in the Renaissance. Recall Donne in "Negative Love": "If that be simply perfectest, / Which can by no way be express'd / But negatives, my love is so" [10–12]. Remembering the once incarnate lover of "Letter V" as the 'positive presence', representing him from without, through tangents, becomes then our poet's way of loving the 'standing negative'.

Cleveland develops his negative praise: "She that affords poor mortals not a *glance* / Of knowledge, but is known by ignorance" [93–4] [emphasis mine]. According to Haffenden, the 'glanced pearls' of "Letter V" carries the suggestion of 'glimpsed' and a secondary one of 'polished'.[154] In both poems, the former sense suggests how fugitive is the glimpse of 'true' form, one available to mind but not to the senses. The implication of 'glance' in "Letter V", however, is primarily physical, as in the mathematician poet's tangents glancing off the lover's body, scintillating his outline—not unlike the Sephardic sense of negative expression one finds later expressed in Derrida's use of "tangentiality", whereby presence is made sensible by a series of glancing blows at "the limit".[155] Or given Empson's interests in oriental philosophy, we may think of Confucius's description of Yao, the old king of Chung-kuo, whose righteous splendour is brought into relief by listing all the articles of ostentation that didn't adorn his noble and austere personage: "His person was not adorned with jewels, and his clothes were without embroidery, simple and without variety", and so Confucius goes on.[156] As the mistress of "Hecatomb" begins to dissolve at the margins of advancing intellect, the poet scrambles to grab onto some connection with reality, "Till bafled Poetry hangs down her head, / She, she it is that doth contain all bliss, / And makes the world but her Periphrasis" [98–100]. Periphrasis is thus used to hold before the reader an outline of vacuum by mentioning all she is not—which becomes the world itself. That is, the 'heap of negatives combin'd' amounts eventually to the world itself, which must be overcome to catch her glimpse. This lament, which we have seen before, for a world left behind in the process of making

[154] Ibid., p. 286.

[155] Jacques Derrida, *On Touching—Jean-Luc Nancy*, trans. by Christine Irizzary (Stanford: Stanford University Press [2000] 2005), p. 136.

[156] Confucius, *Book of History* (rendered and compiled by W.G. Old, 1908), pp. 20–1.

concrete the realm of purer forms, is one of the central aspects of metaphysical poetry that its twentieth century heirs revive.

Cleveland calls his mistress "the *Square Circle*, say / She is the very rule of *Algebra*" [83–4]. Her essence makes a mockery of geometry—the branch of mathematics that conforms most to sense-impressions—and testifies to the rule of algebra, which replaces the concrete with fungible abstract symbols. Empson similarly begins with mathematical periphrases: his lover is not 'locus' but 'envelope'. He uses here an exaggerated mode of poetic comparison, catachresis, which was often employed by Donne. Donne had used the tool to "extend the comparison far beyond rational bounds so that the mind may on its own discover the truth or the 'unspeakableness' of the relation".[157] But the poet of "Letter V" begins in a smug mood; his resort to geometric description doesn't seem a confession of inexpressibility, but rather a haughty scan over his object.

Empson glosses line 1 as follows: "a locus defines a surface by points and an envelope defines it by tangents".[158] 'Locus' seen as points consists with line 11, where the lover becomes but 'grit'. Locus and grit picture the body as seen from the poet's Olympian vantage.[159] A conceptual change in the twentieth century with the popularization of set theory gives 'locus' added layers of meaning. Modern mathematics re-defined 'locus'—which had simply meant 'place'—to a set of points that satisfy a particular condition.[160] Let us draw the locus of points with the condition that they each be at a distance, r, from the origin (0,0) (see Figure 3.1).

Upon connecting the points with this condition, the locus becomes a circle with radius r. The dual sense of locus emerges by taking the circular line as (1) the set of points equidistant from the origin[161] and/or (2) the path traced out by one point such that it remains of equal distance from the central point.[162] In various lines of the poem, either of the senses, of many points or one moving point, is implied: 'grit only' pictures the body as an insignificant point, whose ineffable aura, polished by negative periphrases, gradually glis-

[157] Thomas O. Sloan, 'The Rhetoric in the Poetry of John Donne', *Studies in English Literature, 1500–1900*, 3.1 (1963), 31–44, p. 42.

[158] Empson, *Complete Poems*, ed. Haffenden, p. 285.

[159] And not, as the Gardners seem to think, "her invisible interior essence" (Gardner and Gardner, *God Approached*, p. 149).

[160] Roger Cooke, *The History of Mathematics: A Brief Course* (New Jersey: Wiley [1997] 2013), p. 459.

[161] A.N. Whitehead, *An Introduction to Mathematics* (London: Williams and Norgate, 1911), p. 97.

[162] John Casey, *A Sequel to the First Six Books of the Elements of Euclid Containing an Easy Introduction to Modern Geometry with Numerous Examples* (Dublin: Hodges, Figgis, & Co., 1886), pp. 5–6.

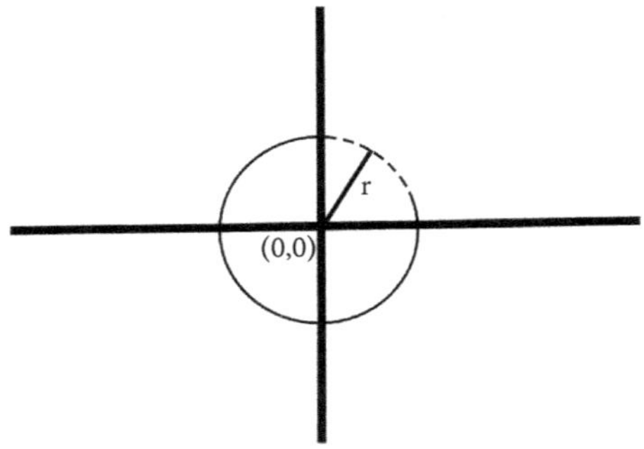

Figure 3.1 Locus of points at distance 'r' from origin.
Source: Image produced by author.

tens to an abstract, other-worldly 'pearl'—and thus, the lonesome hell of the mathematician becomes "darkness visible"; on the other hand, the phrase 'atoms of good form' agrees with the Democritean picture of human being as a collection of points.

Points, atoms, and grit are, of course, what his lover will *not* be praised as. The poet's 'tangent praise' will instead form an envelope, a line or curve that touches all the members of a family of curves or lines.

In Figure 3.2 the envelope is again the circle, but here it is defined as the tangent to each member of a family of lines; that is, the tangent to these criss-crossing lines (defined by some function) happens to be the circle. This poses a seeming complication to the distinction asserted in line 1, because an envelope may just turn out to be a locus. The set of tangents that form an envelope can amount to a locus whose moving point simply has the condition that it trace out the envelope (instead of the condition of maintaining the distance r from the origin which we gave to the last diagram). However, the terms are opposed with such analytic precision in their metaphorical groups that a clear opposition emerges in the poem: declaring, roughly, that "I shall comprehend you as the grazing tangents that imply your form, which outlines an ideal though non-existent image, rather than as the atoms, the points of the loci I can see and feel."

The contrast in line 2 between 'paths of light' and 'atoms of good form' refers to the wave particle duality in quantum theory's conception of light. The poet insists that his mistress appears to him not in just any avatar of light, but as paths, waves, for no point nor particle, nothing vulgarly discrete, can impede the geodesic he is about to trace. These subtexts all introduced in the first stanza contextualize the vision of the poem as emerging from an isolated mind captive to the mathematical world-picture.

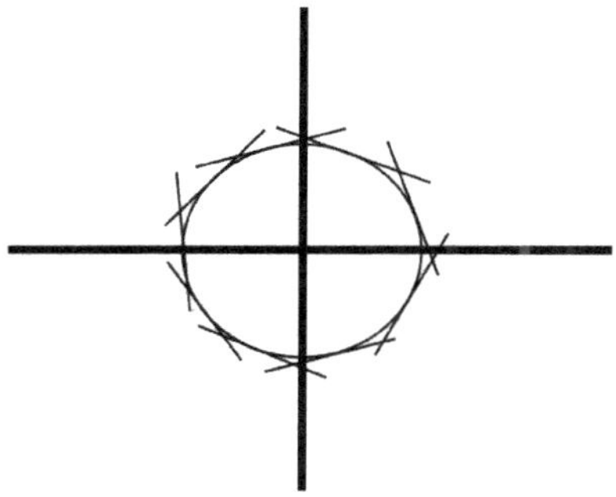

Figure 3.2 Envelope tangent to a family of lines/curves.
Source: Image produced by author.

Empson's habit of using metaphors out of Einstein and Heisenberg resumes Donne's tradition of using the theories of Copernicus and Galileo in his love poems.[163] Association with the often-strained sophistry of Donne has at times led critics to accuse Empson's poems of lacking feeling. In *Renaissance Literature*, Empson said of Rosemond Tuve,

> As I understand her, she treats the Donne line of talk that the idealised woman is a world, or that the two happy lovers are a world, as a straightforward use of the trope amplificatio. [...] I do not think you get anywhere with Donne unless you realise that he felt something different about his repeated metaphor of the separate world; it only stood for a subtle kind of truth, a metaphysical one if you like, and in a way it pretended to be only a trope; but it stood for something so real that he could brood over it again and again.[164]

Here too in the sleepless world of Empson's poems, mathematical images 'pretend to be only a trope' but reveal the poet inching to some 'subtle kind of truth' as he broods 'over it again and again'. Hill marks his "astonishment that Empson's poetry was once considered to possess the 'neutral tone' when his lyrics strike one as being lit by a baroque protean gleam".[165] In Empson, as in Donne, "there is a recreation of thought into feeling"[166] or "felt thought".[167] The adopted metaphysical label allows Empson to refine his wit-induced

[163] Price, 'Empson's Einstein', in Meyer, *Literature and Science*, p. 98.
[164] William Empson, *Essays on Renaissance Literature: Volume 1*, ed. John Haffenden (Cambridge: Cambridge University Press, 1994), p. 5.
[165] Geoffrey Hill, 'The Dream of Reason', *Essays in Criticism*, 14 (1964), 91–101, p. 92.
[166] Eliot, 'Metaphysical Poets', p. 669 (1921).
[167] Herbert Read, *Reason and Romanticism* (London: Faber & Faber, 1926), p. 40.

emotion by modifying convention, as we have seen. He writes that in Donne's poetry,

> The individual praised is the Logos of the virtues he or she typifies, he or she is the abstract idea Beauty or Virtue itself, and therefore constitutes the reality of those qualities in any person who possesses them. The lovers who are a separate planet get part of their dignity from this process, because they epitomise the actual world and are partly its Logos; this is what equips them to be a complete planet.[168]

The youth of "Letter V", however, is further exiled from the liminal state between *morphe* and *eidos* that the Renaissance lady once graced. In loving pure logos, there is 'less (if any) hope' "of getting to bed together".[169] The poem admits no physical, or indeed even intellectual, meeting of lovers: as in "Letter I", one imagines the poet, far from fusing into one planet with his lover, as an isolated star content with himself.[170]

The act of *defining* his peculiar eros in the poem, bordering on the neurotic, develops yet another metaphysical trope.[171] It was attempted by other twentieth-century metaphysicals such as Herbert Read, in his "The Analysis of Love". Bradbrook and Thomas say the word 'definition' in Marvell's "The Definition of Love" would have sounded to its readers as technical pedantry, since the very notion of rigidly circumscribing denotation was alien to Renaissance culture. That "love, the unruliest of the passions, is to receive a definition",[172] might have seemed to the mid-seventeenth century as absurd as passionate praise in geometry would to the twentieth.[173] In *Elizabethan and Metaphysical Imagery*, Tuve points to a similarly jarring exercise in Robert Herrick's two-line poem, "The Definition of Beauty":[174] "Beauty no other thing is, than a beam / Flash'd out between the middle and extreme". Beauty

[168] Empson, *Renaissance*, p. 74.

[169] From Empson's letter to Maxwell-Mahon (Haffenden, *Poems*, p. 286).

[170] Reading given by Katy Price, 'Flame Far Too Hot: William Empson's Non-Euclidean Predicament', *Interdisciplinary Science Reviews*, 30.4 (2005), 312–22.

[171] The Gardners say Marvell's 'The Definition of Love' stages the "human conflict" in 'Letter V' (Gardner and Gardner, p. 147).

[172] M.C. Bradbrook and M.G. Lloyd Thomas, *Andrew Marvell* (Cambridge: Cambridge University Press, 1940), p. 45.

[173] It gradually stopped seeming so with the advancing influence of the Royal Society in the latter half of the seventeenth century; we will recall Thomas Spratt who in his History of the Royal Society (1667), asked for a return to "a close, naked, natural way of speaking [...] bringing all things as near the Mathematical plainness, as they can" (qtd in Reinhold Schiffer, *Oriental Panorama: British Travellers in 19th Century Turkey* (Atlanta: Radopi, 1999), p. 344).

[174] Rosemond Tuve, *Elizabethan and Metaphysical Imagery* (Chicago: Chicago University Press, 1947), p. 302.

draws from the commonly recognizable but expresses itself in ideal perfection. Herrick's may even be a definition of beauty in art: for how is it that the Mary of Michelangelo's pieta is at once maiden and angel, earthly and heavenly Mother? The poet of "Letter V", however, elides the middle, the common, and defines beauty instead from the extremes with 'such tangent praises', and indeed, the results are 'less crashing'.

Empson's parody of convention often seems to take things unnaturally far, but this may not owe only to his unmanaged cleverness. His pyrotechnics, like Donne's, seem designed to demonstrate the barrier to eros, which in the twentieth century was not the region cloud of puritanism, but the cultivated torpor of rationalism. While in "The Definition of Love", "the love which us doth bind / Fate so enviously debars" [29–30], in "Letter V", the mind that perceives only bare 'curves bars even fancy at the gates'. It is not an external fate but the mind itself that has become enemy to eros. The distinction is marked even in the different geometric metaphors that are employed. Marvell uses Euclid's fifth postulate:

> As lines (so loves) oblique may well
> Themselves in every angle greet:
> But ours so truly parallel,
> Though infinite, can never meet. [25–8]

In the Euclidean world, Marvell is able to rely on the dogma that parallel lines never meet to symbolize the eternal company-in-separation of lovers but the twentieth-century poet and his lover, we will recall from "Letter I", don't even enjoy 'space in common'; the poet's imagination is distorted with fantastical shapes in a 'wide-grasping glass in which to gaze alone'. Empson glosses this phrase as giving "the reflection an odd geometry as in non-Euclidean space, so that you can't imagine yourself inside it".[175] All Empson's "Letter" poems end on a desolate note but the almost cosmic loneliness of "Letter V" is actively sought by the poet's stubbornly mathematical mind.

3.2.2 The marble detached

Renaissance conventions are repeatedly ironized in the poem to affect a cultivated mood, sliding from arrogance to loneliness. The change in the poet's

[175] Empson, *Complete Poems*, ed. Haffenden, p. 285.

mood, discernible from his tone, in turn transforms the image of the lover before him. The poem projects a figure in the first four stanzas, which it then retracts in the fifth—the process rehearses for the reader the 'emotional journey', so to put it, of the poet who introverts from eros to logos and back. The effect is achieved via a series of periphrases, 'standing negatives', that modify one another backwards and forward. The word 'gaze' in line 13 has an intentional force which secures our sense of 'glanced'—the negative definitions as vectors 'glancing off' the body of the youth.[176] Taking locus as moving point, we imagined the glossy pearl of line 11 forming around a mote of dust, the lover's body. The lines of the poet's abstracted gaze (recorded in the lines of the poem) *sculpts* the pearl into being, a circumstance which backwardly clarifies the curious phrase of line 6, 'the detached marble'. The ideal and impersonal eyes are as if engraved into existence by the poet's negative periphrases. The transformation of eyes to pearls can also be found in "Letter III", "you heal the blind into a round of pearl" [16]: both poems modify *The Tempest's* "those are pearls that were his eyes"—pearls which intrigue Ariel and the poet of "Letter V" more than the pupiled eyes of people. The Gardners argue that "'detached marble' conveys coldness more than the potentiality of a sculpture in the making".[177] In the 1920s, Hulme had described sculpture as impersonal, and Russell even compared mathematics to sculpture: "mathematics, rightly viewed, possesses not only truth, but supreme beauty—a beauty cold and austere, like that of sculpture".[178] But 'detached marble' in the poem signifies more than the coldness of sculpture, and is likely intended to call up the Afghan Buddhas. Their pupil-less eyes, lacking all speculation, were to Empson the highest symbols of Buddhist impersonality.[179] The unpublished stanza from the manuscripts of the poem contains the line, "those depths in Buddhas".[180] Recalling the discussion of Empson's views on Buddhism in the previous chapter, we might even say that the poet, in the first four stanzas, reaches after the unearthly extreme of the Fire Sermon, where Bhikkhus becomes "weary of the eye", "the visible", "the knowledge of the visible", and "contact of the visible".[181]

[176] Intentional as in Husserl's intentionality, being directed towards something.

[177] Gardner and Gardner, *God Approached*, p. 148.

[178] Bertrand Russell, *Mysticism and Logic: And Other Essays* (London: Allen & Unwin, 1917), p. 60.

[179] William Empson, *The Face of the Buddha*, ed. Rupert Arrowsmith (Oxford: Oxford University Press, 2016), pp. 8–9. Even if Empson had not written this section at the time of writing 'Letter V', this idea is discussed in his main source in *The Face of the Buddha*, which he had definitely read by 1932, namely, Ananda Coomaraswamy, *History of Indian and Indonesian Art* (Mineola: Dover, 1927), p. 23.

[180] Empson, *Complete Poems*, ed. Haffenden, p. 284.

[181] William Empson, *Collected Poems* (New York: Harcourt, 1949), p. 1.

To the young Empson, the Gandhara Buddhas, with their blank, pearly eyes, represented an extreme of spiritual detachment.[182] "Letter V" can be seen as the poet's attempt to compose these statues in words, with 'tangent praise', to erect a 'figureless shape'. The family of curves, the 'envelope', girding the young boy outlines the detached sculpture in line 20, where 'vacuum has your edge, your attitude'. The idea that 'vacuum' defines the boundary of form approaches a contradiction: you are made visible by the invisible (you bar even fancy at the gates). But in the disembodied mind of the poet, his lover has miraculously just flickered into abstract existence. The poet tries to share this *ignis fatuus* with the reader; but only our reason can accept the invitation; we leave our own senses at the gate. The entreaty to disembodied imagination lies in the equation of 'edge' with 'attitude'. The poem is invoking the sense of the term 'attitude' in art criticism to mean "a 'disposition' of a figure in statuary or painting"; the term has been carefully distinguished from 'posture', which invests the pose with psychology. In *Face of the Buddha*, Empson discusses in detail the various meanings of the Buddha's 'mudras'—or "position of hands and arms".[183] These poses, he says, communicate meaning purely through convention which was not conceived as an extension of the individual character being depicted.[184] The 'attitude' of the figure in the poem, in concert with the pupil-less pearls for eyes, thus voids the lover of all individuality. We are further to somehow detach the 'attitude' from the 'marble', or material. It is difficult to see how this is 'praise' of any sort; it is almost like the narrator of *Vanity Fair's* narrator describing the bumbling Dobbin as *decorating* his hands by biting off his nails: "he bit his nails most completely, and nearly to the quick (he was accustomed to ornament his great big hands in this way)".[185] The poem relies on the popular theory of Renaissance Platonism wherein the detached marble can be seen as the pieces of marble removed from the edges of living form, which Michelangelo says is "the task of the sculptor to discover" in irregular blocks of stone. But with vacuum, the negative periphrases, as edge to the poem's figure, we needn't shuffle crumbled refuse off to free the figure it contains: it brings the attitude of a conceptual being, a general persona, into relief: an advent which the poets believes can be ushered satisfactorily, if not entirely, in the abstract language

[182] Empson had a manuscript of the book in 1932 (Arrowsmith, *Buddha*, p. i), around the same time when he was writing 'Letter V' (Empson, *Complete Poems*, ed. Haffenden, p. 283).

[183] Arrowsmith, *Buddha*, p. 7.

[184] Ibid., pp. 8–9; He makes a similar argument about worldliness and the Greek eye of the Gandhara Buddha on p. 66.

[185] W.M. Thackeray, *Vanity Fair: A Novel without a Hero* (New York: Harper & Brothers, 1849), p. 103.

of geometry. The daemon of geometry thus reveals the 'divine states'. The 'glanced pearls', modified by 'detached' and 'attitude' briefly picture, if only to erase, the Buddha's bulbous and impenetrable eye, which 'not for me shall melt back to' the 'small eyes' of the individual his lover once was.

To conceive what cannot be imagined may have been the privilege of the non-Euclidean geometer; but through proleptic and retrodictive modifications, the mathematical poet produces his own geometric eidolon: first a methexis and then its reverse, where matter points to form which, as we will now see, points back to matter at the end of the poem.[186]

3.2.3 Metaphors are lies?

Running parallel to the inversion of 'image' is a reversal in the poem's argument on the nature of poetry. Readers of the poem have struggled to interpret the 'metaphor' of stanza 3; its meaning becomes clear if one takes the tone and attitude of the poet into account. To follow this, we must focus on the advent and adventure of 'you', the addressee, of the poem. The second person is addressed almost as afterthought. The poem begins with no interlocutor until, at last, a nominal other, a 'you', appears in line 8. This prolonged solipsism we will interpret as disdain. Most critics argue that 'they' in 'they are lies' refers to the metaphors.[187] But on an immediate reading, the grammar of the assertion and metrical stress on 'they' promote a sense that instead accuses the anti-metaphors, the periphrases, in the poem of lying. The rival senses are in fact both implied through the mechanics of the poem; but the periphrases only become 'lies' once the shift in tone, from disdain to doubt, has been registered in line 21.

Line 7, 'I can love so for truth, as still for grace' can be interpreted—if 'so' is read as 'thus', as indicating what has come before—as the poet explaining his choice to ignore his lover's physical self to seek a higher truth granted by grace; for a glimpse of the 'divine states' would be, in a Christian sense, a bestowal of Grace. But the syntactic slipperiness of lines 7 and 8 could also suggest a latent sense of the lover being graced by the poet, who exalts the meek creature to the heavenly states of geometry ('for' the sake of his 'grace').

[186] Such as Forrest-Thomson's, that the "best truth is knotted-up in image-complexes" (Veronica Forrest-Thomson, *Poetic Artifice: A Theory of Twentieth-Century Poetry* (New York: St. Martin's Press, 1978), p. 95).

[187] See Empson, *Complete Poems*, ed. Haffenden, p. 285; Gardner and Gardner, *God Approached*, p. 149.

The word 'humility' in line 8 may recall Matthew: "He that shall humble himself shall be exalted".[188] The attribution of Christian humility to his lover allows even a reading of the 'poor grazing misses' of line 23 as the impudent poet styling his misses as lamb (part of the saved flock). The sentiment in line 7—"I will see you as I must despite what you may believe you are"—pronounces the sneer beneath the poet's outward supplication.

The arrogance and indifference toward the individual behind the poet's mathematical lens control the sense of 'you are a metaphor and they are lies'. In the form "you are x" in love poetry, we take x as the metaphor.[189] For example, in the phrase "you are a rose", the 'rose' is a metaphor for 'you', which could mean, "you are beautiful and your beauty pricks me". For interpretive shorthand, let us recast this in Ricardian terms, wherein x is the vehicle whose qualities are mapped on to 'you', the tenor.[190] To the complimented party, it is obvious that the vehicle, the rose, is not in itself significant, but only insofar as it reveals or emphasizes some special characteristic about 'you', the tenor. But in "Letter V", 'you' *is* the metaphor—in other words, the tenor becomes the vehicle. The only purpose of 'you' now is to 'evoke' the qualities of 'truth' that become the poet's real concern. The reduction of the individual to a vehicle thus renders him a mere instrument to the poet's mathematical divination. In like fashion, everything else in the poem of the form 'you are x' is also a metaphor: in 'you are the grit', the grit is the vehicle that suggests the enlightened orbs of the Buddha; 'you are the map only'—'only' is 'merely' here—takes the 'map' as the metaphor, so that your mathematical coordinates transport the poet to the 'divine states'. By deeming his lover a metaphor, the poet manages to love 'what you imply', only 'to exclude' you altogether—just as one discards the vehicle once it has transferred the assigned qualities.

In a metaphor, the vehicle and tenor are not intrinsically alike: the former simply happens to possess a weight and distribution of attributes appropriate to the comparison—that is why the 'knot' is established by chance. In other words, the individual and the form he implies are connected as provisionally as the 'rose' and 'you' in our earlier example. This parallels at the level of language the separation we saw of the figure's 'attitude' from his 'body', at the

[188] Mathew 23:12, KJV.

[189] X is also the metaphor (the pregnant word) in Empson's later account of metaphor in *Complex Words*, pp. 337–9.

[190] I.A. Richards was Empson's supervisor in English at Cambridge and invented the tenor-vehicle scheme to anatomize the metaphor in I.A. Richards, *The Philosophy of Rhetoric* (Oxford: Oxford University Press, 1936), p. 132.

level of image. Because vehicles are in themselves lies, 'pelf', their evocative potential being merely a function of accidental similitude to tenor, the poet chooses instead to describe his object in periphrases—to write around and outside the boy, bringing the invisible truth into relief. In "Letter III", the poet confesses that "my pleasure in the simile thins" [9], and by "Letter V", he declares with Hobbes, Locke, and the natural philosophers of the seventeenth-century that all metaphors are deceptions.[191]

But in stanza 5, the finally achieved metaphysical certainty in his discarnate Idea is suddenly unsettled. The 'yet' of line 17 signals, as a volta in the Renaissance sonnet, a dramatic shift in argument. If your Idea is 'cause unknown', then your divine state is not even connected to the original by causation. If 'you' are merely a metaphorical lie, then the connection between physical being and true form is accidental, not causal—this is confirmed by the collocation "knot *chance*" [emphasis mine], which acts as a temporary definition of 'metaphor' in the poem; even if 'chance' is read as unfurling the 'knot', it does not contradict our reading: metaphors are 'true least' when their parts—vehicle and tenor—are connected by pure chance. With the volta, the poet at last stirs from his dream, conceding that 'cause has at least the form that has been shown'. He begins, in other words, to doubt his own conceit and suspect that the body and its geometry may yet bear some indispensable relation.

Empson notes, "*That has been shown* by the effect of the cause, as in the argument that there must at least be a structure in the external world corresponding to that of our sense-impressions."[192] We will recall from the previous chapter that for Russell too, the common factor between reality and mathematical description was structure, even if the cause for that correspondence was unknowable. Line 18 thus notices something in the world that still corresponds, in structure at least, to the mathematical form with which the poet was solely content until then.

Finding comfort from an unrequited love in the sureness of mathematical vision, one that agreeably reduces the lover to fiction, the poet becomes suddenly clouded with doubt, and his icy attitude begins to thaw. His resolution already began to waver in stanza 4 when he briefly entertained the possibility that in trying to grasp too generally, his lonely gaze may be distorting reality. Empson notes that 'knot chance' has a 'pun with '"not"'.[193] This confirms that

[191] See our discussion of seventeenth-century philosophy's attacks on metaphorical speech in Chapter 1.

[192] Empson, Complete Poems, ed. Haffenden, p. 285. [193] Ibid., p. 285.

lines 9–10 contain a second, contradictory sense, that associations between the metaphor (vehicle) and subject (tenor) are 'not chance'. The punned sense on 'knot' and the syntactic ambiguity in 'they' (whether it refers to poetic metaphor or mathematical periphrasis), is initially dormant and retrodictively activated once the poet registers in stanza 5 an uncertainty about the importance of his vehicle: maybe you, who are a metaphor, and your Form are related by something more than chance; and 'they' that are lies are actually my mathematical periphrases? And insofar as the comparison between poetic metaphor and mathematical periphrasis can be extended to poetry and mathematics in general, as modes of representation that work through momental, sensuous association and sequential, logical connection, the *ex post facto* reading of lines 9–10 demanded by the volta reverses the aspersion cast earlier on poetry.[194] And so we can read the poem after the volta as replying to Hobbes and Locke when it says metaphors are not lies.[195]

The agon staged by the poem between poetic and mathematical reality resolves, in the final stanza, to a 'Principal'. But the reader is yet prevented from easily preferring one or the other by the pun on 'principal'. Line 21 evokes the rule in geometry known as the Principle of Duality. I refer the reader to Section 2.2 for an explanation on how a statement in projective geometry is valid in equal measure whether in the form of lines or points. The subtext of the Principle of Duality, the interconvertibility and equal validity of line and points, in the assertion of line 21 undoes the opposition between lines of outline—envelopes—and body of points—locus—with which the poem began.

The implication of equivalence between lines and points seems in fact to resolve the warring metaphysics of the poem. First, the word 'grant' in line 22 registers the shift in the poet's attitude to his lover. The lover's lines, his formal truth disclosed to the poet by 'grace', is now acknowledged to be granted by the slighted 'you'. As flesh re-emerges from the diffusing outlines of form, the 'lines you grant me' invert through projective transformations from geometric contours to sensible points. As per line 23, the lines are 'paired' with points: in other words, matter and form, idea and instance, in the course of the poem, pin Being to the wall. Even so, the capture, with 'poor grazing

[194] This is why, incidentally, it is incorrect, as the Gardners do, to read metaphors as having "some hidden-logic" (Gardner and Gardner, *God Approached*, p. 149). It isn't the logical, but the sensuous connection in metaphors that is pertinent to the argument at hand.

[195] We will recall the line quoted in the first chapter: "when they [poets] use words metaphorically; that is, in other sense than that they are ordained for; and thereby deceive" (Thomas Hobbes, *Leviathan or The Matter Forme, and Power of a Commonwealth Ecclesiastical and Civil*, annot. G.W.F. Hegel (Amazon Digital Services, 2019)).

misses', is precipitously slant and yields no closure. However unceremoniously the archers of Diocletian arrested St Sebastian, we may venture, his soul remained bound to another. So alas, we mortals, among whom the poet in the end counts himself, may never know what is that tantalizing 'Principal'. As Whitehead declares, the first and last philosophical question remains, "How can concrete fact exhibit entities abstract from itself and yet participated in by its own nature?"[196] But the intellectual retreat from points to lines and the sudden effusion of eros issuing from pointy arrows to sensuous apprehension, is a reversal dramatic enough to suggest that the poet has stirred from his mathematical dream.

3.3 Conclusion

The metaphysical revival of the twentieth century is in some ways the central player in the titular agon. Its fertility allows a poet like Roberts to draw on Renaissance feeling in words and Empson to play with its full arsenal of poetic techne. Both seem to regard the metaphysical style as allowing poets a fuller grasp of reality while submitting the discoveries of mathematical physics to poetic design, if not critique.

Given the paradigm of this work is the agon, it is subject less to the pressures of many works in the literature-and-science or mathematics fields to keep scientific and mathematical images and concepts to the fore of discussion at all times. Understanding the discursive strategies used to undermine the univocal authority of the mathematical world-picture requires a comprehensive analysis of what is being done to words and concepts in the poetic world in order to estimate accurately the place of mathematics in the larger critique. As we have seen in "Letter V", without a careful tailing of moods and attitudes in the poem, we could not catch the reversal in definition of metaphors as forming integral sensuous links between tenor and vehicle: and without that, it would be impossible to say just what is lacking in the metaphor's opponent, mathematical periphrasis. Similarly, the subtle emphases that Roberts places on the image of bronze vectors as opposed to grim quaternions would be obscured if we had not first analyzed the distinction between movement and hardness, and the corresponding symbolism of light and darkness in his poems. Rather than following how his peculiar brand of

[196] A.N. Whitehead, *Process and Reality: An Essay in Cosmology* (Cambridge: At the University Press, 1929), p. 4.

metaphysical poetry invests various aspects of mathematics with different emphases of feeling, we might with a forced promotion of explicitly mathematical material have coloured all of Euclid's world dark.

The mathematical poetry of both Roberts and Empson challenges the mathematical world-picture mainly by bringing to the fore its lack. Roberts concentrates on the semantic fields of poetic diction and hopes, remarkably, to reorient the philological formations of a great passage of European history under the regime of science. If the chief effect of the rise of science has been to atomize and classify the various domains of thought, Roberts responds in his poetry by forcing them back in each other's society. He unifies currents of thought, in the mind's conception of such things as language, landscape, and mathematics, by the strategic development of sensuous significance in words. Empson, on the other hand, makes his most interesting manoeuvres in the process of creating meaning. By demanding of his reader a dizzying focus on words modifying words, forward and backwards, he posits a spectral image of his lover, retracts it, and then resurrects a sexual one, and questions and exalts metaphorical truths, all within the canvas of one poem. The formal elements of verse thus become involved in demonstrating the unique powers of poetry in the domain of 'neat and syllogistic science'.

Despite their steep ambitions, both poets, ultimately, found refuge in, even exalted, but failure and imperfection. In poems like "Nicholas Flamel", "Mona Lisa", and "Letter V", what is ultimately conveyed is done through failure, through the at-times dramatic and other times wistful notes of futility. Failure was indeed central to the fate of love that so greatly preoccupied the poets, for they found the inordinate regard for abstract perfection and rational order in their culture as fundamentally incompatible with the scatter and undulation of erotic emotion.

4
Logos

So far, we have seen how poetry and mathematics can compete as modes over their purchase on reality. In this chapter, our antagonists will no longer compete as instruments merely: a poem was regarded by Laura Riding and Charles Olson as being either true or not: an end in itself. And that end in this chapter we shall term logos.

The term logos has multiple senses that seem distinct; they are, however, related and mutually dependent in the works of Riding and Olson. *The Cambridge Dictionary of Philosophy* says the dominant sense of 'logos' is 'word' and that it has been historically attended by several others: (1) rule, principle, (3) reason, the intellectual part of the soul, (4) measure, proportion.[1]

Both Riding and Olson are concerned primarily, and almost exclusively, with the unit of the word. The sense of a rule or principle in logos, from its uses by the ancient Greeks, into its avatar in Christianity, has often stolen into a higher sense of 'truth'. G.T.W. Patrick, summarizing past translations of Heraclitus, says they used "Logos as the eternal pre-existing law of the identity of being and not-being".[2] This sense will be important in the present chapter, because unlike Empson and Roberts, both Riding and Olson are not abashed to assert a full doctrine of the 'truth of being/not-being'.[3]

The third sense of 'logos', 'reason', is the one upon which the two poets of this chapter most differ. Hobbes said the "Greeks have but one word Logos, for both Speech and Reason; not that they thought there was no Speech without Reason; but no Reasoning without Speech".[4] This tension between reason and speech plays out across the works of the two poets. In a cursory sense,

[1] Robert Audi, ed., 'Logos' in *The Cambridge Dictionary of Philosophy* (Cambridge: Cambridge University Press, 2015), 572–619; sense (2) is the most discursive of the lot: 'proposition, account, explanation, thesis, argument'; and does not pertain to this chapter.

[2] Heraclitus, *The Fragments of the Work of Heraclitus of Ephesus on Nature*, trans., intro. by G.T.W. Patrick (Baltimore: N. Murray, 1889), p. 116 (footnotes).

[3] Carla Billitteri was the first to group Riding and Olson on the criterion of the truth-telling capacity of words in their respective systems (Carla Billiteri, *Language and the Renewal of Society in Walt Whitman, Laura (Riding) Jackson, and Charles Olson: The American Cratylus* (New York: Palgrave Macmillan, 2009)).

[4] Hobbes, *Leviathan*, ch. 4, Subject to Names.

Agon: Poetry's Challenge to the Mathematization of Reality (1920s–1960s). Anirudh Sridhar, Oxford University Press.
 DOI: 10.1093/oso/9780198985457.003.0004

truth is disclosed to Riding in thought, whereas to Olson, in speech; Riding affirms, and Olson denounces the reasoning part of being.

The fourth sense, 'measure', brings us to the topic of this book. The shrinkage of the word 'measure' in modern times to a routine conducted by the instruments of science leaves obscured the richer past of the term. When Protagoras declares man the measure of all things, we may, perversely, think of him as taking man as the chief instrument by which things are gathered about the world. Blake too, as we saw in the opening of this book, regarded the body as the instrument to take the true measure of reality. Both Riding and Olson develop their view of the relationship between word and truth by a corresponding conception of measuring reality in poetry.

4.1 The Non-Platonist Riding

Barbara Adams says, "[Laura] Riding's poetry" is characterized by "the search for a unified identity, an obsession with death and hope of transcendence through art".[5] The term 'transcendence' does not quite capture Riding's view of art. 'Transcendence' was to Riding paradoxically bound to the concreteness of language. She sought a unitary sphere of meaning beyond the fractured world of history: but this elusive totality must be understood not as apart from its parts but rather as the quintessence to which the parts immanently point. The ability to commune between the noumenal and the phenomenal, a power that the Platonic number once was thought to have, lay, to the young Riding, in the poetic word. And so, like some early Christian theologian, while denouncing mathematical Platonism, she borrows the structures of its doctrine to construct an idealist metaphysics of her own logos, the poetic word.

Riding observed through her life poets gradually losing control over the destiny of words and a "shift to the scientific view of language".[6] In the late

[5] Barbara Adams, 'Laura Riding's Autobiographical Poetry: "My Muse as I"', *Concerning Poetry*, 15 (Fall, 1982), 71–87.

[6] Laura (Riding) Jackson and Schuyler B. Jackson, *Rational Meaning: A New Foundation for the Definition of Words and Supplementary Essays*, ed. William Harmon (Charlottesville: University Press of Virginia, 1997), p. 521. This is quoted from an essay, 'The Physical Aspects of Words' whose date of composition is hazy. Joyce Wexler states the chapter in *Rational Meaning*, 'Truth', was written in 1975 (Joyce Wexler, *Laura Riding: A Bibliography* (New York: Garland and Pub, 1981), p. 82). Based on the essays quoted in 'The Physical Aspects of Words' it must have been written around this time. The first bits of *Rational Meaning* were begun in 1948 (Charles Bernstein, 'Introduction to *Rational Meaning*', p. xiii).

1970s, she writes in an essay titled 'Mathematics as an Intellectual Master-Method',

> As the new modern intellectual confidence placed in the quantitative values of mathematics has manifested in attitudes to language, efforts have been made [...] to purge language of its qualitative value-force, and to establish mathematical values, conceived of as universally authenticated basis of logic, as its critical norm. [...] Never before the present era has the seat of logic been placed definitively in mathematics, with language bearing the character of an outlying region of logical activity.[7]

With the gradual elimination of the 'value-force of words', the poetic sense of 'rightness' in words was replaced by the crude binary of 'correctness' offered by mathematical logic.[8] For the young poet, language had been the very central 'region of activity'. Much of her poetry was in fact about language itself. We will in this section read such poems from *Love as Love: Death as Death* (1928), *Though Gently* (1930) and *Poet: A Lying Word* (1933), and attempt to uncover her peculiar and rather extreme philosophy of poetic truth-telling.

She once said "Am I a mystic? No, I am not a mystic, I am Laura";[9] but to anyone who has read the young poet, this will seem but denial. Having once held a unitary vision of language and reality, she abandoned it midway through life, and with her philosophy, Riding also forsook her poetic vocation—her last collection of poems was published in 1938. She wrote in 1944, "I renounced it [poetry] 'as disappointing the hopes it excites as seemingly the way of perfect human utterance, or articulate truth'".[10] Michael Masopust says "few admirers have ever paid poetry the compliment of taking it so seriously, and probably no detractor since Plato's time has taken such pains to discredit it".[11] Peter Temes even entertains the

[7] Riding, *Rational Meaning*, p. 497; The essay's date of composition seems not to be available, but from her discussion of 'contemporary academic linguistics' (pp. 497–9), in which she cites W.W. Bartley's 1974 works, we can tell it was probably written in the late 1970s.

[8] Mark Jacobs has informed me of his interactions with Riding which assured him that she was "very interested and knowledgeable about the subject" of mathematics (Mark Jacobs, interviewed by Anirudh Sridhar, 13/12/2018); also, before enrolling at Cornell University, she would have had to sit basic exams in mathematics ('Chronology', Elizabeth Friedman in Laura Riding, *The Laura (Riding) Jackson Reader*, ed. Elizabeth Friedman (New York: Persea Books, 2005), p. ix).

[9] Laura Riding, *Experts Are Puzzled* (London: Cape, 1930), p. 2.

[10] Laura Riding, *The Failure of Poetry, the Promise of Language*, ed. John Nolan (Ann Arbor: University of Michigan Press, 2007), p. 27.

[11] Michael A. Masopust, 'Laura Riding's Quarrel with Poetry', *South Central Review*, 2.1 (1985), 42–56, p. 42.

possibility that her "rejection redramatized, in larger scale, the central manoeuvre of her best poems: to escape from the role of the object, of the seen and judged"[12]; that the renunciation of her poems would deter their re-presentation and thus preserve them from the depredations of literary critics: if this plausible explanation is indeed the case, due apologies are made for what follows.[13]

When a cabbalistic mood was upon her in youth, it cut across all she wrote, binding the early works together under what we may, for want of a better phrase, call a philosophical system. It will be our task in this section to stitch this metaphysic together, which has yet to be done by literary critics. This might seem a strange task when dealing with the works of a poet, but if any should demand such an undertaking, it is the mysterious young Riding.

While a poet, Riding believed truth can be found only in the lines of a poem. She seems to epitomize this New Critical cliché by elevating the poem to a quasi-religious insularity.[14] Donald Childs argues that Riding's view of a poem as an 'object in itself' was influential in the development of New Critical theory.[15] Her view was, however, not a matter of procedural routine, to be replicated in classrooms, but rather a doctrine that was neither to be argued for nor used. While sharing the New Critical belief in the poem's autarchic power, in *Contemporaries and Snobs* (1928), Riding criticizes modernists of New Critical persuasion, such as Tate and Ransom, for making of the poem a stable object, a well-wrought urn, whose function became but the transmission of facts after a scientific fashion.[16] Unlike the step-by-step routine of an empirical science, to Riding, a poem would cut through the fabric of daily occurrence. "Philosophy is the past explaining the present to the future" and "science is the present explaining the future to the past"; both philosophy and science are, to her, trapped in time, yielding in tandem to the other's theories of phenomena.[17] In her first collection, *The Closed Chaplet*, Riding began thus

[12] Peter S. Temes, '"Code of Silence" Laura (Riding) Jackson and the Refusal to Speak', *PMLA*, 109 (January 1994), 87–99, p. 87.

[13] In my defence, Riding states that "the conversion of nothing into something is the task of criticism" (Laura Riding, *Anarchism Is Not Enough*, ed. Lisa Samuels (Berkeley: University of California Press [1928] 2001), p. 18).

[14] Tom Fisher, 'Reading Renunciation: Laura Riding's Modernism and the End of Poetry', *Journal of Modern Literature*, 33.3 (2010), 1–19, p. 5.

[15] Donald J. Childs, *The Birth of New Criticism: Conflict and Conciliation in the Early Work of William Empson, I.A. Richards, Laura Riding, and Robert Graves* (Montreal: McGill-Queen's University Press, 2013), p. 12.

[16] Laura Riding, *Contemporaries and Snobs*, ed. Laura Heffernan and Jane Malcolm (Tuscaloosa: University of Alabama Press [1928] 2014), p. 115.

[17] Riding, *Puzzled*, p. 37.

to display irrealist inclinations that would intensify through the late 1920s and 1930s.[18] Robert Fitzgerald says of Riding's 1938 *Collected Poems* that,

> The authority, the dignity of truth telling, lost by poetry to science, may gradually be regained. If it is, these poems should one day be a kind of *Principia*. They argue that the art of language is the most fitting instrument with which to press upon full reality and make it known.[19]

And so, the agon resumes.

We will focus mainly on a curious little collection called *Though Gently,* where she sets directly in contention the truth-telling powers of poetry and mathematics. The collection has received next to no critical attention until very recently. Tim Armstrong's essay on the mathematical notions in *Though Gently*, which was my own introduction to the collection, is the only substantial engagement with its 'philosophy'.[20] Its poems do not reappear in the *Collected Poems* and later anthologies: but this is less a reflection of their value in Riding's oeuvre and more a consequence of the fact that each constituent part of *Though Gently* is conjoined to the other to an extent that extracted fragments would in isolation be almost meaningless. If we are to discover what this slim and mysterious volume contains, we must and will undertake the task of viewing it whole, as if it were one long poem.

4.1.1 Word and world

Carla Billiteri characterizes Riding's philosophy of language as Cratylic, referring to *Cratylus*, the Platonic dialogue on the act of naming.[21] Cratylus advances in the dialogue a mysterious belief in which signifiers, to speak anachronistically, emerge physically from their signified. This occult connection allows words in themselves to be judged on a standard of 'correctness'

[18] Laura (Riding) Jackson, *The Close Chaplet*, ed. Mark Jacobs (Nottingham: Nottingham University Press [1926] 2018).

[19] Robert Fitzgerald, review of *The Collected Poems* of Laura Riding by Laura Riding, *Kenyon Review*, 1 (1939), 34.

[20] Tim Armstrong, '"A Transfinite Syntax": Modernism and Mathematics', *Affirmations of the Modern*, 6.1 (2019), pp. 1–29. *Though Gently* is mentioned by Joyce Wexler, who spends two pages on a rather unsatisfactory summary of the collection, as being a negotiation between Riding's private and public thoughts (Joyce Wexler, *Laura Riding's Pursuit of Truth* (Ohio: Ohio University Press, 1979), pp. 63–4).

[21] Billiteri, *Language and the Renewal*, p. 78.

beyond the general sense of rightness in usage. According to Billiteri, the Cratylic position is that the

> reference and meaning of words were established by 'a power more than human' and given to the human race with the specific function of providing essential 'information about things' to those who would use them (*Cratylus* 438c, 435e). In Cratylus's system, the natural world, in its material state of sounds, shapes, and colors, is conceived as intimately connected to a nonhuman language of letters and syllables, each element of a language an emanation of a natural phenomenon carrying a precise, unalterable semantic content.[22]

Billiteri is interested primarily in Riding's later work, *Rational Meaning*, whose composition (1948) began long after Riding had forsaken her poetic vocation. Although (Riding) Jackson—the name she assumed upon shedding her poetic persona[23]—retains in some ways her early lingual mysticism, the older Riding turned to regular human speech, abandoning the perfect poem as a youthful dream. Nevertheless, Billiteri's label Cratylic is a helpful frame for studying her early works.

Riding forwards a 'Cratylic' view first in her 1928 manifesto, *Anarchism is not Enough*. She separates "three historical levels" of words, their "intrinsic sense", "logical" or "applied sense", and

> poetical words, of a misapplied sense, untrue and illogical themselves, but of supposed suggestive power. The most the poet can now do is to take every word he uses through each of these levels, giving it the combined depth of all three, forcing it beyond itself to a death of sense where it is at least safe from the perjuries either of society or poetry.[24]

By thus lauding the misapplied sense, she seems to be reappropriating the slurs of seventeenth century natural philosophy against the lies of metaphors—misapplied senses, we will recall, were Bacon's idols of the marketplace. Riding means also mis-applied here, in that poetic words should not even be applied to things in the world: that their work is beyond mere reference to objects as is the sordid job of scientific statements. Poetic words should clear away all vagueness in language, she says, by evacuating

[22] Ibid., p. 21. [23] Riding, *Failure*, p. 27. [24] Riding, *Anarchism*, p. 12.

all 'sense' but the intrinsic. She hopes thus to protect language from 'scientific treatment', which may be understood as the privileging of the 'logical/applied' senses of words. This was already a threat she sensed in the proto-structuralist works of Ogden and Richards, and Saussure, whom she became acquainted with through the pages of *The Meaning of Meaning*.[25] Her early views on language can be seen as emerging from her disagreements with structuralism.

Ogden and Richards are not usually called structuralists but they were in sympathy with Saussure's seminal act, which they saw "as having for the first time placed linguistics upon a scientific basis".[26] This they say he accomplished by dividing language into two aspects: *Langue* referred to the fixed structures of grammar and syntax, and *parole*, to how language is used in speech, written and vocal.[27]

Fredric Jameson says the "dilemma of linguistics" in the 1920s and 30s is "part of a vaster crisis in the sciences in general: in physics for instance, where the alternation between the wave and particle theories of light begins to cast some doubt on the conception of the atom as a substance".[28] Just as faith in a material substratum beneath symbols of representation was waning in the philosophy of science, he argues, linguistics could not maintain that words are attached to "isolated things" in the world:[29] words were becoming to the scientific linguist plastic vessels holding transferrable meaning. Ogden and Richards aimed their attack against any remnants of cultural Cratylism: "much confusion has been caused by the habit of treating meaning as somehow inherent, or naturally present, in the signs which conventionally serve to convey it".[30] Insofar as Saussure undermined Cratylism, Ogden and Richards were in agreement with him,[31] and sought to encourage this new direction of linguistics. Their relatively minor squabble with Saussure was that he preempts the empirical study of semiology; that "this theory of signs, by

[25] In her introduction to *Anarchism*, Lisa Samuels notes Riding's disenchantment with structural linguistics, especially her intense engagement and disagreement with the conclusions of *The Meaning of Meaning* (p. xlv). We know Riding had read it by the time of writing *Anarchism* (1928).

[26] C.K. Ogden and I.A. Richards, *The Meaning of Meaning: A Study of the Influence of Language upon Thought and of the Science of Symbolism* (New York: Harvest, 1923), p. 4.

[27] Ferdinand de Saussure, *Course in General Linguistics*, trans. R. Harris (Chicago: Open Court Publishing Company [1916] 1982), pp. 9–10, 15.

[28] Fredric Jameson, *The Prison-house of Language: A Critical Account of Structuralism and Russian Formalism* (Princeton: Princeton University Press, 1972), p. 14.

[29] Ibid.

[30] Thomas Sebeok and Jean Umiker-Sebeok, *The Semiotic Sphere* (New York: Plenum, 1986), p. 233.

[31] Ogden and Richards, *Meaning*, pp. 5–6.

neglecting entirely the things for which signs stand, was from the beginning cut off from any contact with scientific methods of investigation".[32] That is, by characterizing the relationship between signifiers and signified in individual use[33] as determined entirely by convention—'applied sense', as Riding puts it—Saussure saw *parole* as subject to changes that are not measurable. In a somewhat circular fashion, Ogden and Richards argued that the question of how words are used should in fact be subject to empirical study, as that will in turn allow for a better use of language.

The Meaning of Meaning is a "confused mixture of philosophy, psychology, ethnology and literature", declares Riding, scoffing at the idea that meaning "is to be known scientifically only through symbols".[34] Rather than a direct passage of meaning between signifier and signified, Ogden and Richards concoct a triangle of reference. The bottom left vertex of their triangle is the "symbol" or the signifying word; the bottom right is the "referent" or meaning; and they introduce at the apex of the triangle a detouring pit stop between sign and referent which they label "thought or reference".[35] Words remain in this triangle Saussurian vehicles for the conveyance of meaning—the only difference is that the thought-process behind reference becomes part of their analysis, psychology thereby smuggling in the 'scientific method' they so desired. As we have seen in *Anarchism*, Riding relegates the logical or applied sense to a mere third of a word's aspect; she thereby denies even the base of the triangle which treats words as mere vehicles. Poetic words, for Riding, in their ideal state, are but themselves, in identity with what is mistakenly separated as their referent.

While Empson might have resisted the vehicular conception of words by charging them with ambiguities, Riding, in *Though Gently*, reaches for the very opposite, a complete semantic cleansing. Billiteri says,

> The piling up of definitions for a single word, she [Riding] vehemently argues, amounts to a radical 'dissolution' of language's 'properties', which she sees as a social problem as well as a linguistic one. As in Plato's *Cratylus*, once the foundational rules of the proper and correct use of words are violated, language becomes mere noise.[36]

[32] Ibid., p. 6.
[33] 'Individual use' is to be distinguished from meaning which "exists in perfection only in the *mass*" [emphasis mine] (Saussure, *Course*, p. 31).
[34] Riding, *Anarchism*, pp. 54–5.
[35] Ogden and Richards, *Meaning*, p. 11.
[36] Billiteri, *Language and the Renewal*, p. 104.

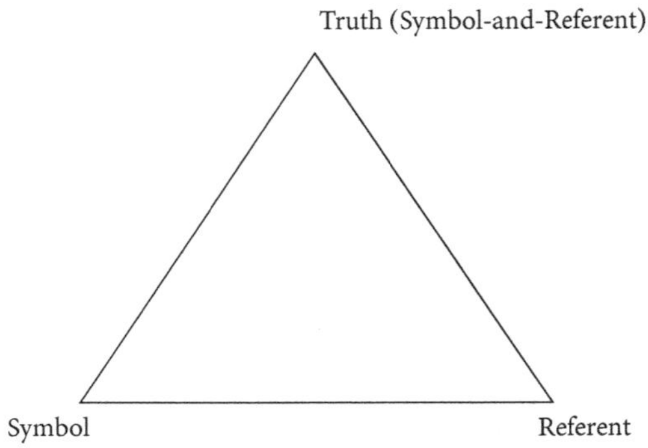

Figure 4.1 Speculative representation of Riding's language theory.
Source: Image produced by author.

Were we to imagine a Cratylic triangle to approximate Riding's theory, it might look like Figure 4.1:

'Truth', in the position it is here afforded serves to underscore her antagonism to the psychologism at the apex of the Ricardian triangle—this is only a placeholder for her complicated theory of language which we will soon elaborate. Ella Ophir says, "in light of the intellectual prestige of science, only truth-value could secure for poetry the authority Riding wanted for it. Any other formulation of its value would, she felt, render it precariously subservient, merely 'expressive' or ornamental".[37] While we have similarly reasoned our way to the shorthand, 'Truth', the term 'truth-value' seems to impose a utilitarianism to her theory that she would have resisted.

4.1.2 Word as word

Paul Auster says of Riding's poetry, "Few poets have ever been able to manipulate abstractions so persuasively. Having been stripped of ornament, reduced to their bare essentials, the poems emerge as a kind of rhetoric, a system of pure argument [...] giving [...] formal pleasure".[38] The presentation of bare essentials before the reader, and the formal mode of argument in Riding's poems, allow us to develop from them something like a 'theory' of language.

[37] Ella Zohar Ophir, 'The Laura Riding Question: Modernism, Poetry, and Truth', *Modern Language Quarterly*, 66.1 (2005), 85–114, p. 89.

[38] Paul Auster, *Collected Prose: Autobiographical Writings, True Stories, Critical Essays, Prefaces and Collaborations with Artists* (London: Faber & Faber, 2003), p. 531.

Let us begin with the following lines from the titular poem of *Love as Love: Death as Death*:

"Death as Death": "To conceive death as death / Is [...], / A blankness fallen among / Images of understanding".[39] The reader is challenged to conceive the idea of death just as it is: the no-longer-being of what is: 'blankness'. An earlier poem, "Dimensions", hints this definition of death: "But death and death is nothing at all".[40] To conceive a word just as the word requires a mental streamlining of its semantic field, holding in a single thought, a word and its meaning sans *difference*—that is, without a routine of tailing associations. But unlike other words which might beget a stream of connotations, it is a happy coincidence that 'blankness' or 'nothingness' aptly illustrates the word 'death': the word's meaning serves as an in-built pause that impedes slippage. The ______ that follows mention of death—both in the poem's line-ending and in the mental conception of 'death' as the advent of nothingness—brings the word and idea, which to Riding are always in union *sub specie aeternitatis*, for a moment, as one to mind in *tempori*. The ambition here can be compared to the effect Beethoven achieves in his late works, by use of fermata against convention. Adorno says of a *Faust II* or *Wanderjahre* that "it leaves only fragments behind, and communicates itself, like a cipher, only through the blank spaces from which it has disengaged itself. Touched by death, the hand of the master sets free the masses of material that he used to form".[41] But of course, the *noesis* of death can only be ephemeral. As soon as will participates, it sees "furnaces / Roar in the ears, then again hell revolves / And the elastic eye holds paradise", thus contaminating with fancy the fleeting purity of abstract linguistic truth.

To force the mind to concentrate on a single word, repetition, as in 'death as death', seems a rather coarse method for a poem to take, but it proves quite effective. It is a strategy repeatedly seen in this volume—for instance, in "love as love", "Earth rounds out Earth", "logic has logic". Her repetitions operate by a curious process that narrowly avoids tautology; the word, in encounter with itself, seems to react in two observable steps: the second iteration brings to mind the denotation, which voids the first of meaning, leaving behind the bare word as residue; next, the evacuated word is replenished with the recently discharged meaning, as if she were springing out old batteries from a

[39] Laura Riding, *Love as Love: Death as Death* (London: Seizin, 1928), p. 1.

[40] This poem appeared in *Fugitive* magazine (1923) but not in her first collection, *The Closed Chaplet* (Laura Riding, 'Dimensions', *The Fugitive* (Aug.–Sept. 1923), p. 124).

[41] Theodor Adorno, 'Late Style in Beethoven', in *Essays on Music*, trans. Susan Gillespie, ed. Richard Leppert (Berkeley: University of California Press, 2002), p. 566.

radio, dusting them off, and replacing them, for a better flow of charge. An example will illustrate the process: "Logic has logic, they remain / Locked in each other's arms" ("World's End"). If 'logic' and 'logic' can be referred to in plural as 'they', they must be distinct. The *word* 'logic' is dismembered from the *concept* of logic by the possessive verb 'has'—it seems odd for a thing to possess itself, so we think the former must refer to something else. But as soon as one stirs from the conjecture provoked by 'has' and realizes that 'logic' does indeed refer to 'logic', the concept is reattached to the word. Both the work of separation and re-attachment is achieved by 'has'; locked in each other's arms', the word logic *has* the concept logic: their connection is thus not adventitious. The word briefly ceases to be a vehicle, and a strange fusion occurs between word and meaning.

Repetition is but one mode by which Riding brings out the relation of language-and-sense. A very early tactic was simply to title a poem, "The Definition of Love", in *The Closed Chaplet* (1926)—the poem, sadly, has none of the poise and playfulness of Marvell's "The Definition of Love". Riding's poem, in its neurotic literalness of purpose, is concerned with actually defining the word 'love'; and this obsession seems to have persisted until 1930, when in *Poems: A Joking Word,* "The Definition of Love" is reworded with the same agenda.[42] Another poem with as literal a purpose but more interesting an effect can be found in *Twenty Poems Less* (1930)[43]—it is titled, straightforwardly, "Meaning"

Meaning words only
Meaning not meaning
[...]
Meaning at last no more to say
Meaning how like nothing is all
Meaning a written a world
Meaning a finished revelation
Meaning less and less

And so she goes on with the mantra, producing proleptic and retrodictive effects across lines, building to a kind of prayer for words. For instance, the cryptic 'meaning not meaning' is backwardly clarified by 'meaning how like nothing is all'; the incantation works by the former erasing meaning to

[42] Laura Riding, *Poems: A Joking Word* (London: Cape, 1930).
[43] Laura Riding, *Twenty Poems Less* (Paris: Hours Press, 1930).

nothing—i.e., a lack of meaning—and the latter making of 'nothing', 'all'—in the way a Buddhist sutra might encourage the supreme elevation of nothingness or Ralph Waldo Emerson's poet who is an invisible eye that is nothing but sees all. It might even recall Donne's poem—"And quickly make that, which was nothing, all"[44]—on how the geographer fills an empty sphere with the entire globe. Meaning is a 'finished revelation' when its *references* becomes 'less and less', until only the word's Cratylic identity, a pregnant nothing, remains.

The meaning-making, or unmaking, procedure of "Death as Death" changes in "The Number", a later poem in *Love as Love*:

> The number is a secret,
> How many elements assemble
> To pronounce *Alive*—
> And leave *Alive* to count places

The poem seems to invite a kind of scientific analysis with the phrase 'how many elements': and simultaneously, it also undermines through its performance the scientific treatment of language. Rather than juxtaposing the word 'Alive' with itself, as in 'death as death', its pronunciation is contrasted with its meaning by first calling up the word as a vocable and then using it as an adjective. The 'elements' that assemble to pronounce any word are the mouth and the mind; but given the scientific context suggested through a possible echo of a doctor pronouncing a patient dead, we might also anatomize the vocable into its two syllabic elements, 'a' and 'live'. The poem means to show how counting elements in the use of a word will say nothing of its meaning; this is sharply felt by the reader who first struggles to fathom the correct way to part '*Alive*' into 'elements' and is then casually enjoined to parse the word in a sentence in line 4. The reader automatically comprehends the meaning of '*Alive*' in line 4, as the phrase 'to leave alive' is readily understood. The process of defamiliarization that the reader was led through in "Death as Death" is reversed in the meaning-making process of this poem by an intensification of familiarization—being alive is, after all, more familiar to us than being dead. In other words, the familiarity of '*Alive*' as used in line 4 is sharpened by the unfamiliarity of its mathematical division in lines 2–3.

[44] Herbert Grierson, *Metaphysical Lyrics and Poems of the Seventeenth Century* (Oxford: Oxford University Press, 1921), p. 11; see discussion of the poem in the review of Grierson by T.S. Eliot, 'The Metaphysical Poets', *TLS*, 669–70 (1921), p. 670.

The poem allows us to distinguish the poetic mode of defamiliarization, the kind of word-alchemy in "Death as Death", from the dissective defamiliarization of 'mathematical lexicographers', as they are dubbed in *Anarchism*.[45] The story proceeds as "The conference adjourned / And the ghosts inaccurate, / Scattering poor memories" [5–7]. The word 'conference' in line 5 brings to mind a collection of professional linguists, where one might find an Ogden or Richards attempting to analyse language by rigorous quantification. As Riding says, "To Mr. Ogden and Mr. Richards, [...] language is this precise mathematical grammar".[46] They, in the poem, are forced to leave 'inaccurate', for how can numbers describe words? Put differently, how can 'ghosts' capture what it is to be 'alive'? The poem thus caricatures scientists of language rending words into morphemes to investigate etymological meaning, 'scattering [the] poor memories' in pieces: a fate, one might venture, common to the explananda of empirical science.[47] The poem says finally, that once quantification is under way, "no sooner known the number, / There is division to prove the whole,/ But never reassembling."

4.1.3 Word and number

Logos, the word as rule, reason, and measure, is the crux of Riding's most serious meditation on the nature of language, *Though Gently*. The word is exalted there in a formally staged dialectic between mathematics and poetry. Unlike the poems we have read thus far, those in this volume unfold in paragraphs of cryptic postulates and pregnant aphorisms (like those of Wittgenstein's *Tractatus*) with rare snatches of verse.

I.M. Parsons, reviewing *Though Gently* in 1931, described its idiom as "a hard, spare utterance, ascetic in its dry intellectuality. But it has, too, something of the confident attraction of a geometrical proposition".[48] David Perkins says, "often they are argument; assertions are made, explained, defended, justified, sometimes questioned and countered".[49] There are, however, no explanations or justifications in the poems despite the argumentative tone of their sentences. *Though Gently* rejects the definitions of terms

[45] Riding, *Anarchism*, p. 23. [46] Ibid., p. 54.

[47] Riding parodies the scientific mind-set in similar terms in *Essays from Epilogue: 1935–1937*, ed. Mark Jacobs (Manchester: Carcanet Press, 2001): "Let us reduce the earth to the smallest conceivable space, so small that it cannot even be called space" (25).

[48] I.M. Parsons, 'Three Seizins', *Spectator*, 147.5396 (1931), p. 739.

[49] David Perkins, *A History of Modern Poetry: Modernism and After* (Cambridge: Harvard University Press, 1987), p. 9.

accorded by convention and prosecutes its argument instead by imposing in places an almost private glossary. The series proceeds by a careful manipulation of meaning around familiar words to arrive finally, if we look carefully, at a definition of poetry.

Parsons characterizes Riding's "method of prose composition" as "an attempt to arrive at conclusiveness, in a series of related statements, by the rigid exclusion of ambiguity or, more precisely, of all irrelevant constructions".[50] Besides ambiguity, the severity of Latinate register in the collection also stunts melopoetic suggestion. As we have seen, Riding believed a poem to, in some mysterious way, convert the phonetic into the phonemic, bringing to surface the oneness between aural structure and reality. Michael Roberts characterized Riding's view of poetry "as the final residue of significance in language, freed from extrinsic decoration".[51] Stating her distress at "the musicification of poetry", Riding believed that poems working only through the interplay of sounds signal a decay of poetry's true purpose.[52]

At the heart of *Though Gently* is a comparison between the Platonic view of number and the Cratylic theory (or so we will initially term it) of words. To stage the argument, Riding erects the figure of Plato as her interlocutor (naming him explicitly in "I Grant You"). Mathematics revealed to Plato the existence of things beyond time, a reality accessible only to *nous*. He casts as "opposites", in the *Seventh Letter,* "every one of the circles which are drawn in geometric exercises", in whatever rude form the clumsy hand manages, and "the Idea of the Circle" that exists in eternal perfection.[53] In place of the bastard circle, for Riding, is the Word and its vulgar misuse by man. In *Meaning of Meaning*, Ogden and Richards cast Heraclitus and Pythagoras to stage a conflict between Cratylism and mathematical Platonism. To Heraclitus—since no man steps in the same river twice—words, by achieving a remarkable stability of reference-to-things amidst the shifting landscapes of time, were tokens, almost, of a possible underlying reality.[54] Pythagoras, on the other hand, held that numbers are the only eternals and can thus reveal the true cosmic order.[55] It becomes the task of this strange collection of poems to challenge the mysticism of mathematicians by advancing a realist theory of language.

50 Parsons, 'Three Seizins', p. 739.

51 Michael Roberts, ed., *Faber Book of Modern Verse* (London: Faber & Faber [1936] 1951), p. 9.

52 Riding, *Anarchism*, p. 32.

53 Plato, *Plato in Twelve Volumes, vol. 7*, trans. R.G. Bury (Cambridge: Harvard University Press, 1966), p. 343a.

54 Ogden and Richards, *Meaning*, p. 32.

55 Ibid., p. 32.

Unlike our prior attempts at close reading which analyse individual poems, this section will be concerned with the collection as a whole, as individual passages cannot be fully understood without reference to formulae and symbolism developed in others. In 1928, Riding, co-authoring with Graves, argued that poem anthologies were bringing about a kind of commodification of poetry, treating them as assortments of flavoured cookies for the puerile purpose of 'popularizing' poetry.[56] It is not surprising then that Riding's collection should be written almost as one long prose poem.

The passage titled "What is there to Believe In" begins with the line "There is a no-sense and a corresponding sense."[57] One may initially read the distinction between 'no-sense' and 'sense' in multiple ways. Trying out the structuralist approach, we may say that a 'sense' corresponds to a thing—as in word → sense → thing. There is also the perceptual 'sense' which similarly mediates two realms, the internal and external. The semiotic and the perceptual 'sense' can be conflated to imply that words come to acquire 'senses' because we attempt to understand the world using sense-perception. Read so, we are prompted to take 'no-sense' to eliminate the cumber of sense: that is, a direct installation of the thing-in-itself, the *ding an sich*, before being tainted by sense-perception which burdens language with 'a *corresponding* sense'. There is also the possibility that she is comparing the sensible and the nonsensical as both being 'There to Believe in'. But the strangeness of the vocable 'no-sense' necessitates a gloss somewhat beyond the realm of pre-existing concepts, such as 'thing-in-itself' a priori, or 'nonsense'. The phrasing of the title 'What is There to Believe In' promises a list of entities whose existence is just 'there', as in readily available for 'belief'. The world of objects outside sense-perception—requiring no-sense—is not readily 'there to believe in', philosophy having got no closer to proving sense-independent reality in Riding's time than when Kant famously declared the failing a scandal. On the other hand, things as they are beyond words (and their senses), *if* taken to exist, announce themselves quite readily—they are simply there, as Dr Johnson rudely reminded us by booting a pebble across the street: they do not require 'belief' either.

Riding is more likely developing an idea that she presents rather clumsily in *Poems: A Joking Word*. "Poems mean doom but doom isn't a meaning. Poems mean but doom is".[58] 'Doom' taken as annihilation allows us to read

[56] Laura Riding and Robert Graves, *A Pamphlet against Anthologies* (Hertfordshire: Garden City Press, 1928), pp. 9–20.

[57] Laura Riding, *Though Gently* (Deya: Seizin, 1930), p. 1.

[58] Riding, *Joking Word*, p. 20.

'poems mean doom' as poems convey 'nothing'—they are outside the illusion of reality in which we recognize 'somethings'. 'Doom' is the only thing that really 'is', and poems mean what *is*, so they refer to no-sense. Since true poetic words do not venture outside themselves to the historical world, the 'no-sense' of these words leads not to *dinge an sich* but the thingliness of words-in-themselves, lending them a substantial reality—a being affirmed by the paradoxical declaration 'there *is*' followed by the hyphenated 'no-sense'. 'No-sense' is the direct object of the verb 'is'—and 'sense', or reference to something in the historical/sensible world, 'corresponds' with no-sense. When merely conveying 'sense', words do not assert this independent existence: as Riding puts it in "Language and Laziness", "it has no reality, it is an empty cipher"[59]—this is the 'logical', Saussurian aspect of words. It is an 'empty cipher' because once it is occupied by its referent, it leaves behind no indelible residue.

The poem then lists a series of things that "correspond": irresponsibility and a corresponding responsibility, a question and a corresponding answer, approximation and corresponding exactness, and so on. This list of corresponding opposites establishes a supposed "equality between sufficient opposites". From its late medieval origins, 'correspond' means "to be in harmony with".[60] But the etymological definition may be a poor guide as it is not clear that 'irresponsibility', whatever be its object, is in harmony with 'responsibility'. By corresponding, somehow, the 'no-sensical words' both subsist and coexist with 'sensical words', despite their being opposites. But, the poem declares, "There is not God because he does not correspond [...] He differs, but he does not differ sufficiently to agree with all that he differs from. There is T and ⊥".[61] 'God' 'does not correspond' because He is fearfully different from any *thing* in our world; though He declares himself in various forms, his avatars do not differ enough to correspond with anything that *is*—in other words, no God has ever simply existed in the manner of temporal beings and things. In *Epilogue*, following an elaborate disquisition on the philological and phonetic genealogy of the word 'God', Riding summarizes the aspect common to all his conceptions and incarnations, namely, that 'God' signifies a remote "something else" that man faces with "a central attitude of perplexity".[62] Poetic words, however, unlike God who does not correspond, correspond between noumena and phenomena.

[59] Riding, *Anarchism*, p. 13.
[60] *OED*.
[61] Riding, *Though Gently*, p. 1.
[62] Riding, *Epilogue*, p. 27.

To explain how they do so, it will prove useful to consult the later Riding's defence against the imputation of Platonism in her youth; it will clarify for us her finer distinctions even if her taxonomy is rather involute. Riding had written in 1928 that "words in their pure use, which I assume to be their poetic use, are denials rather than affirmations of reality. The word *hat*, say, does not create a real hat: it isolates some element in the real hat which is not hat, which is unreal, the hat's self".[63] In 1974, when asked whether this is merely Platonic idealism brought to language, the older woman replied as follows:

> A sympathetic interpreter could point to the reality that [...] was evoked with the word, the realness of 'the hat's self' as distinguished from the realness of hat as 'thing' [...] This was the reality of things as existent to the mind, which knew in terms of words. This is different from the Platonic conception, which involves an endeavour to isolate another reality from the reality of appearance. [...] Plato reduced the numbers of things to idea-forms, and synthesised these into his counter-reality to reality—and had thus a purity abstracted from these. My reality counter to the conventional reality [...] was that which the human self, not extracting its identity from the collective physical or social [...] environment, was in itself, as real of itself, thus unreal by the collective modes of definition.[64]

The analysis of 'hat' here will remind us of her operations on the word 'death' in *Love as Love*. Riding argues that the Theory of the Forms—and the argument also applies to 'God' in *Though Gently*—carves for itself a separate realm from our own, while in words, Riding, to put it reductively, sought to separate their mental and sensual aspects and, as in "Death as Death", demonstrate how pure logos can defy the base corruptions of the lebenswelt.

In *Twenty Poems Less* (1930), which she published immediately after *Though Gently*, there is a poem, "Then Follows", which reads: "Because of being by name a poet, / A creature before man and beyond God".[65] Wexler describes the dilemma as follows: "As poet she faces a more trying task than either [God or man]: she must mediate between the two."[66] God 'does not correspond', but it becomes the task of the poet to sublimate 'sense'—or meaning sullied by history—into 'no-sense'. To posit existence by the phrase 'there is' suggests the subject of existence participates, and thereby corresponds,

[63] Riding, *Anarchism*, pp. 98–9. [64] Ibid., pp. 265–6.
[65] Riding, *Twenty*, p. 12. [66] Wexler, *Truth*, p. 62.

with existence. Thus, 'there is a no-sense' but 'there is not God' (sans hyphen). So whatever T and ⊥ may stand for, they *are*, unlike God, who simply is not.

⊥, or up tack, is a symbol in formal logic that denotes 'false', which we will take for our interpretive purpose as standing for 'negative truth'—a negative that *is*, like Empson's 'standing negative'.[67] To see what it betokens, we are better served looking to another section, "Let", where it is defined.

> Let the sign ⊥ stand for that which all understand and express differently. [...] ⊥ is the exact fulcrum. Let the sign T stand for the interpretive world of leverage. ⊥ is that which is.[68]

'⊥' is an 'exact fulcrum'. It is where language is in union with reality (we will recall the apex of our sketch of Riding's semantic triangle): minds grasp it unaided by sense, but express it varyingly when they speak. The grapheme signifies not only as a sign but also as a diagram. The symbol ⊥ resembles a fulcrum placed upside-down: rather than the teetering balancing of an object placed on a fulcrum, the upside-down image imposes a stability to the connection between word and mind. On the other hand, "T is that which is going on. Therefore T is in immediate opposition to ⊥ but in ultimate reference to ⊥. This is a statement not a hypothesis."[69] T is here to be taken as historical Time, which 'is going on', the world in flux. Unaffected by the variations of time, ⊥ stands opposed to T, but the different expressions of T are in ultimate subordination to the real ⊥ which mind innately comprehends. Referring to temporal truths as 'T truth', she says,

> This is not to be axiomatic. An axiom is a hypothesis adopted as absolute to give authority to all instances covered by it. It is written back from the instances for which authority is required in anticipatory deduction. T truth is necessarily axiomatic: it must be improvised for each immediate set of circumstances.[70]

The phrase 'anticipatory deduction', at first, seems an oxymoron. Customarily, induction induces anticipation, as expectations form on the back of what has been. Deduction is top-down discovery from premises or axioms. Euclidean

[67] In 'Mathematics as an Intellectual Master-Method', Riding suggests that she was familiar with texts on logic in her youth (pp. 597–600). Although none of the texts she mentions use the up-tack, she seems to have been well-read in logic and could have come across the symbol elsewhere.

[68] Riding, *Gently*, p. 2. [69] Ibid., p. 2. [70] Ibid., p. 2.

geometry, for instance, accepts certain axioms and deduces almost all geometric theorems therefrom. This is an instance where Riding's redefining habit is at its most mischievous. 'Axioms' are typically understood as the conditions of mathematical truths that issue from them. In the poem, however, an axiom is said to be 'written back from the instances for which authority is required'. The poem seems to suggest that when the ultimate justification of mathematical truth is questioned, what are tacit, procedural hypotheses about the nature of the world are suddenly posited as 'axioms' to lend the theorems deduced from them final authority—as when the parallel postulate asserted what can only be speculation on the nature of space at the extremes. What was considered beyond question and outside time is thus accused of being infected by history. In contrast to historically contingent axioms, sentences of the poem are declared 'statements', which are 'equal among equals': "This is not to be axiomatic", says the poet of her own poetic sentence, "since it is a statement, not a hypothesis, an equal among equals [...] This is not for the benefit of what follows, for nothing follows."[71] Although the poetic statements are absolute, they do not—by virtue of being 'equals'—claim to be the bases for ensuing statements as do axioms. Once again, it is affirmed that if 'statements' are what follows, and 'nothing follows', then poetic statements are 'nothing'. All types of 'nothing' are in essence 'equal', for differences can only be distinguished by qualities. We might recall "Death as Death", in which "nothing" is defined as "a similarity / without resemblance"; this is because semblance is recognized by sense while 'nothing' is given to pure understanding—like the unutterable yet vivid sense that a child's imagination develops of the word 'hat' over and above the physical hats it has encountered.

By annihilating sense, Riding finds a way to place the truth of poetic statements on a separate plane of abstraction from that of mathematics, whose 'axioms', we have seen, derive ultimately from experience. Falsely declared as absolute for the sake of authority, individual axioms become mere 'dogmas'.

> T truth is necessarily dogmatic: it must first impose its content before it can have validity. This is not to be dogmatic, this does not differ from anything but itself, which is to differ only by degree in what it samely is. This, indeed, is to speak absolutely, since it is not to define the silences of God but the speech of man.[72]

[71] Ibid., p. 2. [72] Ibid., p. 2.

A dogmatic truth 'must first impose its content before it can have validity'; in other words, axioms are synthetic propositions and so cannot 'speak absolutely' as poems can. This much is clear from her early works. But *Though Gently* even attempts to define metaphysically what it means to 'speak absolutely' about 'nothing'. We began by defining the absolute against Platonism: 'statements' are not Platonic in so far as she is not defining 'the silences of God but the speech of man'; the poem unconceals the nature of language positively through its statements and negatively by stating what it is not: neither axiom nor dogma.

Next, she shows us how poetic words relate to "Numbers":

Numbers attempt to arrive at finity through infinity.
According to numbers ⊥ is T+.
Numbers cannot describe totality but only the composition of totality.[73]

We are concerned in this poem with the untranslatability between number and reality, and between number and essence. Because numbers are not entities in the world, they're not bound physically and are thence termed 'infinite'. The futility which the poem points to is the attempt to describe finite entities, sure things in the world, in terms of the non-finite, or borrowing from "The Number", represent what is '*Alive*' with 'ghosts'. The poem states that numbers, incapable even of coming to terms with finite objects in the physical world, attempt to define their essence beyond. That is, while dwelling in the realm of T (the pulses of Time from which counting springs), numbers attempt to arrive at ⊥, the 'essential': they add themselves, the operation symbolized by T+, in the false hope of eventually attaining ⊥.

Tim Armstrong reads these lines as invoking Cantor's theories of infinity, which we discussed in the previous chapter.[74] Cantor discovered, we will recall, infinities of differing sizes: the power set, for instance, is intuitively larger than the set of natural numbers from which it might spring, even though the latter itself is infinite.[75] With the notion of the power set as the context, T+ could be seen as representing man's feeble attempt to count to infinity or count the different infinities, and thereby analyse that which is

[73] Ibid., p. 3. [74] Armstrong, 'Syntax', pp. 8–11.

[75] See ibid., pp. 2–3; chapter 1.2 of Baylee Brits, *Literary Infinities: Number and Narrative in Modern Fiction* (London: Bloomsbury, 2018); and Ivor Grattan-Guinness, *The Search for Mathematical Roots, 1870–1940: Logics, Set Theories and the Foundations of Mathematics from Cantor through Russell to Gödel* (Princeton: Princeton University Press, 2000).

beyond analysis. Armstrong argues that Riding is incorrect to suggest that the infinite essence ⊥ cannot be arrived at by counting:

> If '[a] detail suspends meaning [and] an essential sustains it against suspension' (*TG* 3), then you have the suggestion that meaning is 'suspended' between the realms of number (that is 'detail' or mere facticity) and essence. But that is not quite right, since number, or rather sequences of numbers, can move, at the limit, beyond denumeration and towards finality.[76]

By focussing on the possibilities opened up by Cantor, Armstrong does not allow for the more direct reading of the poem, which is that Riding does not accept the mathematician's definitions of infinity: the poem views the mathematician's 'attempt to arrive at finity through infinity' as a kind of cheap trick. This can be illustrated by reading the second line against another analogy, which it more closely approximates—a much older problem in philosophy than Cantor's, of arriving at the necessary from the contingent: which states that absolute essence or 'totality', ⊥, cannot be attained by a collection of contingent, temporal phenomena, T. We refer to the famous Ontological Argument, in which entities in our world (contingent beings) always require other contingent beings, in an infinite series of partial explanations, to explain their existence—such as, say, a child is explained by her parents who are in turn explained by theirs, and so on; or, as the poet says, "The enumeration of four essences does not lead to the discovery of an essential but of a fifth essence." We are prompted to our analogy by the repetition of the word 'sufficient' thrice in a sentence: "there are sufficient protests of meaning to provoke sufficient details to provoke ultimately a sufficient essential", echoing Leibniz's sufficient reason. An 'existent being'—one that contains within itself sufficient reason for its own existence—must be external to the world; and 'world' can be thought of here as the aggregate of contingent entities. An 'existent being' becomes necessary in this argument as an explanation for the phenomenal world, or reality.[77] While the necessary being in the Argument is God, Riding has already distinguished ⊥ from God. The word 'God', she says in *Twenty Poems Less*, is only for the small-minded that are trapped in convention: "For a poor understanding / It ["The Gentle Truth"] may all be reduced / To a

[76] Armstrong, 'Syntax', p. 10.

[77] Those interested in a clear summary of the argument can see Bertrand Russell, *Why I Am Not a Christian* (London: George Allen & Unwin, 1957), p. 127.

simple word: God!"[78] We will instead apply the Argument to the 'essence' disclosed in the poetic word.

In her use of repetition in *Love as Love,* we observed how Riding endows words in her poems with reason, in themselves, for themselves, without reference to their conventional meaning (which is determined by T [historical time]). If $\perp$ is the poetic word, we may paraphrase the second line as saying $T_1+T_2+T_3+\ldots T_n \neq \perp$ or that the series $T_1, T_2, T_3, \ldots$ does not culminate in $\perp$. Thus, even if lines 1–2 recall Cantor's efforts at defining infinity, the poem must be read as baulking at such attempts, as the essence to Riding is entirely different from what mathematicians take it to be. This is also the only reading that excuses the paradoxical assertion in line 6 that "Numbers are life, imitation, or analogy and they lead only to further numbers. / Numbers are detail". The absurd phrasing must not be confused as suggesting numbers are '*Alive*', as in actively being—they are, after all, 'ghosts'. The word 'life' must instead be taken in the context of its listed synonyms: 'imitation' and 'analogy'. The line suggests that numbers are sedimented, as is 'life' or the "myth", which is "the unreal, literary, psychologically organised self-world [;] collective-real: its existence depends on a belief in reality, though in reality as a myth"[79]. Numbers as life is not too much of a semantic stretch for a poet who has made axioms historical. It then becomes the task of the poetic word to remove us from T time, the numbered, serialized world of myth. In 1974, Riding attempts to clarify some of these ideas:

> My theme [was] the poet as one who does not live, think, speak, within the frame of a <u>concept</u> of reality; and it is poetry as of non-conceptual substance; and it is the poem as just itself. That is, I am letting the poetic function meet my sense of an ultimate fullness, perfection, of speaking, an <u>end</u> that words imply [...] So I say [...] 'A poem is nothing'.[80]

The equation between 'fullness', 'perfection' and 'nothing' in this passage is a hint for us to take 'essence', 'totality', and 'nothing' in *Though Gently* as all differently referring to the same thing. Line 3 states that 'numbers cannot describe totality but only the composition of totality'. Reality is thus beyond the grasp of numbers, which can only list its components.

[78] Riding, *Twenty*, p. 16.
[79] Riding, *Anarchism*, p. 100.
[80] Ibid. (Appendix I), p. 256.

In a later poem in *Though Gently*, "I Grant You", Riding states that "For Plato [...] poetry was an arrogant experiment in perfection":[81] Plato believed poets deluded in thinking their work perfect when in fact poetry (and all art) was but 'imitation' of the shadowy objects of phenomena, which were in turn 'imitations' of perfect objects, eidoi, whose celestial splendour it was the vocation of the mathematician to discover. By defining numbers as 'imitation', Riding reverses Plato's charge.[82] Plato excommunicated poets from his Republic for entrenching our imprisonment in the cave. In the *Phaedo*, Socrates says to Cebes,

> [W]ould you not be cautious of affirming that the addition of one to one, or the division of one, is the cause of two? And you would loudly asseverate that you know of no way in which anything comes into existence except by participation in its own proper essence, and consequently, as far as you know, the only cause of two is the participation in duality—this is the way to make two, and the participation in one is the way to make one.

Arithmetic is, argues Socrates, meaningless without an ideal conception of number. The poem "Numbers", repudiating this transcendental dogma, confines numbers to their own realm, a kind of mathematical purgatory where they are trapped to 'lead only to further numbers'. Poetry, on the other hand, is said to disclose "the possibility of planes of thought beyond common reach", as "the inspiration of the poet was divine".[83]

Returning to *Though Gently*, we have the lines,

> A detail suspends meaning; an essential sustains it against
>   suspension.
> There is ultimately only one essential; [...]
> Detail is measured in number, the essential in degree.
> But such degree is expressed numerically, to allow compatibility
> between detail and essential.[84]

We will recall that 'numbers are detail': while numbers 'suspend meaning', the poetic word is an 'essential' that 'sustains' or preserves meaning 'against suspension'. The next line—'Detail is measured in number, the essential in

[81] Riding, *Though Gently*, p. 18.
[82] I refer to the famous Book X of *The Republic*, wherein Socrates declares poets and artists as belonging to the imitative tribe.
[83] Riding, *Though Gently*, p. 18. [84] Ibid., p. 3.

degree'—I believe, allows us to attempt—scandalizing the iconoclastic Riding, had she known—a rude illustration of that unspeakable nothing which poetic words ultimately are. Before we understand the use of the word 'degree' here, let us harken to its earlier use: it was stated that poetic statements 'differ only by degree in what it samely is'. God thus mirrors poetic statements, because He 'differs, but he does not differ sufficiently to agree with all that he differs from'. I have characterized God as *mirroring*—occupying a different dimension, as it were, from—poetic statements or ⊥ because they, like the Bodhisattva, deign to re-enter the world (correspond) after attaining Heaven while God does not.

Now let us collect all we have learnt about the essential, ⊥: it is (1) indivisible, (2) differs but is the same, (3) differs only in degrees, and (4) God mirrors it. From these postulates, the image of ⊥ that emerges is a sphere: a unitary Parmenidean sphere of existence which differs when approached from its various angles or degrees of rotation but is in essence the same all over—the ideal continuous object of topology, even. The sphere also calls up sacred images from antiquity, from Plato, who likened God's design of the universe to a sphere in the *Timaeus*, to Byzantine art which depicts God as a circle, to Nicholas of Cusa, who defined God as an "infinite circle whose centre is everywhere and whose circumference is nowhere".[85] Indeed, the reflection of a sphere would be a circle in the mirror, which would then occupy a separate dimension like the god of *Though Gently*. The sphere must have been for the poet a metaphor-in-absentia for the world (not the oblate spheroid of astronomy but world seen as total reality). In a true poem, as per Riding's definitions, language illumines regions of this unitary sphere of existence which 'differs only by degree in what it samely is'—it is the common quarry from which vocables are extracted: vocables that when rightly received in the lines of poetry, reveal the essence of *prima materia*.

But we have left out in our list of facts about ⊥ the main descriptor, namely, that it is 'nothing', 'doom', 'no-sense'. Nothing can be truly said about the sphere as a whole—pace the $4/3\pi r^3$ of geometry—because it does not have a gradation of attributes: it is a 'similarity without resemblance'.[86] Nothing, we have shown, also equates to 'totality': this equivalence computes only if we see

[85] Plato, *Timaeus and Critias*, trans. Desmond Lee (London: Penguin Books, 1977), p. 9; Katherine M. Dunbabin, *Mosaics of the Greek and Roman World* (Cambridge: Cambridge University Press, 1999), p. 95; For Cusano's quote, see Anthony Levi, *Renaissance and Reformation: The Intellectual Genesis* (New Haven: Yale University Press, 2002), p. 151.

[86] As a consequence, everything said in the last two pages, including the image of the sphere, must be taken as mere analogy. It is only to prepare a reception of the poems in *Though Gently*.

both 'nothing' and 'totality' as sharing their one articulable attribute, namely, being *not something*. The existence of 'nothing/totality' can, says Riding, only be displayed or unveiled in the course of poetry. It therefore becomes the task of the poetic word to shuttle between detail and essence, T and ⊥, conveying the institutionalized reader from her imprisonment in history to a purer region of abstract linguistic truth. Just so, the reader is primed to understand correctly the question at the heart of Laura Riding's pursuit of truth: "An image of meaning is eternally suspended? / Devilishly contrived to stand eternally between detail and the essential?"[87]

4.1.4 Loss of certainty

In a later collection, *Lying Word*, we see that Riding's faith in poetic words rescuing a society in the thrall of number seems to have lapsed. The disappointed mystic turns her conservative animus to guarding poetry, instead, from society—perhaps for posterity, perhaps for eternity. The following lines are from the poem "Come, Words, Away",[88]

> But I know a way to soothe
> The whirl of you when speech blasphemes
> Against the silent half of language.

Regular speech constantly offends the silent half, the word itself as known to mind, by shouting words in loose and promiscuous ways. The poet has had enough:

> Come, words, away to where
> The meaning is not thickened
> With the voice's fretting substance.

One might expect the renaissance turn of phrase in 'come, away' in a Herrick poem, beseeching his mistress to elope, or even in Shakespeare, asking death to "come away, come away". Although the word is the poet's amour here, the typical context remains of a tryst and its ruin in vulgar society. The

[87] From "The Sphinx" (Riding, *Though Gently*, p. 4).
[88] Laura Riding, *Poet: A Lying Word* (London: A. Barker, 1933).

Aristophanic union between mind and word *in ovum* is being gnawed away by speakers of those words, as if their mouths were marauding scavengers.

Her Manichaeism as a poet becomes steadily more severe with age; she seems to shun the world altogether (in the opening lines of *Epilogue*) so her words can be roused from the stigma of history:

> Now time has reached the flurrying curtain-fall
> That wakens thought from historied reverie
> And gives the word to uninfected discourse.[89]

The crafty nonce 'historied', echoing sullied, presents words in their historical sense, utilized in discourse, as infected by sin. The line recalls Sidney, who in his *Defence of Poesie,* describes the tragedy of those who labour with words: "Sith our erected wit maketh us know what perfection is, and yet our infected will keepeth us from reaching unto it".[90] But unlike Sidney, Riding, in her youth, seems to have believed redemption from the corruption of discourse possible.

Despite an eventual loss of faith, Riding continued to develop the concepts set forth in *Though Gently* throughout her later, better known, poems. In "Unless Infinity is only Time",[91] the totality unattainable to numbers counting to infinity is seen through the lens of time, where Essence becomes "the indescribable moment of enough" [5]. The vertiginous equivalence between totality and nothingness in *Though Gently* is transferred in this poem to eternity and the moment. She argues in *Epilogue* that "the scientific universe is [...] the repetition of a same circumstance [...] composed of a succession of instantaneous events; and so long as there is at least one such event the succession is for itself infinite".[92] The indescribable moment, the instant in time, like we saw with Empson's calculus metaphor, is for Riding too a doctrinal point, an Essence, unapproachable to science. The scientific universe rends time into "the moment upon moment upon moment that bears down, / In mathematical spite" [19–20]. And yet, T+ ≠ ⊥. But readers of *Though Gently* will remember that it had once been the poet's vocation, now foregone burden, to reveal, ever and again, the "unnumbered one", the 'indescribable moment of enough', to the arithmetical procession of history if only to follow

[89] Riding, *Epilogue*, p. 1.
[90] Philip Sidney, *The Defence of Poesie* (London: Printed for William Ponsonby, 1595), B4.
[91] Riding, *Lying Word*, p. 17.
[92] Riding, *Epilogue*, pp. 24–5.

Blake's commandment to poets "To see a World in a Grain of Sand / [...] Hold Infinity in the palm of your hand / And Eternity in an hour".[93]

Riding is the first among the poets we have read in this book to retract poetry from the agon for *descriptive* authority. She fashions the poetic word as participle to the inviolable one of reality, from which it communes with—rather breaches—the illusory fabric of culture: a fabric that is parted, sifted, counted, and thereby described by number. *Descriptive* authority, then, does not secure access to truth, the ultimate object of the agon. We move now to another American poet who attempted to retire poetry from the competition over description; but who, unlike Riding, found meaning almost entirely in the saying and speaking of poetry. Although Charles Olson was more a public figure than the recluse Riding, we will sense in his writing, both in register and style, an equal uncanniness, a pervading feeling that verses are indited as a kind of incantation to rouse the reader from some ancient slumber.

4.2 The Non-Euclidean Olson

As removed the word was from the foibles of man and history to Riding, so much was it marked by these to Olson. Logos, word and measure, was for Olson from and of the human body; the body, an instrument to commune with reality, but also vulnerable to corruption with the advent of mass culture.

> O my people, where shall you find it, how, where, where shall
> you listen
> When all is become billboards, when, all, even silence, is
> spray-gunned?[94]

Enervated by the loudness of modern mass culture, the body is no more fit, says the poet, to hear poetry aright. It was Olson's hope that through the mechanics of reading his poems, the body of old that was once in company with nature would somehow be resurrected.

The idea that the visual, auditory, and semantic detritus of mass culture was disturbing the aptness of language, what was prophesied in the early works of Riding, is a sustained preoccupation of Olson's poetry.

[93] McGann discusses the influence of Blake's Platonic ideas of language on Riding in Jerome McGann, 'Laura (Riding) Jackson and the Literal Truth', *Critical Inquiry*, 18 (1992), 454–73, p. 468.

[94] Charles Olson, *The Maximus Poems*, ed. George F. Butterick (Berkeley: University of California Press, 1983), p. 6.

Colored pictures
[...]: dirty
Postcards
  And words, words, words
All over everything
          No eyes or ears left
To do their own doings (all
Invaded, appropriated, outraged, all senses[95]

There is an assumption behind Olson's poetic experiments that the transfiguration of the commonplace under late capitalism was triggering a synesthetic derangement of the inner life and atrophying sensuous comprehension in and through words. To recover the connection between world, body, and language—what will serve as a preliminary definition of Olson's 'field'—was an ambition that lasted his career as poet and critic.

Olson is sympathetic with Lawrence, Yeats, and other critics of modern statecraft: "to him [Dostoevsky]", he says, "only a hideous levelling of man can come when revolution establishes itself 'on the elements of science and reason'. What he sensed and what we know is that the modern revolutionary state denies the dignity and the value of individual human personality."[96] Olson also channels the later Auden's attacks on the levelling—under the guise of equality—tendency of social statistics:[97] "history", says Olson, should at least be conceived as "events and laws, not this dreadful beast, some average and statistic".[98] These recognizable critiques of the intrusion of number into the social world seem to cast Olson in a sequel, supporting role in modernist poetry's agon of the 1920s and '30s. But unlike Empson and Roberts, Olson, with Riding, is a metaphysician, and is interested in the ontology of language more than its descriptive powers. And unlike Riding, he does not view the discursive empire of mathematics as an entirely hostile power. Peter Middleton says Olson "thought of physics as opportunity, not as a source of

[95] Olson, *Maximus*, p. 17.

[96] Charles Olson, *Collected Prose*, eds, Donald Allen and Benjamin Friedlander (Berkeley: University of California Press, 1997), p. 129.

[97] See Anirudh Sridhar, 'Auden and the Infernal Sciences', *Essays in Criticism* 71.4 (2022): 431–453.

[98] 'A Bibliography on America for Ed Dorn' (Olson, *Prose*, p. 297). Olson admires Auden for dwelling on Melville's anti-rationalism in *The Enchafèd Flood* (in 'The Materials and Weights of Herman Melville' (Olson, *Prose*, p. 113). Olson's critique of sociology seems the facsimile of Auden's 'The Kingdom of Number', in his differentiations between "the individual" and "single and plural", dubbing the latter "King Numbers" (Olson, *Prose*, p. 297); see "The Virgin and the Dynamo", in W. H. Auden, *The Dyer's Hand and Other Essays* (New York: Random House, 1962), pp. 61–71.

envy".[99] He participated in the agon but through appropriation as much as denunciation. The classical mathematical world-picture was still his antagonist; but modern mathematics would be competing on his side.

Nineteenth-century mathematics becomes central to Olson's construction of what he calls "the Real"—the prize, as we have seen, in the agon of the faculties.[100] It should be noted, however, that Olson is prone to wayward extrapolations from mathematical theory. He admits to an audience, once, "I should like immediately to disburden myself of any idea on your part that I have any adequate knowledge of mathematics & geometry."[101] While it would test the patience of a mathematically literate reader to take what Olson says about mathematics too seriously, we will see that Olson's (ab-)uses of mathematics provide a kind of precise, diagrammatic analogue to his metaphysics. We will at times have to guess at what Olson understood from the mathematical ideas he brandishes—less amusing perhaps than picturing windmills through Don Quixote's eyes—but given his musings on mathematics do clarify his poetry, and his poetry is so strange and striking, it is a task worth undertaking.

Olson found in modern mathematics a liberation from the spent dualisms of ancient Greece. The problem facing the Western mind, says Olson in "Human Universe", in a characteristically sweeping manner, is that "[w]e have lived in a generalizing time, at least since 450 B.C".[102] All thought has been subject either to Plato's "logos, and the reason necessary to it" or Aristotle's "logic and classification".[103] Modern mathematics, for Olson, would disentangle these oppressive ancient divisions, between subject and object, object and motion, form and content, the individual and the collective.

Olson's basic set of concepts is introduced in his first manifesto, "Projective Verse" (1950), and developed throughout the following decade. Roughly in the period of fifteen years, from his first book in 1947 (*Call me Ishmael*) to his final manifesto in 1962 ("Proprioception"), Olson completely articulates his puzzle of existence: bringing a notion of the Real, the correct function of language, and its relation to human biology, all, in a plausible manner, together—they are, respectively, the broad subjects of the following sections. An elucidation of the 'non-Euclidean relationship'—with due apologies for the phrasing—between world, language, and body will, we will see, allow us to understand what Olson was attempting with his 'projective' verse.

[99] Peter Middleton, *Physics Envy: American Poetry and Science in the Cold War and after* (Chicago: University of Chicago Press, 2016), p. 10.

[100] 'Equal to the Real' (Olson, *Prose*, p. 120).

[101] Mark Byers, *Charles Olson and American Modernism: The Practice of the Self* (Oxford: Oxford University Press, 2018), p. 68.

[102] Olson, *Prose*, p. 155.

[103] Olson, *Prose*, pp. 155, 156.

4.2.1 Non-Euclidean reality

> ya, selva oscura, but hell now
> is not exterior, is not to be got out of, is
> the coat of your own self, the beasts
> emblazoned on you[104]

So declares the poet, "In Cold Hell in Thicket", lost in that dark wood where Dante wandered midway through life, unsure what his new situation portends. It appears he knows not what he is, even, as the surrounding thicket seems to be enveloping him like a bodily coat, the encircling beasts but apparitions on his unfurling self. This sense of a self sans the lyrical interference of the ego, coterminous with ambient space, is a thing Olson rears and moulds through his works. The poet asks, "And who / can turn this total thing, invert / and let the ragged sleeves be seen / by any bitch or common character?"[105] Olson perhaps regarded himself a prophet tiding the new revelation: of the inversion of this 'total thing', of self become object and world become body. But to see what ushered in this new reality, a space that our bodies will come to call self, we must turn to the dizzying mathematics of topology.

Olson saw the advent of non-Euclidean geometry as the seminal episode of modern mathematics. Kimmelman says Olson draws from "Weyl, Riemann, Carl Gauss, Nikolai Lobachevsky and János Bolyai".[106] Olson scribbled in one of his notebooks what appears to be a shorthand for the history of non-Euclidean geometry:

Euclidean—parabolic
hyperbolic—Bolyai-Lobatchewsky
spherical
elliptic—Riemann—Cayley[107]

Among his heroes, Bernhard Riemann seems to have held a special significance. Riemann comes before the other nineteenth-century figures—ones more familiar to modernist poets such as Bergson and James—who challenged the Cartesian picture of space as the observable and measurable backdrop to existence. Olson finds himself gladly thrust into his

[104] Olson, *Poems*, p. 158. [105] Ibid.
[106] Burt Kimmelman, "'Equal, That Is, To The Real Itself': The new physics, Charles Olson, and avant-garde poetics." *DQR Studies in Literature*, 47 (2011), 641–67, p. 661.
[107] Byers, *Modernism*, p. 67; Charles Olson, 'Verse & Geometry plus E. P.', I Box 49:37, CORC.

surroundings when apprehending the world of Riemannian geometry, as he does when reading *Moby Dick*, with whose writing, he notes, Riemann's theories happily coincide.

> It took thirty-one years (Melville's age when he wrote *Moby-Dick*) for the German mathematician Riemann to define the real as men have since exploited it: he distinguished two kinds of manifold, the discrete (which would be the old system, and it includes discourse, language as it had been since Socrates) and, what he took to be more true, the continuous.[108]

Although the difference between the discrete and the continuous is important in Riemann's works, Olson has allowed his imagination here to run unchecked: he seems to suggest that the implications of Riemann's work were intended toward language and culture—quite fantastic. But even through the haze of conjecture, the distinction between the discrete and continuous seems to signal something clear to Olson. 'The discrete' in Olson's system (Figure 4.2) represents one side of a duality. He uses the term in two ways: first of space as discrete from events, as stable backdrop to the drama of human affairs. The other sense is of objects as separate from one another. A function such as $y = x^2$ (where x belongs to the natural numbers and is between 0 and 3)

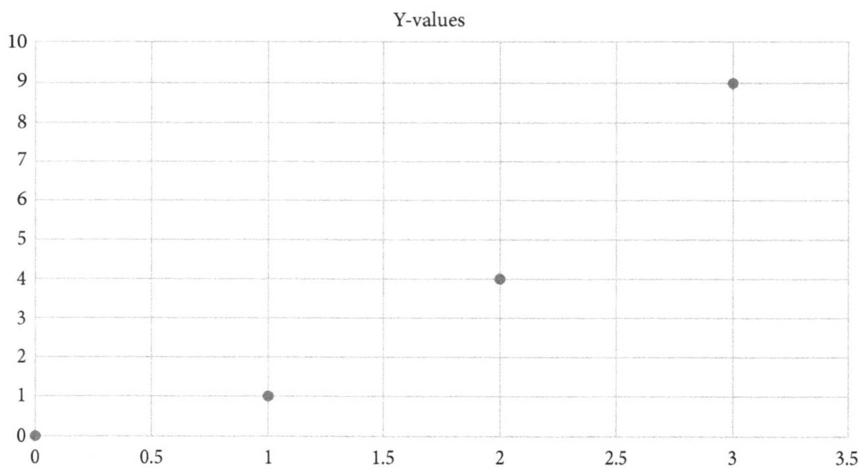

Figure 4.2 Discrete picture of function ($y = x^2$).
Source: Image produced by author.

[108] 'Equal to the Real' (Olson, *Prose*, p. 120).

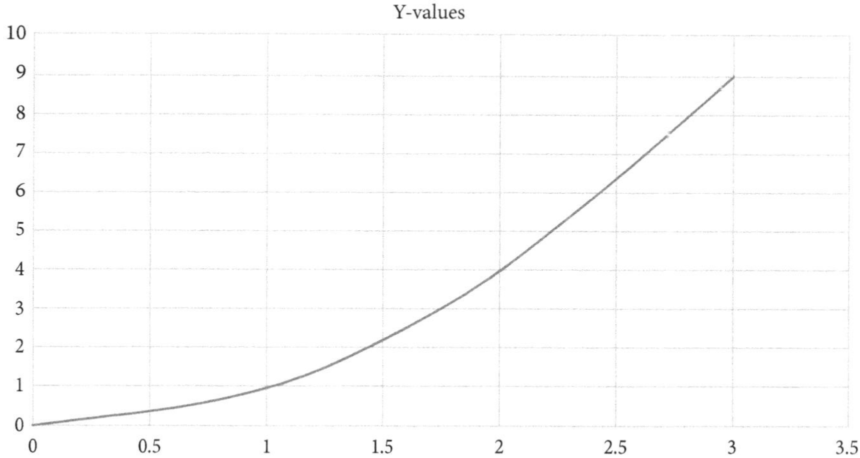

Figure 4.3 Continuous picture of function ($y = x^2$).
Source: Image produced by author.

would be discrete because yields individual values that are distinct from one another.

On the other hand, the same function $y = x^2$ (where x belongs to the *real* numbers and is between 0 and 3) would be continuous (Figure 4.3) because yields the values of the previous function along with all the numbers in between.

"Discrete magnitudes", says Riemann, rely "upon the postulate that certain given things are to be regarded as equivalent; quantity is accomplished in the case of discrete magnitudes by counting".[109] One can immediately see from Figures 4.2 and 4.3 why the discrete mode would suffice for and be suitable to such activities as counting, classifying, taxonomizing, all of which Olson seconds to the *ancien regime*. In a continuous conception of things, however, no individual locus has unique value: there is only neighbourhood, proximity, and deviation.

Discrete mathematics is described, in the texts on mathematical philosophy—especially Hermann Weyl's *Philosophy of Mathematics and Natural Science*[110]—that Olson read in the 1950s, as a kind of cultural artifice. Weyl shows the pre-Socratics to have been circumspect about converting the

[109] Bernhard Riemann, *On the Hypotheses which Lie as the Bases of Geometry*, ed. Juergen Jost (Switzerland: Springer International, 2016), p. 32.

[110] Ralph Maud, *Charles Olson's Reading: A Biography* (Carbondale: Southern Illinois University Press, 1996), p. 45.

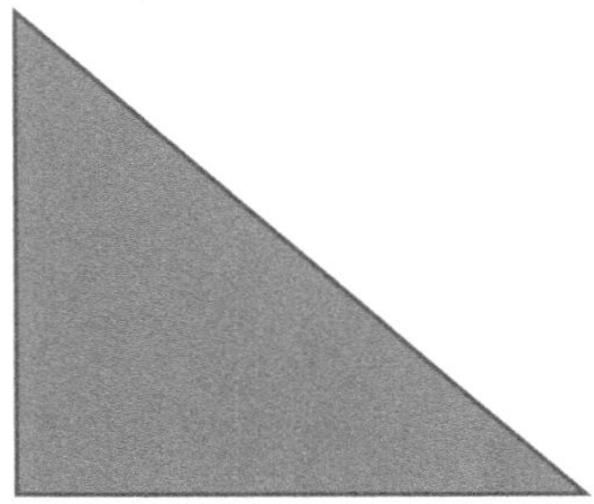

Figure 4.4 Continuous triangular region.
Source: Image produced by author.

continuous to the discrete; they "had been deterred from this step", he says, "because they took the discovery of the irrational seriously".[111] An irrational is a number which cannot be represented as a fraction.

The triangle can be thought of as a continuous region (Figure 4.4). If we take the height and base as a unit of 1 each, the hypotenuse, by the theorem of Pythagoras, would be $\sqrt{2}$. $\sqrt{2}$ cannot be represented as a whole number or fraction, and thus, the continuous region would resist representation in discrete terms. As more such miscreants as $\sqrt{2}$ were discovered, further gaps began to appear in the number line: an embarrassment that is evident in the resort to ellipsis, with $\sqrt{2}$ equalling 1.41421..., and π equalling 3.14159..., and so on. In the nineteenth century, however, these gaps were filled, as it were, in the process of giving rigorous definition to the real number line.[112] A real number would thereafter be the limit of a sequence of rational numbers (numbers that can be represented as fractions), each subsequent rational number getting nearer and nearer the elusive irrational, not unlike tangents approaching the slope of a point on the curve in calculus.[113]

Weyl says, "Geometry proper, whose ideal structures could only be carried with exactness into a real continuum after this has been spun over with a subdivision net of a fineness increasing *ad infinitum*."[114] The fineness of sequential detail with which the numbers could be described extended the powers of geometry in the nineteenth century. Working in the new geometry typically requires the use of sets. Sets can be open and closed. When the

[111] Hermann Weyl, *Philosophy of Mathematics and Natural Science*, transl. by Olaf Helmer (Princeton: Princeton University Press, 1949), p. 68.
[112] Terrence Tao, *Analysis I* (Delhi: Hindustan Book Agency [2006] 2022), p. 102.
[113] It would be more accurate to say that real numbers were limits of Cauchy sequences, so that each rational in its neighbourhood would be ϵ-close to it, where $\epsilon>0$. But elaborating on this would take us far off course (Tao, *Analysis*, p. 102).
[114] Weyl, *Philosophy*, p. 91.

Figure 4.5 Open interval.
Source: Image produced by author.

boundary values are included in the set, it is called a closed set; and when not included, an open set. For instance, the set of the values of x+y where $5<=x+y<=8$ would be a closed set because includes 5 and 8; whereas the set of $5<x+y<8$ would be an open set because excludes 5 and 8 The notion of a topology on a manifold was defined using open subsets.[115] And it is the 'openness' of these sets which forms the conceptual foundation for Olson's later construction of the "OPEN poem"[116]—of which hereafter.

We might think of the hollow bubbles in Figure 4.5 as the open set ($\sqrt{2}$, π), where neither $\sqrt{2}$ nor π themselves belong in the set; but the members of the set are infinite and approach nearer and nearer to the irrational, never-reached termini in either direction.[117] But the use of open subsets is not limited to describing irrationals. Say a point travels from 2 to 4 on the number line. In topology, the real numbers 2.0 and 4.0 would be unreachable in the open set because they are limits, not boundaries: the topological set would extend infinitely, say, from…2.0001, 2.0002, 2.0003…and 3.9997, 3.9998, 3.9999,…in either direction.[118] Now if we extrapolate the number line to higher-dimensional contexts, we can imagine regions of space seamlessly opening into one another.

For instance, the boundary of the circle in Figure 4.6 represents the closed set of points [x,y] that satisfy the equation of the circle $x^2 + y^2 = r^2$; but the shaded area within is an open set, which contains all points (x,y) that satisfy the equation $x^2 + y^2 < r^2$. The shaded circle, we will observe, blends into its limit; the region of space extends infinitely and infinitesimally towards the bounding other. This is one reason why Olson regarded topology, the study of manifolds, as being 'more true'. By moving from classical geometry to topology, Riemann had, to Olson, liberated the Western mind from "the exact death quantity does offer, if it is numbers, and extension".[119]

[115] Hermann Weyl, *The Concept of a Riemann Surface*, trans. Gerald R. MacLane (New York: Dover [1913], 2009), p. 18.

[116] 'Projective Verse' (Olson, *Prose*, p. 240).

[117] There are actually infinitely many irrational numbers within the open set of ($\sqrt{2}$, π), but we are ignoring these and considering only the rational numbers within for exposition.

[118] These numbers are to be used mnemonically and discarded thereafter: for instance, there is a proof in mathematics that makes '3.999…' actually equal to '4', which would make of it the very boundary that the open set precludes.

[119] 'Equal to the Real' (Olson, *Prose*, p. 121).

Figure 4.6 Open disk.
Source: Image produced by author.

For Olson, the discrete's main mischief is that it suffers consciousness to become attached to arbitrary placeholders, concreting boundaries that occlude the messy thicket between. Take Olson's poem, "The Kingfishers":[120]

> The message is
> A discrete or continuous sequence of measurable events
> distributed in time
> is the birth of air, is
> the birth of water, is
> a state between
> the origin and
> the end, between
> birth and the beginning of
> another fetid nest[121]

The poem celebrates the Heraclitean flux defeating our abstractions.[122] What lies in between the arbitrary 'origin' and 'end' is what the poet must exhume. His task is to invest the continuous manifold of reality with significance alien to a Western consciousness assembled under the logos of Socrates. We will note the positive spin on the word 'fetid', which, in modifying 'nest', comes to

[120] George Butterick says Olson borrowed the following lines about the discrete and the continuous from Norbert Wiener's *Cybernetics* (George F. Butterick, *A Guide to The Maximus Poems of Charles Olson* (Berkeley: University of California Press, 1978), p. 632).
[121] Olson, *Poems*, p. 90.
[122] Earlier in the poem, Olson quotes directly from Heraclitus's famous 12th Fragment (Olson, *Poems*, p. 89).

effect a sensory rewiring in the reader. Those among us who have fared through a swamp will recall the plashy, mingling, unsteady discomfort of the surroundings, unrelieved by discrete rock or lea—the bodily memory of that discomfort is picked out and nestled with the tangled 'nest', our new home. Olson's poetry, we will see, often seeks thus to deploy words to somatically and subliminally register the ideas supposedly set in motion by nineteenth-century geometry.

The new version of reality, for Olson, demanded that we set aside the old book in which man had catalogued reality like a stamp collector. "Taking it in towards writing, the discrete, for example, wasn't any longer a good enough base for discourse: classification was exposed as mere taxonomy."[123] Prose, says Olson, should free itself from "naturalism",[124] from the rigidities of discrete mimesis, by incorporating in language the outré features of topology. Weyl says,

> Space is infinite not only in the sense that it never comes to an end; but at every place it is, so to speak, inwardly infinite, inasmuch as a point can only be fixed step- by-step by a process of subdivision which progresses *ad infinitum.* This is in contrast with the resting and complete existence that intuition ascribes to space. The 'open' character is communicated by the continuous space and the continuously graded qualities to the things of the external world. A real thing can never be given adequately, its 'inner horizon' is unfolded by an infinitely continued process of ever new and more exact experiences [...]. For this reason it is impossible to posit the real thing as existing, closed and complete in itself.[125]

The infinite rash that space had newly acquired meant that objects within would also be transmogrified in unseemly ways; they could no longer sit elegantly closed and completed, as Weyl puts it. Two objects would not be deemed alike or different because of the bulk they boasted on the Cartesian plane—this notion of things seemed almost childish innocence to non-Euclidean man. For Olson, all things were to keep pace with the wayward tantrums of the numbers. It was the task of writers to display the world 'topologically', so to speak, to reject stable representations, where objects in a scene are named and arranged as if they were "numerable", capturable as discrete entities, and to instead "stay in the condition of things".[126] There is

[123] 'Equal to the Real' (Olson, *Prose*, p. 122). [124] Ibid., p. 121.
[125] Weyl, *Philosophy*, p. 41. [126] 'Equal' (Olson, *Prose*, p. 120).

no "inert fact" for the writer's pen to capture: the "character of the external world is that", like open sets, "all things extend out" into a "reality without interruption".[127]

Expressing the new reality would require writers to find the right measure of equivalence between word and world. But to understand how this was to be attempted, we must see what it means for one thing to be the same as another, what congruence came to mean in Olson's new literary landscape. Olson says,

> Congruence, which had been the measure of the space a solid fills in two of its positions, became a point-by-point mapping power of such flexibility that anything which stays the same, no matter where it goes and into whatever varying conditions (it can suffer deformation), it can be followed, and, if it is art, led, including, what is so important to prose, such physical quantities as velocity, force and field strength.[128]

The changing signification of congruence in the nineteenth century represents to Olson almost as dramatic a shift in thought as that between the discrete and the continuous. With the new sense of congruence, Olson thought the idea of shape and identity that had become reified in the Western imagination could somehow be deranged. In the Euclidean world-picture, if the portions of space filled by two solids, and by implication, their internal angles, are the same, they are said to be congruent. An object's identity is thus determined by the extent of space it occupies. For instance, the two triangles in Figure 4.7 are deemed congruent on a flat plane.[129]

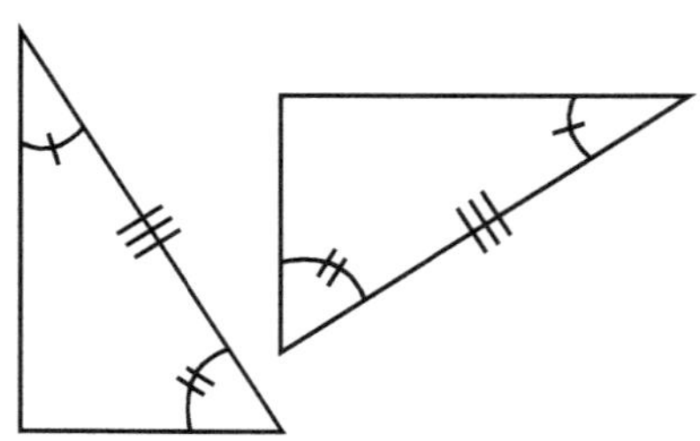

Figure 4.7 Congruent triangles in Euclidean space.
Source: Image produced by author.

[127] Ibid., p. 122. [128] Ibid., p. 123.
[129] DeMcadams, Image distributed under Creative Commons 4.0 licence.

The old Euclidean picture assumes, *mutatis mutandis*, that the "congruent mappings express an intrinsic structure of space itself; a structure stamped by space on all spatial objects".[130] The nature of space postulated in and through this process is that "all points in space are objectively alike".[131] Topology liberates space from these dogmas: it allows congruence to obtain between things that otherwise seem different to the detached observer; and this was owing to the 'point-by-point mapping power', as Olson calls it, of continuous mathematics.

Topology thus became for Olson the germ of non-representational art. An image in Olson's new art must be truly congruent with reality; it must show what the representational image under the Euclidean world did not, the tendency of things to mingle: and for that, the word must itself mingle, or as Olson would term it, "project" into ambient space.[132] Weyl presents a vivid analogy to display the difference:

> Euclidean space may be compared to a crystal, built up of uniform unchangeable atoms in the regular and rigid unchangeable arrangement of a lattice; Riemannian space to a liquid, consisting of the same indiscernible unchangeable atoms, whose arrangement and orientation, however, are mobile and yielding to forces acting upon them.[133]

To imagine Riemannian space as liquid will allow us to compare the old idea of congruence with the new, which, we will recall, Olson had described as being able to 'suffer deformation'. Plotnitsky says,

> insofar as one deforms a given figure continuously (i.e. insofar as one does not separate points previously connected and, conversely, does not connect points previously separated) the resulting figure is considered the same. Thus, all spheres, of whatever size and however deformed, are topologically equivalent. They are, however, topologically distinct from tori.[134]

130 Weyl, *Mathematics*, p. 79. 131 Ibid., p. 71.

132 It is, however, rather unfair to levy the entire blame on Euclid: Erwin Panofsky has shown that a mistranslation of Euclid's 8th postulate in Renaissance texts led to a stiff and unreal linearity becoming entrenched as our idea of perspective, as opposed to the curvy sections of our experiential visual field, as was intended by the Greeks (Erwin Panofsky, *Perspective as Symbolic Form*, trans. Christopher S. Wood (New York: Zone Books [1924] 1991), p. 35).

133 Weyl, *Mathematics*, p. 88.

134 Arkady Plotnitsky, 'Manifolds: On the Concept of Space in Riemann and Deleuze' in *Virtual Mathematics: The logic of difference*, ed. Simon B. Duffy (Manchester: Clinamen Press, 2006):187–208, pp. 191–2.

In this way, bizarrely, to a topologist, a doughnut and a coffee mug are the same. Topological congruence may be obtained from shapes grotesquely varying in appearance or size, as long as they have the same number of interruptions: thus, a pyramid, cube, tetrahedron, icosahedron are all topologically equivalent to a continuous sphere, but not to a doughnut or coffee mug, both of which have a hole in them, which in turn are not equivalent to a figure eight or handcuff, and so on.[135]

Olson's implicit assumption is that our conception of things is beholden to our characterizations of them. For instance, most of us are schooled to distinguish polygons by counting the number of vertices, edges, and faces (Figure 4.8).[136]

But if we characterize polygons by their Euler characteristic X, which is the Vertices – Edges + Faces (X = V – E + F), we find that they all have the same characteristic, namely, 2. A tetrahedron's characteristic would be 4 – 6 + 4, for instance, a cube's would be 8 – 12 + 6, a dodecahedron's would be 20 – 30 + 12, all of which are simply 2.[137] In topology, a sphere too can be shown to have an Euler characteristic of 2.[138] For most geometric objects that we can imagine, the Euler characteristic uniquely determines congruence; and it depends only on the number of interruptions (a doughnut—or even a cube with a hole in it—would have an Euler characteristic of 0, for instance).

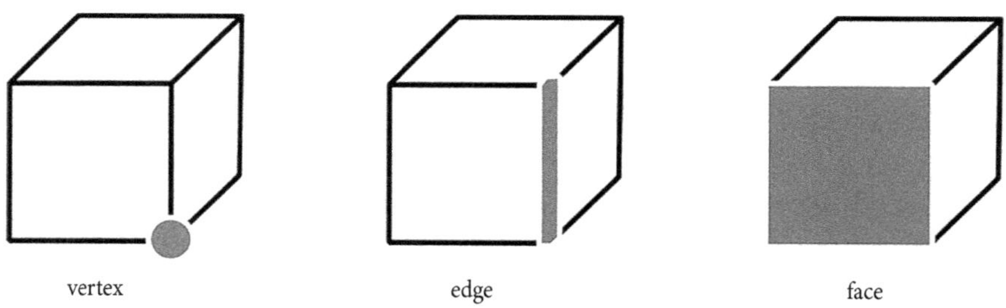

Figure 4.8 Euler characteristic is V – E + F.
Source: This file is made available under the Creative Commons CC0 1.0 Universal Public Domain Dedication, https://commons.wikimedia.org/wiki/File:Vertex_edge_face.svg.

[135] Sue Goodman, *Beginning Topology* (Providence: American Mathematical Society, 2009), pp. 36–7.
[136] This file is made available under the Creative Commons CC0 1.0 Universal Public Domain Dedication, https://commons.wikimedia.org/wiki/File:Vertex_edge_face.svg.
[137] David Richeson, *Euler's Gem: The Polyhedron Formula and the Birth of Topology* (Princeton; Princeton University Press, 2019), pp. 260–6.
[138] Goodman, *Topology*, p. 66.

Olson's poem, "The Praises", caricatures the Platonic idea that the universe is a copy of the perfect and divine shape, the dodecahedron.

> Observing
> That there are five solid figures, the Master [...]
> Concluded that
> The Sphere of the Universe arose from
> The dodecahedron.[139]

We might imagine Olson replying to the master that in topology, the dodecahedron and sphere are the same.[140] If the sphere of physical reality 'arose from' the dodecahedron—for the latter "approaches the sphere most nearly" in form, and so is "associated with the spherical heaven" and the "spherical earth"[141]—then topology to Olson initiates a kind of methexis-in-reverse that rejoins the 'Real' with the actual, suturing the ancient gap rent by Socrates.

My mathematician friend struck upon an apology for the seemingly pathological showings of topological congruence: "when one object is transformed into a seemingly radically different object, for everyone standing on the surface it does not appear, looking around them, that anything has changed".[142] This description would have no doubt delighted Olson. Olson regards topology as 'more true' because finds congruence from a hypothetically situated level: he rather loosely extends this to mean from the level of experience. But if we take his claims about mathematics as metaphors for his poetics, we can see how it gave Olson a language to render man "midst of it all", "folded in", "a thing among things".[143] He gleefully finds in topology an art that jettisons size in favour of a one-to-one relation of elastic significance. It became for Olson the task of poets to harness the 'point-by-point mapping power' of Riemannian congruence in their comprehension of reality.[144]

Brendan Gillot says there are but two points of consensus on Olson's poetics, namely, "the openness of the 'field'" and the "corporeality of future

[139] Charles Olson, *Collected Poems—Excluding the Maximus Poems* (Berkeley: University of California Press [1987] 1997), p. 97.

[140] We are using the Euler characteristic for ease of exposition. This is not entirely equivalent to homeomorphism in topology, and does not represent how homeomorphism is defined. It is important to note that if two objects are homeomorphic, they have the same Euler characteristic. The converse does not necessarily hold, but it does for most three-dimensional surfaces that can be envisaged.

[141] Plato, *Timaeus*, p. 77.

[142] Adam Jones (Oxford University), interviewed by Anirudh Sridhar, 03/08/2019.

[143] 'Equal' (Olson, *Prose*, p. 123).

[144] Ibid.

verse".[145] But the consensus has not offered us an adequate explanation of what terms like 'openness' and 'corporeality' mean and how the 'field' is to mediate between word, body, and world. We can understand these oft-repeated terms by attending to Olson's interests in topology. Non-representational, projective art would for Olson operate through the 'field', wherein art would be reanimated from its inert, subordinate role in mimesis and subjected to the 'velocity' and 'force' that attend the objects of experience. Poets were to convert "the large area of the whole poem into the FIELD [...] where all the syllables and all the lines must be managed in their relations to each other" with the express purpose of achieving continuity through congruent mappings with reality.[146] It is generally agreed among critics that the closest analogue to Olson's idea of the poetic field is the vector field, and its uses in physics, where physicality and movement—the central features of phenomena to Olson—can be mathematically represented—for vectors, unlike points, have direction and magnitude.[147] Figure 4.9 illustrates a vector field to help imagination along:[148]

Given words in a poem act upon others, applying perlocutionary forces on each other, it is useful to think of the poetic field as a system of masses on a manifold (reality), where each vocable is a vector with directed magnitude: and the mutual doings of words as functions that project from one point to another in specific directions; and proleptic and retrodictive modifications of words within a poem as creating desired and fortuitous reorientations of meaning.[149] To clarify Olson's field analogy, it may be useful to think of traditional, naturalistic images, sentences that are pictures of facts, if you will, as scalar functions which return stable values in every invocation. And so, vector functions, which take vectors and return further vectors in the field, would symbolize the restless workings of Olson's projective verse. The projective "image, therefore, is vector", declares Olson.[150]

This is one significant way in which Olson differs from the other poets we have discussed. He was not interested in using poetry to rescue phenomena

[145] Brendan C. Gillott, 'Charles Olson's "Projective Verse" and the Inscription of the Breath', *Humanities*, 7.4 (2018), 1–20, p. 2.

[146] 'Projective Verse' (Olson, *Prose*, p. 243).

[147] Rosmarie Waldrop, 'Charles Olson: Process and Relationship', *Twentieth Century Literature*, 23.4 (1977), 467–86; Tim Armstrong, 'Poetry and Science', in *A Companion to Twentieth-century Poetry*, ed. Neil Roberts (Oxford: Blackwell Publishing, 2001), p. 84; Don Byrd, 'The Possibility of Measure in Olson's Maximus', *Boundary*, 2, 2 1,2 (1973/74), 39–54, p. 50.

[148] Image in the public domain: https://commons.wikimedia.org/wiki/File:VectorField.svg.

[149] The following paper by a mathematician explains how the analogy can be understood in detail: Hans Mathews, 'Words and Picturables', *Phalanx*, 3 (2009), 1–22, p. 5.

[150] 'Projective Verse' (Olson, *Prose*, p. 252).

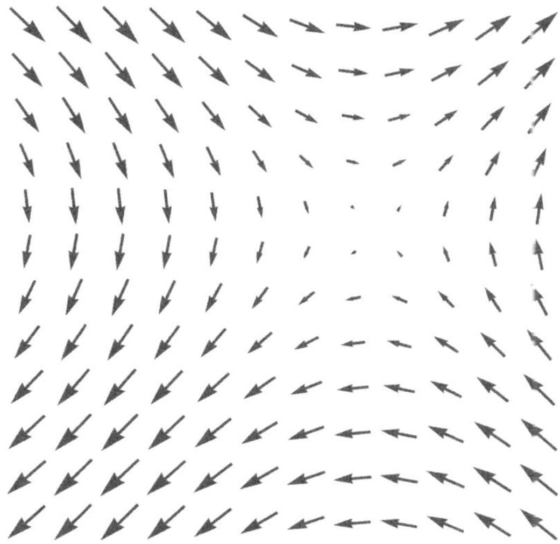

Figure 4.9 Visual sample of vector field.
Source: Image in the public domain: https://commons.wikimedia.org/wiki/File:VectorField.svg

from mathematical distortion. Recalling Eddington's elephant tumbling down the hill from previous chapters, we may reimagine the elephant under the influence of gravity and friction as a vector quantity on a manifold.[151] After the physicist has abstracted the scene, Olson would say that 'velocity, force and field strength' are all that remain. But Olson seems pleased with what Eddington bemoans: for to Olson, there really is no discrete elephant and hill—both being fictions of an imagination captured by Euclidean propaganda—but only a complex of mass, force, and velocity.

The central idea in Olson's poetics is that movement from the external world to the body to imagination, escorted by the word, was to be in continuum. Any congruence to reality attempted in language could thus not be restricted to inert representation; the function of words was not to capture reality but to project into the process, the field, of meaning-transfer. He says in his manifesto that "a poem is energy transferred from where the poet got it (he will have some several causations), by way of the poem itself to, all the way over to, the reader".[152] In "Projective Verse" (1950), Olson declared that the new poetics must involve a new "stance toward reality".[153] In subsequent

[151] In physics after Riemann and the real numbers, vector fields come to be defined on manifolds. See Weyl's chapter on 'Metrical Continuum' in *Space-Time-Matter* trans. Henry L. Brose (New York: Dover, [1918] 1952), pp. 163–65; and Weyl, *Riemann*, p. 63.

[152] Ibid., p. 240.

[153] Olson, *Prose*, p. 246.

essays, we discover that the pillars of this new architecture of reality were Riemannian continuity and congruence, whose features verse would embody once reincarnated in the field.

4.2.2 Non-Euclidean language

Two inventions of nineteenth-century geometry—in continuity and congruence—were to Olson blows fatal to the world-picture of Euclid, and with it, he hoped, naturalistic art. We move now to the other great strike of the nineteenth century, one that was smote in words, not numbers. Central to Olson's theory is that the new 'stance towards reality' was experienced first in the primordial journey of the Pequod.

Moby Dick became to Olson the spring and testament of his non-Euclidean theory, so to speak, of language. In *Call me Ishmael*, Olson first records his idea that Melville saw character in its awful form, not as subject but "OBJECT in MOTION".[154] The importance of the continuous was something "Melville blindly knew", he says.[155] The reader can sense it from "his approach to physicality" and "his address to human character as necessary human force".[156] Unlike the omniscient narrators of Victorian fiction, Melville did not paint a sophisticated picture of the inner lives of his characters; though psychologically opaque, his characters communicate their inward fury through outward impression of presence. Olson might say they were vectors in the field of action. And *Moby Dick*'s field of action was not set in the diachronic arc of Victorian narrative: it was, to borrow Auerbach's description of Homer's *Odyssey*: "a procession of phenomena tak[ing] place in the foreground—that is, in a local and temporal present".[157] Olson sees "modern events" as "reduced" due to their framing within the fiction of mechanical causation: "seen only as the eye of the needle and the camel left out".[158] In Olson's Maximus poems, the poet says "cause / is not the equal of, the error of, act".[159] The mechanical construction of cause-and-effect, for Olson, occludes the entire tissue of connected, continuous, events before and after the discrete stop-points salient to science. Melville, however, jettisons plot, a structure proceeding by cause and effect—one episode being the rational trigger for the

[154] Olson, *Prose*, p. 63.
[155] 'The Materials and Weights of Herman Melville' (Olson, *Prose*, p. 116).
[156] 'Materials and Weights' (Olson, *Prose*, p. 116).
[157] Erich Auerbach, *Mimesis: Representation of Reality in Western Thought*, trans. Willard B. Trask (Princeton: Princeton University Press [1953] 2003), p. 52; Olson at many points in his essays compares Melville to Homer.
[158] Ibid., p. 118. [159] Olson, *Maximus*, p. 97.

next—and takes the reader through his many ekphrases of the leviathan in an oppressive foreground upon which the fiery Pequod, freighted with savages, hurtles into the tempestuous night.

Olson develops the notion of continuity between character and action through another distinction between physicality (somatic experience) and physiology (anatomical picturing).

> Melville's knowingness of object and motion, those factors of a thing which declare what we call its physicality (and do not mean physiology). 'The Tail' is as lovely an evidence as any other of Melville's ability to go *inside* a thing, and from its motion and his to show and to know, not its essence along [...] but its *dimension*, that part of a thing which ideality—by its ideal, its World Forms or its Perfections—tended to diminish.[160]

The word dimension here is not to be understood in the Euclidean sense of extension, but how the magnitude of the object imposes on the reader. I believe the following lines from the chapter, "The Tail", will prove illustrative:

> This peaking of the whale's flukes is perhaps the grandest sight to be seen in all animated nature. Out of the bottomless profundities the gigantic tail seems spasmodically snatching at the highest heaven. So in dreams have I seen majestic Satan thrusting forth his tormented colossal claw from the flame Baltic of Hell.[161]

At once the unseen dimension, 'bottomless profundities', visible aspect, 'grandest sight', kinetic of soma, 'spasmodically snatching', and cadence from power to pathos, 'thrusting forth his tormented'—the word 'forth' insisting on mobility, keeping the image from being petrified to picture—all surge as energy from beholder to reader. Melville has found in words a congruent mapping of "that quality of any particular thing or event which comes in any one of our consciousness; how it comes in on us with a force peculiar to itself and to ourself".[162] A lesser writer might have sought a static symbol for the tail, a tree, say, or Platonic solid—not so, Melville: with Satan's thrusting claw, he breaches the "given physicality and moves from its essence into its kinetic".[163] Olson just so finds some deep resonance between Riemannian

160 'Materials and Weights' (Olson, *Prose*, p. 117).
161 Herman Melville, *Moby Dick; or, The Whale* (New York: Harper & Brothers, 1851), p. 420.
162 'Materials and Weights' (Olson, *Prose*, p. 117). 163 Ibid., p. 117.

geometry and "Melville's non-Euclidean penetrations of reality", as together they free Western man from the shadow of the Forms.[164]

> It is my experience that only some such sense of form as the topological includes, able to discriminate and get in between the vague types of form morphology offers and the ideal structures of geometry proper, explains Melville's unique ability to reveal the very large (such a thing as his whale, or himself on whiteness, or Ahab's monomania) by the small.[165]

Michael Jonik takes Olson to mean that Melville literally exhibits the geometric developments of his time. "Melville explicitly invokes a non-Euclidean form as he carefully anatomizes the 'curious internal structure' of the whale's head: 'Regarding the Sperm Whale's head as a solid oblong, you may, on an inclined plane, sideways divide it into two quoins.'"[166] This is an example of taking Olson's analogies literally, and how doing so defeats any power they may retain through their often loose applications.

When Olson finds continuity and congruence 'of form as the topological includes' in Melville's prose, it is never a question of what Melville says about the whale but *how* he says it. A simple dichotomy from "Human Universe" will make this clear: language, says Olson, needs to be much more than "the act of thought *about* the instant", it needs to be "the act of the instant".[167] Language must be able to recreate the immediacy of experience in its very articulation, to fill the void of "discrimination (logos)" between word and world with a resounding "shout (tongue)".[168] None of this is understood by finding literal geometric shapes in Melville's description of the whale's head—to say of the whale's head that it appears oblong is but scalar representation. One might in fact say that when Ishmael is running his readers through his many autopsies of the whale, the prose is least like a vector field. Olson is most in awe of Melville's prose when it imbues the scene with continuity between the outward and inward, where character is not disclosed by the omniscient narrator exposing to the reader its thoughts and feelings, but is itself extended as object onto the scene: this of course manifests most and best in Captain Ahab, who is not person but presence—even in absence.

[164] Ibid., pp. 117–18. [165] 'Equal to the Real' (ibid., pp. 122–3).
[166] Michael Jonik, *Herman Melville and the Politics of the Inhuman* (Cambridge: Cambridge University Press, 2018), p. 53; To be fair to Jonik, his interest is primarily in Melville, not Olson.
[167] 'Human Universe' in Olson, *Prose*, p. 156. [168] Ibid., p. 155.

The passage I will quote now will be excused for its length by its being one of the great achievements in prose in the history of literature:

> The hatch, removed from the top of the works, now afforded a wide hearth in front of them. Standing on this were the Tartarean shapes of the pagan harpooneers, always the whale-ship's stokers. With huge pronged poles they pitched hissing masses of blubber into the scalding pots, or stirred up the fires beneath, till the snaky flames darted, curling, out of the doors to catch them by the feet. The smoke rolled away in sullen heaps. To every pitch of the ship there was a pitch of the boiling oil, which seemed all eagerness to leap into their faces. Opposite the mouth of the works, on the further side of the wide wooden hearth, was the windlass. This served for a sea-sofa. Here lounged the watch, when not otherwise employed, looking into the red heat of the fire, till their eyes felt scorched in their heads. Their tawny features, now all begrimed with smoke and sweat, their matted beards, and the contrasting barbaric brilliancy of their teeth, all these were strangely revealed in the capricious emblazonings of the works. As they narrated to each other their unholy adventures, their tales of terror told in words of mirth; as their uncivilized laughter forked upwards out of them, like the flames from the furnace; as to and fro, in their front, the harpooneers wildly gesticulated with their huge pronged forks and dippers; as the wind howled on, and the sea leaped, and the ship groaned and dived, and yet steadfastly shot her red hell further and further into the blackness of the sea and the night, and scornfully champed the white bone in her mouth, and viciously spat round her on all sides; then the rushing Pequod, freighted with savages, and laden with fire, and burning a corpse, and plunging into that blackness of darkness, seemed the material counterpart of her monomaniac commander's soul.[169]

It would be improper to explain. I will trust that enough has been said in previous sections to make the significance of these words to Olson clear and move swiftly on.

For Olson, the continuity of character with world corresponds to the continuity between language and the reader—they are achieving congruence in and across different dimensions within the same manifold, as it were. This is made clear in his manifesto, 'Projective Verse': the poem, says Olson, should create in its field a language congruent—in the topological sense—to the

[169] Melville, *Moby Dick*, p. 483.

poet's original impressions and release, "at all points energy at least the equivalent of the energy which propelled him in the first place".[170] The work of words is to transfer embodied experience to the reading body by bombarding it with equivalent impressions, unlike descriptive, naturalistic prose, which attempts remote representational fidelity. This requires the retelling of events to be projective in the sense of being projectile rather than spatially accurate, as of renaissance perspective.

Olson says, "the point is also the overall 'space' of Moby Dick. That space, and those of which it is made up, have the properties of projective space (otherwise they should all come out more familiar, and round, because they would stay Euclidean)."[171] An excursus into projective geometry will help us understand how Olson situates language-and-reader vis-à-vis his character-world continuum. The developments of eighteenth- and

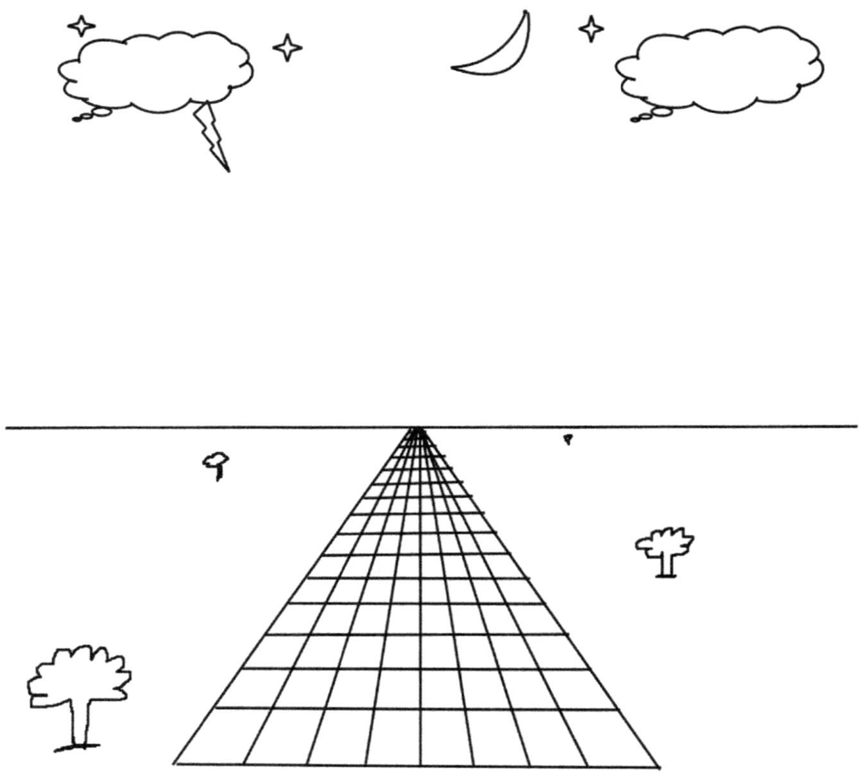

Figure 4.10 Renaissance courtyard.
Source: Image produced by author.

[170] 'Projective Verse' (ibid., p. 240).

[171] 'Equal to the Real' (ibid., p. 123).

nineteenth-century projective geometry solve, in some ways, the renaissance problems of the horizon, the fictional line at infinity where parallel lines seem bound to meet.[172] We may imagine a courtyard in a renaissance sketch but comically extended to stretch out before us to the horizon (Figure 4.10).

The parallel lines seem bound to meet—as they do when standing on a railways track, say—at the mythical infinity bar of the horizon, where the horizontal lines of the grid too, increasingly pinched together, appear to converge to a point. Now if we were to imagine a shape drawn on this grid, say, a parabola, we find that, owing to the seeming convergence at the horizon, it folds into an ellipse. That is, Figure 4.11 becomes Figure 4.12 in our facsimile of the renaissance canvas.

But nineteenth-century geometry, equipped with the real number line, possessed a 'point-by-point mapping power' that allowed the horizon, the line at infinity, to be represented and thereby brought, as it were, into the realm of the discernible.[173] By imagining a higher-dimensional projective space (see Figure 4.13), we can draw the parabola on a Z = 1 plane, which would be the plane parallel to the two-dimensional XY plane. We are simply

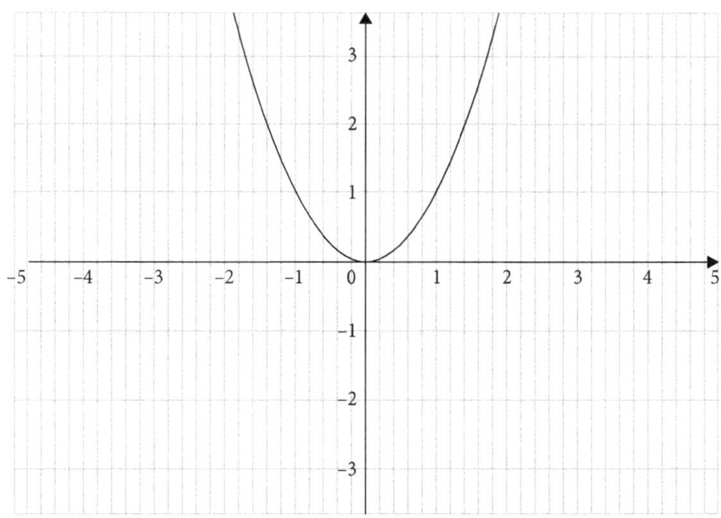

Figure 4.11 Parabola.
Source: Graph generated by author.

[172] See Jeremy Gray, *Worlds out of Nothing: A Course in the History of Geometry in the 19th Century* (London: Springer, 2007), p. 29.
[173] Gray, *Worlds*, pp. 76–7.

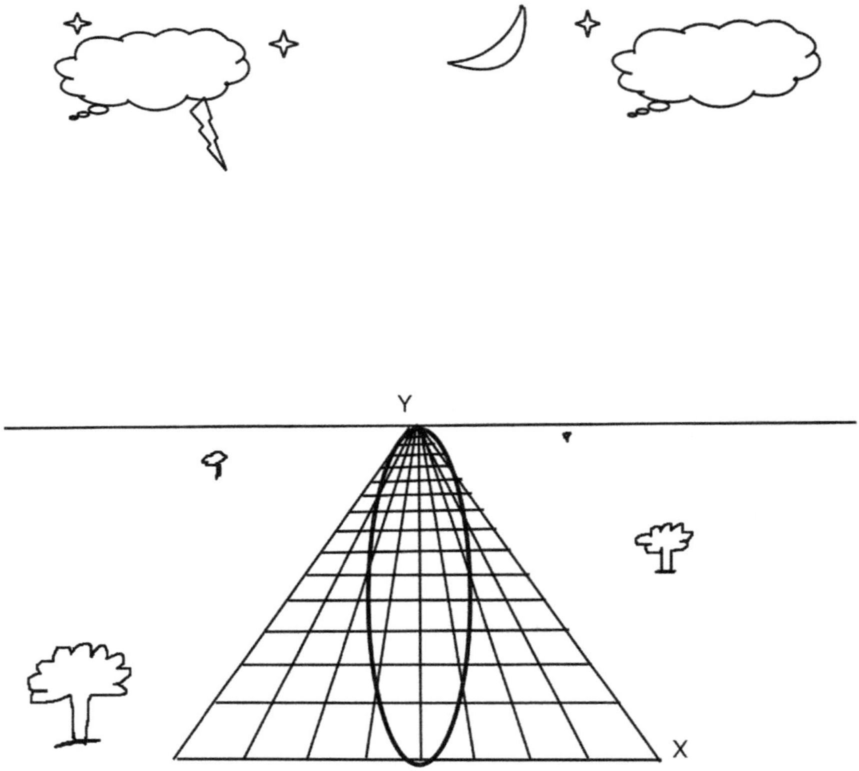

Figure 4.12 Parabola becomes ellipse at the horizon.
Source: Image produced by author.

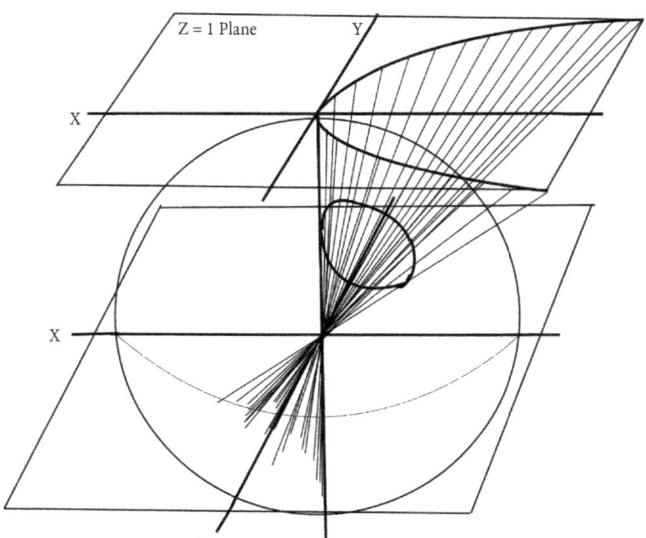

Figure 4.13 Transformation of ellipse in projective space.
Source: Image produced by author.

suspending a plate on top of another so that they are parallel and drawing the parabola on the upper plate. Any line piercing the bottom plate will eventually intersect with the upper plate, save those lines actually along the bottom plate, which will remain parallel to the Z = 1 upper plate. Now let us think of the bottom plate as the erstwhile horizon of the Renaissance canvas. We can imagine the Z = 1 plane resting on a hollow sphere whose centre is the origin (0,0,0). We are only interested in the sphere's upper hemisphere here, because the base of the hemisphere, the XY-plane, as we have said, is the old horizon, where the image terminates, but never reaches. Now the upper hemisphere can be thought of as the eye looking upwards, through the Z = 1 plane at the real-world object beyond that is being imaged. Now if we draw lines from the parabola on Z = 1 through the origin, we find that these lines all meet the sphere at various points to form the anticipated ellipse (whose limit, we have said, is the XY plane or the horizon).[174]

To understand Olson's use of 'projective space', we have imagined the point of view, the region of experience, as it were, as being below the Z = 1 plane, within the upper hemisphere. We notice then that the ellipse is formed around the eye. Thus, if Renaissance perspective attempted to bring the abstract, aerial view of the geometer to a sideways, more situated view, nineteenth-century projective space could pitch the viewer further into the midst of it all, where the point at infinity becomes, as it were, the point of view.

Olson offers as example of *Moby Dick*'s projective space Ishmael's description of "the delivery of Tashtego from the whale's head".[175] The reference here is to the events that transpire once Tashtego mounts the suspended head of the whale and slices it open to draw oil, and slipping on the spermaceti, plunges into the freshly opened gash, as the head of the whale is tossed back into the sea. As the head is descending in the water, Tashtego spinning in its cranium, Queequeg dives for the rescue:

> Now, how had this noble rescue been accomplished? Why, diving after the slowly descending head, Queequeg with his keen sword had made side lunges near its bottom, so as to scuttle a large hole there; then dropping his sword, had thrust his long arm far inwards and upwards, and so hauled out our poor Tash by the head. He averred, that upon first thrusting in for him, a leg was presented; but well knowing that that was not as it ought to be, and

[174] This exposition has been adapted from Gray, as cited above, and also a lecture by the mathematician Norman Wildberger, 'Projective Space', *Math History*, 2011.

[175] 'Equal to the Real' (Olson, *Prose*, p. 123).

> might occasion great trouble;—he had thrust back the leg, and by a dexterous heave and toss, had wrought a somerset upon the Indian; so that with the next trial, he came forth in the good old way—head foremost. As for the great head itself, that was doing as well as could be expected.[176]

The reader is first disoriented from an aerial perspective over events by the directional derangement in the phrase 'side lunges near its bottom'—a lunge being a forward movement, reoriented by 'side' and 'bottom'. Rather than describing the scene from starboard, we might say, Melville thrusts the reader into the projective space of the cranium: a field instinct with motion: where Queequeg is fumbling for Tashtego's head from without as the latter is rotating within. A terse and simple phrase, 'a leg was presented', suffices to reset the reader's perspective, placing it right amid the action (in our projective geometry analogy, within the upper hemisphere). A more tortured phrase like 'wrought a somerset upon the Indian' is required to dissolve the characters into the field of action. A thing is 'wrought' by physical effort. The word seems to exist only in the past tense, as evidence of effort exhibited in the finished object ('wrought' being too distant in usage from the quotidian 'work'). The unusual use of 'wrought' in the place where an active verb should have been seems to detach the work done from the agency of the doer—Queequeg's effort is now part of, exhibited in, the changing motions of Tashtego. A summersault, on the other hand, is individually achieved; but again, the unusual preposition 'upon' works to prevent agency from being transferred back to the helplessly swirling Indian. As when calculating the mean of two different numbers yields a number in between that is neither, 'wrought', 'somerset', and 'upon' work together to recede both personalities from the scene, leaving behind only a complex of mass, effort, and motion.

Olson goes on to explain his projective space analogy by saying Melville "makes things stand out" in a "homogeneous" fashion.[177] In our last diagram, the (X, Y, Z) coordinates of the ellipse described are considered homogeneous vis-à-vis the coordinates of the parabola.[178] As the values of X, Y, and Z are scaled higher and lower on the upper plate, the lines translating the parabola below move around the sphere, tracing out the corresponding homogeneous coordinates of the ellipse. Olson might say that Melville's prose translates the movements of characters in the scene homogeneously to the reading imagination, its experiments with case and tense

[176] Melville, *Moby Dick*, pp. 382–3. [177] 'Equal to the Real' (ibid., p. 123).
[178] Gray, *Worlds*, p. 160; Weyl, *Mathematics*, p. 69.

making events in the invisible cranium stand out to embodied perspective (the upper hemisphere in our diagram). There is also the common sense of 'homogeneous' to mean alike; as in all aspects of the scene strike the reader with equal salience, unlike the privileged perspective of the observer in naturalistic prose or renaissance painting, the latter of which, we might venture, distributes the furniture of a scene according to the rules of Euclidean geometry. The rescue is confusing: while the cranium is downwardly sinking, Tashtego is rapidly rotating, and Queequeg is upwardly groping; but since Melville writes from the field of events, from the situation of activity, we are able to sense proprioceptively the weighty descent of the head, the nimble rotation of the Indian and the powerful strokes of the cannibal, simultaneously and in equal measure.

4.2.3 Non-Euclidean body

> The chain of memory is resurrection [...]
> The vector of space is resurrection [...]
> The being of man is resurrection [...]
> Direction—a directed magnitude—is resurrection [...][179]

We may infer from this series of propositions in a nameless poem that 'The vector of space' = 'direction – a directed magnitude' = 'the being of man' = 'the chain of memory'. Part of this equation, between the vector space, vectors, and human being, should already seem familiar. We have seen how Melville projected the directed magnitude of character-and-space onto an embodied reading perspective in the field of his prose. We will now proceed to see if these ideas help us better understand Olson's own poetry, and its fantastic ambition to agitate the body's 'resurrection' through the syllable, word, and line.

In the Maximus poems, there is a constant feeling of urgency; the body is not only to be resurrected from its millennial slumber under Logos but, we will recall, saved from falling worse to the mass culture of late capitalism. The workings of Olson's purportedly salvific verse in *Maximus* are based on the notion he develops in his essays of verbal 'energy-transfer'. Olson says there are two registers of the 'Real' in the body: "the acquisitions of his ear

[179] Olson, *Poems*, pp. 372–4.

and the pressures of his breath".[180] In an ideal poetic process, the union of ear and mind births the syllable, and the breath demarcates the line. Congruence between the experience of the writer and the reader is achieved by energy-transfer, through the hearing and saying of the syllable: "the HEAD, by way of the EAR, to the SYLLABLE".[181] We have seen that the mapping from source to recipient must in verse be continuous (in the Riemannian sense of not stopping at countable points), which in "Projective Verse", Olson thusly describes: "ONE PERCEPTION MUST IMMEDIATELY AND DIRECTLY LEAD TO A FURTHER PERCEPTION".[182] Olson later explains in his final manifesto, "Proprioception", how this constant bombardment of perceptions, one directly leading to another, should impact the body. He defines proprioception as "the data of depth sensibility / the 'body' of us as object which spontaneously or of its own order produces experience of, 'depth'".[183] The primacy of proprioception, the sensation of depth, in our recognition of space and substance affirms to Olson the idea that 'movement or action' is more primal to human experience than matter and form. A babe born will innately sense depth, but will acquire language, and the ability to identify entities in discrete noun-concepts, later in its development. In the Maximus poems, the poet laments the indoctrination of language that makes us forget this primitive experience:

> one loves only form,
> And form only comes
> Into existence when
> The thing is born
> Born of yourself[184]

'Born of yourself' in that 'form', the well-defined object, the thing, is a creation of the self, something learnt.[185]

In "Projective Verse", Olson describes the "sentence as first act of nature, as lightning, as passage of force from subject to object".[186] The projective sentence, for Olson, should reproduce proprioception, not conception. One way it does so, as we have seen from Queequeg's rescue, is by resetting perspective

[180] 'Projective Verse' (Olson, *Prose*, p. 241). [181] Ibid., p. 242. [182] Ibid., p. 240.
[183] 'Proprioception' (ibid., p. 181). [184] Olson, *Maximus*, p. 7.
[185] David Herd has pointed out that this sequence of lines on 'form' is copied from a letter Olson had written in 1950 to Frances Boldereff, on the difficulty of popularizing his ideas in "Projective Verse" (David Herd, *Contemporary Olson* (Manchester: Manchester University Press, 2016), p. 154).
[186] 'Projective Verse' (Olson, *Prose*, p. 244).

to the midst of the action. But beyond perspective, the sound of projective language must excite the fundamental proprioceptive sense which mankind perceives by "SENSIBILITY WITHIN THE ORGANISM BY MOVEMENT OF ITS OWN TISSUES".[187] The movement of tissue can be scored, as it were, in a projective verse that will "get on with it, keep moving, keep in, speed, the nerves, their speed, the perceptions, theirs, the acts".[188] The buffeting of bodily nervure among the elements, what we call experience, can thus be stored in and released from the body of language, its sound.

In "Human Universe" (1951), we will recall, Olson made the important distinction between language that is an 'act of the instant' and 'act of thought about the instant': the former is Real, continuous language because undistinguished from experience, while the latter keeps the body at bay from mental cognition to identify discrete things: thought fashioned the concept, and the concept engendered the object. The essay "Proprioception" confirms the metaphysic we have so far been stitching, namely, that the 'Real' is pure movement and conformations of matter are but epiphenomena of action. These ideas seem to be an elaboration in poetics of Whitehead's *Process and Reality*,[189] which Olson read a decade prior.[190]

Whitehead gifted to Olson a vision of the universe as a great symphony of activity whereof such abstractions as subject and object are merely the changing notes of wilful endurance. What we perceive as discrete objects became in Whitehead's philosophy merely periods of cooperation among primary modalities of activity. Despite appearances, argued Whitehead, time will eventually spread discord amidst the most abiding patterns of solidarity, precipitating disaggregation and destruction—so it comes to pass that even mountains fall.[191] With process replacing matter as the foundation of reality, the erstwhile secondary qualities of Locke, smell, colour, and all that became to Olson ingredient in the fetid nest of life, could once again be promoted. Unlike mountains, the colour red (a favourite among modernist philosophers),

[187] 'Proprioception' (ibid., p. 181). [188] 'Projective Verse' (ibid., p. 240).

[189] Whitehead's influences on Olson is a well-trodden topic in literary criticism. See Shachar Bram, *Charles Olson and Alfred North Whitehead: An Essay on Poetry*, trans. Batya Stein (Lewisburg: Bucknell University Press, 2004); Jeremy Campbell, 'Observer and Object, Reader and Text: Some Parallel Themes in Modern Science and Literature', in *Beyond the Two Cultures: Essays on Science, Technology, and Literature*, eds Joseph W. Slade and Judith Yaross Lee (Ames: Iowa State University Press, 1990), p. 23.

[190] He read *Process and Reality* around 1955 (having already been acquainted with Whitehead's philosophy through *Adventures of Ideas*) (Robert von Hallberg, *Charles Olson: The Scholar's Art* (Cambridge: Harvard University Press, 1978, p. 83).

[191] Chapters 4 and 5 of *Science and the Modern World* might be the most straightforward explication of these ideas (A.N. Whitehead, *Science and the Modern World* (New York: The Free Press [1925] 1967)).

although flitting in and out of phenomena, returns just the same every time, for all time: "colour" thus became to Whitehead an "eternal. It haunts time like a spirit."[192]

But eternal though secondary qualities were, eternality still remained a quality; the substance, as it were, was to Whitehead un-random activity, process. "It is nonsense to ask if the colour red is real. The colour red is ingredient in the process of realization. The realities of nature are the prehensions in nature, that is to say, the events in nature".[193] If these prehensions (that is, the immediate grasp, the 'proprioception' of the way of things as manifest, or as Blake might put it, life's essential conspiracy between Energy and Reason) are primary in nature, events replace objects as the foundation of reality. "Nature", says Whitehead, "is a process. As in the case of everything directly exhibited in sense-awareness, there can be *no explanation* of this characteristic of nature. All that can be done is to use language which may speculatively demonstrate it."[194] Perhaps Olson took this as a singular opportunity in the agon: for he argued that it would be poetic language that will ultimately demonstrate nature as process.

But not any poetry could achieve this feat. Olson maintains that the abstracting legacy of logos has left us with a language better equipped to describe matter, the epiphenomenon, than primordial process. So long as the role of language is to represent, there will never be continuity. Instead of a ceaseless stream from writer to reader, closed verse is a discrete record of 'thought about the instant': derivative abstraction rather than integrated action. Olson thus proposes a new mechanism: an event in the world will foment a premonitory rumbling in the bodily nervure; this will register as sound in the cochlear labyrinth; being weighed to the motive thrust of the source, the syllable will minister the transubstantiation of experience to language. The whole process then unfolds in reverse as the reader reads: or to better illustrate, we can read the equation from the unnamed poem in reverse: as 'the chain of memory' = 'the being of man' = 'direction—a directed magnitude' = 'the vector of space'. Memory is where the process begins, for when reading, syllable becomes associated with sensation from the storage of past experience. This is how, Olson's wished, bodies would ideally participate in the poetic process. His ideal reader seems to be a prelapsarian being before the noise and flash of modern America and its omnipresent 'billboards',

[192] Whitehead, *Modern World*, p. 87. [193] Ibid., p. 72.

[194] A.N. Whitehead, *The Concept of Nature* (Cambridge: Cambridge University Press [1920] 1964), p. 54.

'spray-guns', and 'coloured pictures', and outside the occidental tradition after Socrates. This lost body, he hoped, would be resurrected through the primitive and hymn-like utterances of his poems.[195]

The Maximus poems are divided into 'Letters'; their ambition is to bio-mechanically attune the body to its surroundings. "Letter 9" is an ironic play on Whitman's song of himself,

> I measure my song,
> measure the sources of my song,
> measure me, measure
> my forces[196]

By channelling the seminal lines of America's original bard, Olson announces his great ambition. Maximus will convert the grand ego of Whitman's 'Song of Myself' to object. By splitting his self into subject, 'I', and object, 'me', the poem declares that his conscious self writing the 'song' will measure his existing self experiencing nature[197]—the 'sources of his song'. The source experience is 'measured' by its impressions on himself, his body. The poet will then release as much force with his words as is congruent to the experience registered in bodily memory: the result is his measured song. Madeleine Cooper says in Olson, the "self [is] defined proprioceptively as the nexus of physical exchange between subject and object".[198] In Letter 9, the two 'measures' flanking the object 'me' in the third line shows the resurrection of the Whitmanian ego "I" as 'object'—harkening to his conclusion in "Projective Verse" that "man is himself an object".[199] This shift is brought about by a kind of two-way measurement: of body against nature and body against language. So, the poet, before writing, must finely calibrate his body, his device for the registry of movement, to reality. This comes from living—in the Ruskinian sense of doing—or as Olson himself puts it in "Tyrian Business" ("Letter 8"), "felicity/ resulting from a life of activity".[200]

The Maximus poems unfold as a series of concentrated anecdotes of fishing and other activities, from the daily life of a Bay Stater from Gloucester. These terse sketches have the energy of a Kafka short or a Williams verse. Olson

[195] Olson finds these ideal beings among the contemporary Mayans, who sit comfortably connected to one another in dense busses, jostling together in rhythmic union ('Human Universe', Olson, *Prose*, p. 159).

[196] Olson, *Maximus*, p. 48.

[197] This is thus not a critique of the act of measuring itself, as Herd suggests (Herd, *Olson*, p. 244).

[198] Madeleine J. Cooper, 'Finding a Centre: The Poetry of Charles Olson' (PhD Dissertation: University of Nottingham, 1977), p. 225.

[199] 'Projective Verse' (Olson, *Prose*, p. 247).

[200] Olson, *Maximus*, p. 42.

dedicates a poem on the difficulty of translating life to language, "Red Mallows", to WCW, likely in homage to "The Red Wheelbarrow". The poem, reflecting on itself, rejects beauty in favour experience as the true subject of poetry:

> Form
> Descends
>
> And the inhabitation
> —the eyes, and the ears,
> [...] nose
> About—it is they
>
> Who also
> come up.[201]

Consistent with this promise to WCW, the Maximus poems are written with the purpose of communicating 'inhabitation', not 'form'; they are designed to force the senses to go 'nosing about' the world.

Olson's overwhelming certitude that motion and force form the essence of reality—what he learns from *Moby Dick*—is reflected in his many experiments with grammar. One such is the constant repetition of verbs. Take, for instance, "Letter 1":

> O kill kill kill kill kill
> [...]
> in! in! the bow-sprit, bird, the beak
> in, the bend is, in goes in the form
> that which you make, what holds, which is
> the law of object, [...] what
> the force can throw up, can, right now hereinafter erect,
> the mast, the mast, the tender
> mast![202]

The whole scene pulsates with what Olson's mentor, F.O. Matthiessen, called "verbs of action",[203] as the poet attempts to release "primitive energies latent

[201] Olson, *Poems*, p. 307. [202] Olson, *Maximus*, p. 8.

[203] F.O. Matthiessen, *American Renaissance: Art and Expression in the Age of Emerson and Whitman* (Oxford: Oxford University Press, 1968 [1941]), p. 430. Sherman Paul says there is in this passage an "estrangement from the familiar world to the misuse of language"—the particular 'misuses' we are about to discuss, I believe, are key to understanding what the poems seek to do (Sherman Paul, *Olson's Push: Origin, Black Mountain, and Recent American Poetry* (Baton Rouge: Louisiana State University Press, 1978), p. 127).

in words".[204] The use of the spatial preposition 'in', in a slightly archaic manner as a truncated phrasal verb, seems to animate even space; that is, the 'inside' is both an area and a directed action. The energy of the anecdote erases any even fleeting representational images from mind, as the 'law of object' makes object itself unavailable for picturing: what one *is* is instead 'thrown up by force', by the melee of becoming. Throughout the Maximus poems, Olson repeats his verbs and rations his nouns as a means to frustrate objectification. For instance, in "Letter 10" is a line, "It was fishing was first".[205] By such a simple manoeuvre, a quotidian act like 'fishing' can be made to acquire an independent mystery: had the sentence began, 'Fishing was...'—fishing would have become a recognizable class of pastime, which the 'chain of memory' would proceed to concretize. The interest of the sentence would then have been in the object, as in, fishing was what? As it stands, however, the object was a doing, 'it was fishing', and the doing was primary.

This experimentation with grammar is motivated by a commandment that Olson draws from the Vedas, and states as two equations:[206]

> There may be no more names than there are objects
> There can be no more verbs than there are actions.[207]

Besides the Vedas, Olson's edict on nouns has been traced to Fenellosa, who in *The Chinese Written Character as a Medium for Poetry*, says "a true noun, an isolated thing, does not exist in nature. Things are only terminal points [...] of actions".[208] This passage, which Olson underlined in his copy,[209] seems a linguistic parallel to Olson's understanding of Whitehead's 'process'. If all things that we—through the hallucinatory spell of language—take to be discrete, countable entities are epiphenomena reified for comprehension from underlying actions, restless subterranean forces, then to dispel the illusion, the most pedestrian of sentences would need to be supplied with many

[204] Matthiessen, *Renaissance*, p. 423; Mathiessen uses this phrase to describe Melville's language. Olson and Matthiessen exchanged letters, throughout their careers, wherein Melville is often discussed (Olson, *Letters*, p. 26).

[205] Olson, *Maximus*, p. 48.

[206] Butterick traces these lines to a 'Syllabary for a Dancer', in which Olson says, the "quotation you have offered me from the Vedas, was it—that there may be no more names than there are objects. That will cover nouns adequately, and we can do the whole job by adding a like statement to cover verbs: there can be no more verbs than there are actions in the human universe" (Butterick, *Guide*, p. 60).

[207] 'Tryian Business' (Olson, *Maximus*, p. 40).

[208] Ernest Fenollosa, *The Chinese Written Character as a Medium for Poetry*, ed. Ezra Pound (London: Stanley Nott [1919] 1936), p. 6, qtd. in Waldrop, 'Olson', p. 469; Waldrop cites the title of Fenollosa's essay incorrectly, as *Notes on the Chinese Written Character*.

[209] Waldrop, 'Olson', p. 469.

additional verbs: verbs that would unsettle the concretion of discrete imagery. There can be many verbs because action is all there is. At times, Olson even avoids reaching an object in his poetic sentences: "a hollow muscular organ which, by contracting vigorously, keeps up the / (to have a heart".[210] Or in the same letter, further on, "the seedling of morning" is described as "to move, the problems (after the night's presences) the first hours of/ he had noticed [...]".[211] In both instances, the sentence stops short of a noun, circumventing its object by slipping into a new sentence. Instead of 'heart' or 'morning', the lines end abruptly, to be extracted into action, 'to *have* a heart' and 'he *had* noticed'. There are, in other words, as many names (or nouns) as exist objects—the 'muscular organ' need not be reified into another noun in the same line.

Even when the poet occasionally settles on a noun, it is often phonetically charged to carry the sensation of movement. We have seen an example in "Letter 1" in his liberal use of 'mast', wherein the undulation of facial muscles in saying 'mast', from what phoneticians call the bilabial nasal ('ma') to the sibilant ('st'), seems to conjure the flapping of fabric. But there are more powerful nouns that Olson repeats, such as 'whorl' and 'wharf', throughout the poems.[212] Both begin with velar fricatives ('wh') that by constricting the flow of air, release a sharp breath, which is then gradually arrested by the alveolar caress of the rhotic ('rl' or 'rf'). Olson wants the energy in sounding 'whorl' and 'wharf' to modify their sense, to convey petals or ships as *being* arranged around a point—their formation as coming into being. In the final section of "Tyrian Business", the poet first provides a "definition": "the crooked timbers/ scarfed together to form the lower part of the compound rib", and then attaches to it a noun, or what "we call 'em'", namely, "futtocks".[213] The term 'futtock' should do, in theory, to signify the middle frames of a ship, with the even flow between the vowels, 'u' and 'o', expressing the imaginary row of straight planks. However, the main feature of the ship in the definition provided is that it is ribbed with *crooked* timbers. These, the poet avers, are better expressed by the undulation of descending chin to ascending lip in the saying of "fylfot",[214] as an American would. The *sound* (not the dictionary sense) of fylfot is what "she look like",[215] because planks on a ship cannot realistically be arranged as swastikas.

[210] 'Tryian Business' (Olson, *Maximus*, p. 40).
[211] Olson, *Maximus*, p. 41.
[212] Ibid., pp. 11, 26, 32, 40, 363, 467 (and likely more).
[214] Ibid., p. 44.
[215] Ibid., p. 44.

In “Place; & Names”, Olson distinguishes between ‘name’ and ‘noun’. Proper nouns of “cities or person” are inadequate “to the order of creation of anything except names—including possibly mathematics (?)”. The crucial difference he seeks to draw from inherited nouns and the new names that the poet will christen things with is that “names / be as parts of the body, common, & capable / therefore of having cells which can decant / total experience”[216]—‘muscular organ’ thus is preferred to ‘heart’. In the act of naming creation, which right poets inherit from Adam,[217] it was important to Olson that the appellations emboss the impression of the named object on ‘parts of the body’ through their syllabic dance on the tympanum. The name must contain the “Story” of experience, which “is if not superior / at least equal to ultimate mathematical language”[218]—by which we can be sure, he means topology.

There is an interesting and adventurous extension of the one-to-one equation Olson demands between names and objects. This is in his deliberate confounding of the singular and plural. The following lines are from “Letter 3”: “When he came / there were three hundred sail could fill the harbour, / if they were all in, as for the Races”.[219] Although there are three hundred of them, their noun is singular, ‘sail’, suggesting how in a ‘Race’, the scale and size of the scene is better conveyed viewing the sails as one sprawling, continuous cloud of movement. But in the very next section of “Letter 3”, there is another sequence: “I speak to any of you, not to you all, to no group [...] / Only a man or a girl who hear a word”.[220] Here we are forestalled from conceiving a collection of individuals as one entity—‘no group’! An individual, ‘a man or a girl’, is then given a plural verb ‘hear’. Here the solecism functions not to communicate magnitude, or the sublime, but to erode the belief that an individual is a closed entity: to posit instead Olson’s metaphysics of being as process, a whorl of invisible happenings, never coalescing to the discrete identity of an individual ‘I’, but always continuously dissolving in the course of events: character as a collection of vectors in a manifold. “Letter 3” ends with an appeal to the alienated subjects of Gloucester: “Isolated person in Gloucester, Massachusetts, I, Maximus, address you / you islands / of men and girls”.[221] The lines seem almost cast as a spell, to resurrect the

[216] Olson, *Prose*, p. 200.

[217] Catherine Stimpson makes a similar point about Olson’s command that there should be ‘no more names than there are objects’, when she says, “since each man creates speech anew, each event will be its own name” (Catherine Stimpson, ‘Charles Olson: Preliminary Images’, in *Early Postmodernism: Foundational Essays*, ed. Paul A. Bové (Durham: Duke University Press, 1995), p. 143) although we must read Olson as specifically referring to poets, not ‘each man’.

[218] Olson, *Prose*, p. 201. [219] Olson, *Maximus*, p. 14.

[220] Ibid., p. 15. [221] Ibid., p. 16.

individuated and alienated subjects, and invite them once more into the field of events, to rise unto Donne's sermon, that "no man is an island, / Entire of itself / Every man is a piece of the continent, / A part of the main".[222]

The being that is slowly assembled through Olson's selective deployment of plurality emerges finally in "Letter 27", when Maximus says,

> No Greek will be able
> To discriminate my body
>   An American
> is a complex of occasions,
> themselves a geometry
> of spatial nature.[223]

The lasting influence of Euclid located, or circumscribed, the individual to a discrete portion of space, from which Olson hoped to liberate the American, casting 'them' into Riemann's 'spatial nature'.[224] The rather impudent description of the American as a geometry of spatial nature is an invitation to what Husserl called *rückfragen*, which I will liberally translate to 'back-questing', the anamnetic recovery of the sense of being as a 'complex of occasions'.

Olson's poems are thus formal and grammatical experiments in how far verse can be made to work in accordance with the metaphysics he drew from the new geometry. If nineteenth-century geometry could influence culture as Cartesian geometry had Enlightenment thinking, the shape of the new reality—and individuals in it—would come to seem buffeted by events like Weyl's liquid in a lurching vehicle. And it was the poet, experient of this feverish existence, who would bring the revelation to all.

> The branches made against the sky are not of use [...]
> who has to raise (Who puts this on, this damning of his flesh?)
> he can, but how far [...] can he raise the thickets of
> this wilderness?

[222] John Donne, *The Works of John Donne, vol III*, ed. Henry Alford (London: John W. Parker, 1839), pp. 574–5; David Herd alternatively suggests that Olson here echoes "Melville's equation of 'isolatoes' with 'Islanders'" (Herd, *Olson*, p. 182), but Melville forms a third equation with the phrase "continent of men" (Melville, *Moby Dick*, p. 133), which shows that Melville himself has read his Donne well.

[223] Olson, *Maximus*, pp. 184–5.

[224] He lifts the phrase from Whitehead: "the human body is indubitably a complex of occasions which are part of spatial nature" in A.N. Whitehead, *Adventures of Ideas* (Cambridge: Cambridge University Press, 1948), p. 91; Bram, *Olson and Whitehead*, p. 54.

[...]
How can he make out, he asks,
of this low eye-view,
size?[225]

One gets the impression that the reality we find ourselves in once the comforting illusion of Euclidean space dissolves is a confused one. Reality no longer obliges understanding by organizing itself into discrete objects sitting pretty for the observing; for counting and taxonomizing. How far can the 'thickets of this new wilderness' be raised to consciousness? Only the poet's experiments in language—how the experiments fare in a world with 'words, words, words / All over everything'—can tell.

"How can he change", the poets asks, these "silvered knivings / these awkwardnesses? / How can he make these blood-points into panels?"[226] It remains to be seen if these 'blood-points'—a useful phrase to render the fleshed points, the vectors, organic and alive, unlike the dead points of the Cartesian plane—can be shown; whether a tractable expression can be found for what the body is in actuality: "a wagon, for a sleigh, for the beak of, the running sides of / a vessel fit for moving". Once found, the words would be carried off right away in a wagon, a sleigh, a beak, like the nouns-in-absentia of the Maximus poems extracted into verbs—for a reality made continuous is a terrible thing: only to be revealed fleetingly, 'kniving' the ego, disintegrating the individual 'I' in its passing. But once we stop examining space from the heavenly vantage of the Euclidean geometer, and see from the level of experience, propriocieve from 'this low-eye view', how will we know 'size', how perceive space?

In hell it is not easy
to know the traceries, the markings
(the canals, the pits, the mountings by which space
declares herself[227]

The visible insignia of a place, the tangible markings, such as 'traceries', 'canals', and 'pits' by which space declares itself to a mind in the Cartesian spell is overwhelmed by sensuous perception from below, the 'low-eye view'.

[225] From 'In Cold Hell, in Thicket' in Olson, *Poems*, p. 156.
[226] Olson, *Poems*, p. 157. [227] Ibid., p. 157.

Hallberg says, "Romantic poets could identify with the landscape by an extension of feeling; something colder than sympathy is here advocated."[228] But the poem uses the features of landscape as a sign of something more general, of objects on Euclidean space, visible, discriminable, from the 'high-eye-view' of the Romantic wandering lonely as a cloud over the vales and hills. The 'low-eye view' is in topology the vantage from which shapes of varying sizes and forms—the doughnut and coffee mug, say—appear the same. It is a view without a 'sight', whose darkness is made visible, sensible rather—as in the sulphurous gloom of Hell oppressing the stupefied angels of *Paradise Lost*—in and through the physicality of words.

"Space", in Olson's new hell is "arched, as she is, the sister/awkward stars drawn for teats to pleasure him, the brother / who lies in stasis under her".[229] We are prompted to take 'sensuous apprehension' as the obverse of 'knowing the markings' by the feral suggestion in the word 'teats'. The jump from tangible markings to teats pleasuring him, from 'to know' to 'lying under', is an almost clinical feat in Olson's cold hell; differently achieved from the native somatic vigour with which Milton's words illumine his hot hell. We might remember Satan steering his flight "aloft, incumbent on the dusky air":[230] how dusky charges the air with a density which backwardly imbues incumbent with the proprioceptive potency to communicate the weighty effort of flight to the poem's featherless, bipedal readers; and then compare the intentional 'awkwardnesses' with which Olson's 'space' involves body, 'the vessel fit for moving', through the sudden shift in the verse from the discrete to the continuous (from canals and pits to being arched under stars drawn for teats). A germinal version of such an effect can be found in Flaubert's *Salammbo*:[231]

> The plain, which was wholly bounded by mountains, expanded around them. Here and there a palm tree leaned over a sand hill, and pines and oaks flecked the sides of the precipices: sometimes the rain of a storm would hang from the sky like a long scarf, while the country everywhere was still covered with azure and serenity; then a warm wind would drive before it tornadoes of dust, and a stream would descend in cascades from the heights of Sicca, where, with its roofing of gold on its columns of brass, rose the temple of the Carthaginian Venus, the mistress of the land. She seemed to

[228] Hallberg, *Olson*, p. 146.

[229] Olson, *Poems*, p. 157.

[230] From Book 1 of *Paradise Lost*.

[231] Olson and Robert Duncan discuss Flaubert on many occasions. See *An Open Map: The Correspondence of Robert Duncan and Charles Olson*, eds Dale Smith and Robert Bertholf (New Mexico: University of New Mexico Press, 2017).

> fill it with her soul. In such convulsions of the soil, such alternations of temperature, and such plays of light would she manifest the extravagance of her might with the beauty of her eternal smile. The mountains at their summits were crescent-shaped; others were like women's bosoms presenting their swelling breasts, and the Barbarians felt a heaviness that was full of delight weighing down their fatigues.[232]

The scene begins in an innocuous naturalistic description naming individual trees in the landscape. But once Venus is introduced as mistress of the land, landscape suddenly becomes an extension of her presence, revealing in its glittering and sultry air her glamorous sensuality and in its undulations and swellings her lissom and buxom form, which weigh in and onto the Barbarians trogging on her open surface.

Olson welcomes the new reality not with a celebratory ekphrasis: he wonders what happiness it brooks, how the reluctant might escape, "how / shall he convert this underbrush, how turn this unbidden place / how trace and arch again / the necessary goddess?"[233] Unlike *Salammbo*, Olson's poem does not rely on personification (with the introduction of the goddess) to shift gears from the discrete to the continuous: it works by inviting at various points different aspects of the bodily sensorium: while 'canals' and 'pits' are—though already sensuously laden—physical objects readily available for mental picturing, lying 'in stasis under' 'awkward stars drawn for teats' is as opaque to imagination as it is available to bodily proprioception.

Amid the seemingly random, at times crazed, sallies of striking sentences, it is easy to miss the simple idea that runs through Olson's writings: that reality, language, and body should be in mystical yet precise communion. Analogy to non-Euclidean geometry has proven to us crucial in gathering this idea together. The poet was bidden no longer to seek just visual conformity but somatic impact on the reader congruent to his original experience. In the new art, the human organism is treated as an open system, like the 'coat of your own self', breathing in the 'Real' and dissolving into 'open verse'. We have discretely identified the poet, the word, and reader in the continuum only for critical exposition: these points should not be gathered into some theory of cognitive linguistics but seen rather as limits on the open set, say, or wrinkles of significance in the continuous whole of Olson's new Real. The

[232] Gustave Flaubert, *Salammbo: A Realistic Romance of Ancient Carthage*, trans. J.S. Chartres (London: Gibbings &co., 1899), p. 29.

[233] Olson, *Poems*, p. 158.

'man' and 'girl' referred to in plural sufficiently indicate this. Just as two open sets bleed into one another as they approach the same limit, the 'Real' merges with the poet, dissolves into the syllable and line, rising again, just so, into the reading organism.

After all, "Limits / are what any of us / are inside of".[234]

4.3 Conclusion

Spinoza once declared "truth" to be its own sign.[235] To the poets in this chapter, the word became its own sign. It is rare to find such extensive similarity and intensive difference in two philosophies of poetry. The reason for this, I believe, is that the interests of the poets is concentrated on the basic unit of the word. This can of course also be said of both Empson and Roberts. After all, Empson's *Complex Words* (1951) is perhaps the most extraordinary meditation on the meaning-making process of words in the period—written indeed in the same decade that saw the *Philosophical Investigations* (1953) and *How to Do Things with Words* (1955). Roberts, similarly, is interested in how words record and facilitate the historical merger and fission of mental faculties. What differentiates Riding and Olson from the rest is their interest in words as words, above and beyond anything to which they refer. The study of semantic fields or ambiguity was for them but secondary.

The protagonists of this chapter are thus distinguished from the others in their having a philosophy of language: that is, in their departure from the realm of reference to the metaphysical question of what language is. In considering the two principal senses of logos, 'word' and 'truth', both say it is fundamentally a mistake—that has become historically entrenched—to view words as proceeding world and referring back to it. It was not the task of words to represent reality. To Riding, language precedes reality, or history, as experienced, whereas to Olson, language is a part of experienced reality: the word participates with the saying organism in reality's acts of being. This belief in the ontological primacy of the word must also be seen as a rhetorical manoeuvre in the agon against mathematical representation. What can be a truer mode than poetry when the poetic word itself becomes part of the world?

[234] Olson, *Maximus*, p. 21.

[235] The actual statement was, "truth is the standard of [...] itself" (Benedictus Spinoza, *Ethics*, ed. and trans. Edwin Curley. 1677. London: Penguin, 1996, p. 43s), but this implies that it imposes its own criterion for itself and is thus its own sign.

Although both elevated language to or above what was understood as its source, their philosophies were antithetical. Olson's conception of human being as a conduit in the process by which reality issues would seem barbaric to the rational Riding, who, at least in her youth, held fast to a disembodied vision of perfect identity between meaning and essence—where Olson insisted on moving "from its essence into its kinetic".[236] In her manifesto "A Prophecy or a Plea", Riding expressed her distaste for the old "definition [in which] man is but a stream of passage between the source that is life and the outlet that is poetry".[237] To sweep aside the vulgar mouths that in speaking, offend the intellection of words, was a lasting wish even of the older Riding. We thus have in Riding and Olson the most extreme partisans of mind and body.

Despite irreconcilable differences—or perhaps because they were prone to irreconcilable differences—they are both convinced in having finally breached millennial delusions. They offer views of language that seem to cut through the sedimentation of poor past uses based in errant assumptions of its nature. Given such grand ambitions, we might question the discursive strategies employed by both poets. As regards Riding, the dissociation of language from historical experience is a curious manoeuvre. If it is the case that history after the Enlightenment had proven disappointing to Riding, and she did not see a realistic way forward wherein language could be rescued from its 'mathematical', as she terms it, reification, it might have seemed sensible to carve for words a separate and higher realm whereof poets are masters. But language, regardless of the mental process that assimilates vocabulary, is learnt in concert with other individuals in the course of daily life: it happens squarely within the historical-social world which Riding would sooner do away with. It seems rather unlikely that Cratylus will find many votaries besides the hermetic Riding in a bustling modernity. As regards the restless fury of projective verse, there seems a possibility of an organic response among readers that agrees with the aims of Olson. A legate in some ways of Hopkins and the young Rimbaud, and their attempts to pulsate verse with movement, Olson forces the reader to alertly anticipate and even experience action. The persistent circumvention of the discrete—at times, even of nouns—in his poems creates an intriguing reading experience regardless of whether it proceeds to restore some primordial mode of being. But we are forced to ask when wading through the Maximus poems whether in fact the

[236] 'Materials and Weights' (Olson, *Prose*, p. 117).

[237] Riding, *Chaplet*, pp. 51–2.

surrender to bodily experience is not better facilitated through traditional rhythm, as in Rimbaud, or even when caught in the sensual music of Yeats. It seems at times that the tortured syntax requires too much the intervention of mind to sort and analyse the poem's doings for any prior or proprioceptive state to gain sway. In the last, despite the grand metaphysical aims of both writers, their verse may only ever be remembered as experimental.

5
Aesthetics

We have argued in this book that there was a tendency natural to poets of the modernist period to resist the mathematical picture of the world because it had, through intentional and impersonal forces, undermined the poetic world-picture in value, effacing its power to an alarming degree from public imagination. As a consequence, we have shown how modernist poetry, particularly those poetics and poems that contended with the staring fact of mathematization, is best seen as bringing to culmination a three-hundred-year agon between rival world-pictures: that poetry's bold foray into mathematics in the period can be understood as the last serious literary challenge to the univocal mathematization of reality.

It might help to exhibit the fitness of the agonistic paradigm if we, somewhat as mathematicians do when seeking proof, assume, and work back from, its contrary. We will in this chapter explore instances of consilience in the modernist period between poetry and mathematics with the hope of demonstrating the fitness of the agonistic paradigm. We will explore such concurrence both in terms of modes and the issuing world-pictures.

While we have so far regarded mathematics as a unified force, one great antagonist, we will in this chapter treat the artefacts of mathematics and the mathematization of reality separately. The mathematical world-picture, we said, imposes a certain univocal character on the nature of things: things that can be understood differently if approached from alternate lenses—light becomes photon, say, and colour becomes wavelength in the mathematical world-picture because it is only certain properties of these phenomena that yield to mensuration, whereas both may signify differently and more in art. But if we take the objects that occupy the mathematical world-picture not as standing for or representing the commonly shared objects of experience but as artefacts in themselves of the labour of mathematicians, we are better placed to understand why there has long been among poets a deep aesthetic fascination with mathematics—a heightened form of the commonplace exclamation of beauty in a proof or elegance in a theorem.

Agon: Poetry's Challenge to the Mathematization of Reality (1920s–1960s). Anirudh Sridhar, Oxford University Press.
 DOI: 10.1093/oso/9780198985457.003.0005

5.1 The Beauty of Geometry

An enchantment with pure shapes and numbers, even a feeling of aesthetic kinship, stole over some modernist poets who wrote of mathematics as a gateway to the true and the beautiful. These poets can be described as mathematical Platonists. We will leave it to A.E. Mackay to define what we mean by the term. She writes,

> The region of pure reason is a calm,
> White moonlit night,
> All earthly forms appear
> Etched black against the light
> Mysterious, clear.
> Pure mathematics are a mirror
> Reflecting abstract truth for ever
> Through regions free from error[1]

This could serve as a manifesto for Platonism. The 'earthly forms appear', are cleared of detail, and their backlit outlines waver into regularity. Whatever is earthly about earth gives way to an essence purely abstract, unburdened by circumstance. To the poet seeking timeless forms beneath an irregular and impious history, to be "for one brief moment in eternity",[2] mathematical truth seems an ideal condition.

Edna St. Vincent Millay is another poet who affirms the truth and beauty of geometry but she does so more in resignation.

> Euclid alone has looked on Beauty bare.
> Let all who prate of Beauty hold their peace,
> And lay them prone upon the earth and cease
> To ponder on themselves, the while they stare
> At nothing, intricately drawn nowhere
> In shapes of shifting lineage[3]

The attraction for poetry in mathematics has perhaps never been summarized so aptly as here in her modernist resignation: that fallen art—imagination

[1] A. E. Mackay, *The Garden of the Gods, and Other Poems* (London: Grant Richards, 1931), pp. 14–15.

[2] Ibid., p. 14.

[3] Edna St Vincent Millay, *Collected Poems*, ed. Norma Millay (New York, Harper, 1956), p. 605 (from *The Harp-Weaver and Other Poems*, 1923).

grasping after organic form—can no longer hope for perfection. Allan Burns summarizes the poem as seeking "the elusive nature of beauty"[4]—beauty seems to reside now in that other world, to whose perfect forms the mathematician alone has access through the ghostly objects of his trade: those nothings, drawn nowhere.

The poem appreciates geometry in terms more familiar to modern aesthetics than the Academy: as *useless* beauty. This can be discerned by tailing the tone attending the phrase 'beauty bare'. In a previous sonnet, the poet had said, "Love is not blind. I see with single eye / Your ugliness and other women's grace. / I know the imperfection of your face".[5] Love does not in the poet exalt her lover's imperfections to oblivion as in troubadour praise.[6] She seems to grasp conceptually the 'blindness of love', but feels no empathy for poets 'who prate of beauty' in an amorous fog: "Well I know / What is this beauty men are babbling of; / I wonder only why they prize it so".[7] Incapable of an elevated appreciation of sexual beauty, the poet says, "Still will I harvest beauty where it grows: / In coloured fungus and the spotted fog [...] In empty tins; and in some spongy log / Whence headlong leaps the oozy emerald frog".[8] When the sense of beauty is not naturally perceived but must be sought after, there is little that seems to reliably direct the eye—a frustration Dorothea Brooke and even Isobel Archer shared when stood before Florentine masterpieces with insufficient aesthetic education. In "Inert Perfection" (from a previous collection), the poet recommends a disinterested appreciation of beauty: if beauty "be bound by Function, that it be / Less than Perfection". Continuing in *The Harp-Weaver*, the poet laments, "Pity me that the heart is slow to learn / What the swift mind beholds at every turn".[9] It is rare that a poet should try so earnestly to emote a lack of feeling. One not given to spontaneous overflows of powerful emotion perhaps becomes content in more abstract pursuits. It is sympathy with this gradual retreat into mind that adds pathos to the line, 'Euclid alone has looked on beauty bare'. The stresses lie on 'lone' and 'bare'. While she glimpses the non-existent shapes of geometry with

[4] Allan Burns, *Thematic Guide to American Poetry* (Westport: Greenwood Press, 2002), p. 4.

[5] Millay, *Poems*, p. 586 (Collected as Sonnet xxvi).

[6] Or take Isabel Archer, for instance, who debates whether she ought not fix upon the outward flaws of a supplicating gentleman: "She had reminded herself more than once that this was a frivolous objection to a person of his importance; and then she had amended the rebuke by saying that it would be a frivolous objection only if she were in love with him. She was not in love with him and therefore might criticise his small defects as well as his great" (Henry James, *Portrait of a Lady* (Raleigh, Freebook Publisher [1881] 2020), p. 86).

[7] Millay, *Poems*, p. 586. [8] Ibid., p. 603 (Collected as Sonnet xliii).

[9] Ibid., p. 589 (Collected as Sonnet xxix).

the rapture of an Actaeon espying Diana bare, when that beauty is enfleshed, it retains little of interest to the intelligent poet.

Walter Turner, a member of the twentieth century pastoral movement calling itself Georgian, was so captured by mathematics that he, like Descartes, issued the passport of existence to those things only that conform to the ideal objects of the imagination. Yeats regarded Turner "as the first poet to read a mathematical equation; a musical score, a book of verse, with an equal understanding; he seems to ride in an observation balloon, blue heaven above, earth beneath an abstract pattern".[10] He finds in the phrase 'earth beneath an abstract pattern' an accurate description of Turner's attitude, which involved a love of sensuous beauty yet with an undertone that their ultimate justification lies in their disclosure of some higher form that can be grasped by mind alone.

Turner's long poem called *The Seven Days of the Sun* is a rather gloomy series that exalts mathematics to pave the poet's retreat from phenomena. The poet has "felt the whiteness of a lily / Upon my palate; / And the solidity of their slender curves / Like a beautiful mathematical proposition / In my brain".[11] The beauty of the world lies in the subject's reordering of impressions. Features we may attribute to phenomena are actually abstracted and patterned by the sorting mind before they are conceived and made real.

> When I look upon a beautiful body
> And rapidly make an *Abstract*
> I tingle with pleasure at the deviations and aberrations
> Of the *Real*.
> I become alive through a series of shocks.
>
> My heart thunders as I race along this asymptote ever-
> Lasting
> Which is my life among other bodies.
>
> [...]
>
> The curve of misery is asymmetrical
> I would close joy in a perfect circle.[12]

[10] W.B. Yeats, *Later Essays*, ed. William H. O'Donnell (London: Charles Scribner's Sons, 1994), p. 462.

[11] W.J. Turner, *The Seven Days of the Sun: A Dramatic Poem* (London: Chatto & Windus, 1925), p. 26.

[12] Ibid., p. 29.

Like the nominal boy of Empson's "Letter V", who was a mote of dust that suggested ideal Form, phenomena seem to Turner 'deviations' and 'aberrations' from the '*Abstract Real*'. But the mortal coil unwinds as an 'asymptote' towards the world—as in Empson's "The World's End", mind, readily possessing the objects of geometry, will never know world, just as the asymptote will never merge with its limit. The unattainable asymmetries of the world thus become objects of desire and misery to the reasoning mind, and the resigned Platonist encloses 'joy in a perfect circle'.

Marianne Moore's poetry ventures beyond the aesthetic affirmations of mathematical Platonism to its moral doctrines. In a poem titled "Icosasphere" (after the platonic solid icosahedron), the poet says,

> the birds nesting in the merged green density,
> weave little bits of string and moths and feathers
> [...]
> in parabolic concentric curves'
> and, working for concavity, leave spherical feats
> of rare efficiency[13]

The weaving birds, symbolizing the natural world here, are an unblemished good. Their goodness is demonstrated by their achieving amid the messy mud of earth perfect geometric forms. The poem declares the "Icosasphere" the "summit of economy" as it is "geometrically/neat". Linda Leavell says, "like other Moore poems, 'The Icosasphere' equates geometric perfection with moral perfection".[14] "Whereas" human beings, says the poet, "through lack of integration, / avid for someone's fortune", were "slain" and "committed perjury".[15] 'Integration' in the poem refers to the birds merging with nature like Empson's magnolia. This union seems to be lost to a flawed humanity that perjures and is slain, it appears, in some Old Testament measure of retribution. To Augustine, animals did what nature bade and were thus part of God's good design. Man sought what was not his 'fortune' (to be understood in the poem also as fate), dissembled to gain what was not given: but in striving so, has also achieved all that he has: and this made man the only "evil good",[16] and the Fall thus fortunate. Against the sordid conduct of mankind,

[13] Marianne Moore, *Collected Poems* (New York: Macmillan, 1951), p. 142.

[14] Linda Leavell, 'Kirkwood and Kindergarten: A Modernist's Childhood', in *Critics and Poets on Marianne Moore: 'A Right Good Salvo of Barks'*, eds Linda Leavell, Cristianne Miller, and Robin G. Schulze (Lewisburg: Bucknell University Press, 2005), p. 36.

[15] Moore, *Poems*, p. 143.

[16] Augustine, *Confessions*, p. 75.

just so, the poem sets the aspiring harmony of nature and the symmetrical justice of geometry.

Despite such a discernible strain of mathematical Platonism among poets in the period, the agonistic paradigm is not undermined because the modernist struggle was against *mathematization*, not the objects of mathematics. But let us see if the paradigm can also produce a positive account for the aesthetic affinity between poetry and pure mathematics. World-pictures attain victory over others, we have said, not by successfully comprehending all the facts of phenomena but through the rhetorical pursuit of authority. That is, the process of establishing authority will involve colouring their mode of discourse to advantage. We have witnessed such manoeuvres in seventeenth-century philosophy, with Bacon and the Royal Society exalting the virtues of science against what they painted as a deceptively metaphorical English. We also noted the separation of the academic disciplines of mathematics and physics in the nineteenth century that led to the emergence of what has been termed 'modernist mathematics', which included new branches of mathematics that seemed autonomous from the physical world, such as formalism, analysis, and non-Euclidean geometry, among others. The severance of pure mathematics from the physical world naturally excited some anxiety among mathematicians. Doubts as to the origins of mathematics were expressed at the time by Russell: "mathematics may be defined as the subject in which we never know what we are talking about, nor whether what we are saying is true".[17] Once the physical world was no longer the foundation for mathematical knowledge, we noted in the introduction the various attempts to find a secure grounding for mathematical truths: there emerged logic, as the basis of mathematics, posited first by Gottlob Frege in 1884, and championed later by Russell. Hilbert attempted to bring all mathematics into an internally consistent, self-contained system called formalism; Kronecker, as we have seen, tried to rein mathematics in from the infinities in the clouds to a tractable 'finitism', which laid the foundation for Brouwer to locate the origin of mathematics in intuition.[18] Why should it be, we may ask, that a 'philosophy of pure

[17] Bertrand Russell, 'Recent Work on the Principles of Mathematics', *International Monthly* (1901). Reprinted in *Bertrand Russell: His Works, vol. 3: Towards the 'Principles of Mathematics' 1900–02* (New York: Routledge, 1994), p. 84.

[18] "Three traditionally important views on the nature of mathematics: logicism, intuitionism, and formalism" (Paul Benacerraf and Hilary Putnam, ed., *Philosophy of Mathematics: Selected Readings* (Cambridge: Cambridge University Press [1964] 1983), p. 1)). See also the collection of essays in Part I. The Foundations of Mathematics, pp. 41–65; also see the three chapters (5. Logicism, 6. Formalism, and 7. Intuitionism) on "Mathematics and Its Foundation", in David Bostock, *Philosophy of Mathematics: An Introduction* (Oxford: Wiley-Blackwell, 2009).

mathematics' should emerge in the late nineteenth century, where mathematicians of the previous centuries, for the most part, had simply got on with the business of doing mathematics? The agonistic paradigm would argue that attempts to find the source and ground the truth of mathematics were defensive moves in the wider competition of the faculties: as ways of fastening mathematics to secular foundations in an age that would no longer tolerate Pythagorean mysticism. It is difficult to explain in any other way such baffling attempts as Russel's, for instance, of deriving the numbers—the surest objects to mind—from something as treacherous and slippery as logic.

In this vein, there was another trend in mathematical philosophy which might set the context for the modernist trope of exalting mathematics in poetry. This is the attempt in the late nineteenth century to characterize pure mathematics in the language of aesthetics: this can be seen as the first rhetorical move of a new entrant in the agon of faculties—rather, of the most ancient of players newly aliened from the physical world with which it had so long been strangely united.

The art historian Wilhelm Worringer observes at the beginning of the twentieth century that, "we frequently find the, at first sight, astonishing idea [...] that mathematics is the highest art form".[19] The alliance seems astonishing to Worringer in the age of Maxwell and Einstein but Baylee Brits points out that mathematics historically allied with the arts much more than the practical activities of scientists.[20] R.G. Collingwood even argued that in *The Republic*, by describing art as imitation of imitation, Plato had unwittingly created a 'field' of aesthetics;[21] although designated to a lower plane, art was ontologically separate from the world, as much so as geometry. It is, however, the legacy of German aesthetics that perhaps best explains the attraction for modern mathematicians. Andrea Henderson says, "Kant's aesthetic form sounds remarkably like the Victorian ideal of 'pure mathematics'".[22] The notion that appreciation of art is *disinterested* and that art is itself useless is a legacy of Kant's third critique. Kant distinguished judgement of beauty from personal interests of the individual and separated the uses to which a work of art is put from the artwork itself, which, he maintained, appeals without

[19] Wilhelm Worringer, *Abstraction and Empathy: A Contribution to the Psychology of Style*, trans. Michael Bullock (Chicago: Ivan R. Dee [1908] 1997), p. 19.

[20] Brits, Baylee, *Literary Infinities: Number and Narrative in Modern Fiction* (London: Bloomsbury, 2018), pp. 2–3.

[21] R.G. Collingwood, 'Plato's Philosophy of Art', *Mind*, 34 (1925), 154–72.

[22] Andrea K. Henderson, *Algebraic Art: Mathematical Formalism and Victorian Culture* (Oxford: Oxford University Press, 2018), p. 11.

agenda; has "purposiveness without purpose".[23] The separation of judgement from earthly interest is one of the reasons why aesthetics was a fertile discourse in which to ground a discipline—pure mathematics—recently freed from physical reality.

Kant's critique straddled the continental and British traditions of aesthetics: while David Hume defined taste subjectively as the accretion of aesthetic experience, Baumgarten made the aesthetic deducible from objective principles of beauty.[24] Kant combined the two into a theory of beauty as subjective judgement about which one can nonetheless argue and seek agreement.[25] Poincaré's analysis of aesthetics in mathematics mirrors the Kantian theory. Milena Ivanova states,

> Aesthetic judgements are not simply emotional responses, differing between individuals with different tastes and preferences. Nor are they objective, since they do not refer to or reflect an objective property of a theory. It is reasonable to suppose that for Poincaré aesthetic judgements are objective in that there is intersubjective agreement between beings like us who share the same intellectual capacities.[26]

Poincaré, in a chapter on "Mathematical Discovery", claims that this "harmony" between subjective judgement and intersubjective agreement is "a satisfaction of our aesthetic requirements".[27] He argues that the ultimately useless nature of mathematical theory is its salvation in an age that saw number as weapon for utilitarianism: a mathematician does not study his subject "because it is useful; he studies it because he delights in it, and he delights in it because it is beautiful".[28] He says, pace Russell, "the distinguishing feature of the mathematical mind is not logical but aesthetic".[29] He even borrows the notorious creed of Bohemian decadence, *l'art pour l'art*, in his attempt to

[23] Immanuel Kant, *Critique of Judgment*, trans. Werner S. Pluhar (Cambridge: Hackett Publishing [1790] 1987), p. 73.

[24] David Hume, *Four Dissertations* (London: A. Millar in the Strand, 1757); Alexander Gottlieb Baumgarten, *Aesthetica* (Hildesheim: G. Olms [1750] 1961).

[25] Kant, *Judgement*, p. 100.

[26] Milena Ivanova, 'Poincaré's Aesthetics of Science', *Synthese*, 194.7 (2017), 2581–94, p. 2588.

[27] Henri Poincaré, *Science and Method*, trans. Francis Maitland (London: Thomas Maitland and Sons [1908] 1914) p. 38.

[28] Henri Poincaré, *The Value of Science: Essential Writings of Henri Poincaré*, ed. Stephen Gould (New York: Modern Library [1908] 2001), p. 368.

[29] Qtd in Tommy Dreyfus and Theodore Eisenberg, 'On the Aesthetics of Mathematical Thought, for the Learning of Mathematics', *EPDF*, 6.1 (1986), 2–10, p. 2.

restyle the mathematician rake: "intellectual beauty is sufficient unto itself, and it is for its sake".[30]

Mathematical beauty could of course not draw from sensual pleasure; like Millay, Poincaré found it in pure form: symmetry and order were to be its ornaments:

> Of course I do not here speak of that beauty which strikes the senses, the beauty of qualities and of appearances; not that I undervalue such beauty, far from it, but [...] I mean that profounder beauty which comes from the harmonious order of the parts and which a pure intelligence can grasp.[31]

Russell would similarly distinguish the forms of beauty to mathematics' advantage—he employs that other carefully devised term of aesthetics, the sublime. Through the works of Kant, Burke, and Schopenhauer, among others, beauty had come to be associated with fineness while the sublime, with the majestic and awe-inspiring. The latter knifed the ego most, as to behold the tempests of Turner or the infinities of mathematics was to have the realization of the self's insignificance come crashing with joy. Russell says,

> Mathematics, rightly viewed, possesses not only truth, but supreme beauty—a beauty cold and austere, like that of sculpture, without appeal to any part of our weaker nature, without the gorgeous trappings of painting or music, yet sublimely pure, and capable of a stern perfection such as only the greatest art can show. The true spirit of delight, the exaltation, the sense of being more than Man, which is the touchstone of the highest excellence, is to be found in mathematics as surely as poetry.[32]

He is developing for mathematics an old, morally charged, aesthetic distinction between the gorgeous and the austere, which, for instance, expresses itself in Hellenistic narratives of the decline of the arts from the golden age, noticing the laboured sinewiness of their sculpted athletes,[33] or in the Benedictine exaltations of abstract symmetry in their invectives against the extravagances of the early Romanesque.[34] Russel's idea that mathematics was a cold beauty unlike the hot nudes of the canvas was influential in mathematics

[30] Henri Poincaré, *The Choice of Facts*, trans. G.B. Halsted, *Monist*, 19.2 (1909), 231–39, p. 237.
[31] Poincaré, *Facts*, p. 237.
[32] Bertrand Russell, *Mysticism and Logic: And Other Essays* (London: Allen & Unwin, 1917), p. 60.
[33] Cyril Robinson, *A History of Greece* (London: Methuen and co., 1929), p. 280.
[34] Bernard of Clairvaux, *Apology (Apologia)*, 1125.

surreptitiously scaling the aesthetic ladder in the twentieth century. G.H. Hardy, who popularized and tried to establish mathematical aestheticism as a branch of philosophical inquiry, says, "mathematics exercises a coldly impersonal attraction".[35] Hardy argues with the aesthete that what is of ultimate value is never of immediate service: "Is mathematics 'useful', directly useful, as other sciences such as chemistry and physiology are? This is not an altogether easy or uncontroversial question, and I shall ultimately say No."[36]

Hardy's puts forth three criteria for the judgement of mathematical beauty: "economy", "unexpectedness", and "inevitability".[37] Even the poets who we have seen resisting the mathematical world-picture would aestheticize mathematics in similar ways. As regards 'economy', Roberts said,

> The true mathematician instinctively aims at rigour and economy of argument [...] These preferences are ultimate, they cannot be explained, but they are widespread, and the solution of a problem which satisfies these preferences gives an intense satisfaction to the solver: a satisfaction which, mathematicians claim, is indistinguishable from that produced by certain works of art.[38]

Eliot even says mathematical logic has made language a more economical medium for poets: he praises "how much the work of logicians has done to make of English a language in which it is possible to think clearly and exactly on any subject. The *Principia Mathematica* are perhaps a greater contribution to our language than they are to mathematics".[39] Empson dwells on the inward tension between inevitability and unexpectedness: "the aesthetic value of a mathematical process lies in a [...] perpetual slight surprise, which on the next moment's consideration is turned to a richer acceptance, was what Aristotle found most fundamental to exalted beauty. This pleasure is inherent in the method of mathematics"[40]—as in most things, Empson ultimately values paradox: and the aesthetic value of mathematics for Empson is in the contrary experience of unexpectedness arising from the recognition of inevitability. With the divergence of pure mathematics from physics, poets began thus to accept the move of mathematics toward aesthetics; they even seem happy to concede to the gambit of mathematicians to insinuate their art

[35] G.H. Hardy, *A Mathematician's Apology* (Cambridge: Cambridge University Press, 1940), p. 14.
[36] Ibid., p. 8. [37] Ibid., p. 29. [38] Roberts, *Critique*, p. 107.
[39] T.S. Eliot (1927), "A Commentary", *The Monthly Criterion*, 6(4), 289–91, p. 291.
[40] From Empson's diaries (1925), recorded in John Haffenden, *William Empson: Among the Mandarins* (Oxford: Oxford University Press, 2005), p. 105.

atop Parnassus as the fairest of them all; after all, art's insecurity lay not in its charms but its claim to truth.

Accepting the association of mathematics with aesthetics was not without benefits in the agon. The mathematicians who grounded their subject in aesthetics were also making the useful equation of beauty with truth. Hardy believed his three aesthetic criteria to be of chief importance when assessing the truth of a theorem; "The mathematician's patterns, like the painter's or the poet's must be beautiful; the ideas like the colours or the words, must fit together in a harmonious way. Beauty is the first test: there is no permanent place in the world for ugly mathematics."[41] With inevitability, unexpectedness, economy, and now the criterion of 'rightness', a mathematical theorem could be judged by a Ruskinian who held " *there is but one right way of doing any given thing required of an artist*".[42]

The glossing of pure mathematics in the language of aesthetics at the turn of the century seems to have won ground in the agon as the manoeuvre even begins to attract the mathematical theories of physics. Herbert Read says, "scientists like Mach, Poincaré, and Karl Pearson have agreed in defining science as the economy of thought. But this is also a good working definition of art."[43] J.W.N. Sullivan, like Hardy with mathematics, nominates beauty as the chief characteristic of a good scientific theory: "The measure of success of a scientific theory is, in fact, a measure of its aesthetic value [...] The measure in which science falls short of art is the measure in which it is incomplete as science."[44] We will note that the term 'theory' in science referred at the time to the mathematical theories of physics only and was yet to be generously extended to the narrative descriptions of biology.

Mathematical physics even followed the mathematician's use of aesthetic criteria in assessing the validity of theory—undoubtedly a promising development in the agon for art. Read says, "with Professor Eddington, 'I cannot reject the hope that theory [and, I would add, art] is by slow stages leading us nearer to the truth of things'."[45] Heisenberg suggests that Nature, in the evermore elegant ways in which it submits to representation—as in relativity theory after classical mechanics—seems to offer beauty as a sign of approaching

[41] Hardy, *Apology*, p. 14.

[42] John Ruskin, *The Two Paths: Being Lectures on Art and Its Application to Decoration and Manufacture Delivered in 1858–59*, ed. Christine Roth (West Lafayette: Parlor Press, 2004), p. 32.

[43] Herbert Read, 'Readers and Writers', *The New Age*, 22 (December 1921), 67–8, p. 68.

[44] J.W.N. Sullivan, 'The Justification of the Scientific Method', *Athenaeum*, 4644 (1919), 274–5, p. 275; to justify his hypothesis, Sullivan quotes Poincaré (p. 275).

[45] Read, 'Readers', p. 68; Eddington quote taken from *Space, Time and Gravitation: An Outline of the General Relativity Theory* (London: Cambridge University Press, 1920), p. 25.

truth: "[i]f nature leads us to mathematical forms of great simplicity and beauty we cannot help thinking that they are 'true', that they reveal a genuine feature of nature".[46] Paul Dirac reserved a crucial epistemic role for beauty in accepting a scientific theory as true: "one has a great confidence in the theory arising from its great beauty, quite independent of its detailed successes".[47] Weyl's position was between these tendencies—of beauty as truth or as an indicator of it: he says, "my work always tried to unite the true with the beautiful, but when I had to choose one or the other, I usually chose the beautiful".[48] The popular appeal of aesthetic criteria in judging scientific theory can thus be viewed as a brief success for the partnership between poetry and mathematics in their agon against physics (we are wilfully personifying the disciplines to suggest that this episode of the agon took place between faculties more than individuals like Heisenberg or Poincaré, who may have regarded themselves as much mathematicians as scientists or aesthetes). One might imagine mathematics, channeling Mackay or Millay, saying to science, "our objects might be circling regions of pure calm, mirrors reflecting beauty bare, but your highest achievements are but species of our works, and it is our beauty, in the end, that determines your truth". We already find this air of superiority in the magnolia-like insolence with which Hilbert, presenting twenty-three important problems in mathematics to the International Congress of Mathematicians in 1900, lists as problem number six the "mathematical treatment of the axioms of physics".[49] Cleaning up the remnants of nineteenth-century mechanical models, and beautifying the foundations of physics, is but one minor task among the many loftier callings of mathematics.

The aesthetic reimagining of mathematics in its so-called modernist phase thus appears too sustained and deliberate to be ignored. Mathematics became useless, sublime, for its own sake, unexpected, and even right. If indeed it was the case that the new mathematics had renounced earthly concerns, it does not follow that it must be beautiful or sublime, that it must be useless in the way that natural beauty, and later art, was thought to be.[50] This tells us that

[46] Werner Heisenberg, *Physics and Philosophy: The Revolution in Modern Science* (London: Allen & Unwin, 1959), p. 68.

[47] Paul Dirac, 'The Excellence of Einstein's Theory of Gravitation', in *Einstein: The First Hundred Years*, eds Maurice Goldsmith, Alan Mackay, and James Woudhuysen (Oxford: Pergamon Press [1980] 2013), 41–6, p. 44.

[48] Qtd in S. Chandrashekar, *Truth and Beauty: Aesthetics and Motivation in Science* (Chicago: University of Chicago Press, Chicago, 1987), p. 52.

[49] David Hilbert (1902). 'Mathematical Problems', *Bulletin of the American Mathematical Society*, 8.10, 437–79.

[50] Henderson argues that Victorian mathematicians modified Kant's aestheticism by grounding their metaphysics outside the Euclidean world, but the idea of disinterested appreciation does not depend on a Euclidean premise (Henderson, *Algebraic Art*, p. 43).

the envy and subterfuge of the agon was not unidirectional: that the many-sidedness of the agon was rather intricate, and that there emerged interesting alliances—not wholly unsuccessful ones—during the contest.

5.2 The Mathematical Mode

A deeper form of concurrence between mathematics and poetry—than the outward beauty of their world-pictures—is that between the poetic and mathematical modes. In our introduction, we said mode precedes and engenders a world-picture: that while world-pictures might have overlapping characteristics, their modes are always distinct, and that, in fact, it is the irreducibility of this distinction that makes the uniqueness of discourses inviolable (while they may yet be conquerable). There have, for instance, always been differences in the organons of mathematics and physics but they were considered one because their objects were for long alike. But in the nineteenth century, after non-Euclidean geometry departed from the Newtonian world-picture, the discourses eventually split, making it plain that it was method that was actually integral in them. The split has in fact abided till our day despite the theories of physics becoming more and more purely mathematical in content and appearance in the last century.[51] Given the agon of the faculties is premised on the uniqueness modes of understanding and representation, we must consider if attempts to conciliate the poetic and mathematical modes threaten the paradigm and its account of cross-disciplinary tension.

We have so far narrated events in the period of the 1920s–60s when poets resisted most fiercely the mathematization of reality. But if we look to the 1910s, to movements such as Futurism and Vorticism, we find an unabashed affirmation of science and its works; and a willingness to retrain art in accordance with its rule.[52] The founder of Futurism, Filippo Marinetti, wrote a manifesto in 1914 called 'Geometric and Mechanical Splendour and the Numerical Sensibility', which I can only describe as an attempt to remake poetic art in the image of the machine.[53] Its surrender to the mathematical world-picture is unparalleled among the artistic movements of the time.

[51] See Roger Penrose, *Fashion, Faith, and Fantasy in the New Physics of the Universe* (Princeton: Princeton University Press, 2017), ch. 1.

[52] See Matteo D'Ambrosio, 'A "Poetry of Numbers"? Khlebnikov, Jakobson, Marinetti, Schwitters, Kostelanetz', *Cherchez Le Text: Proceedings of the Elo 2013 Conference*, Section 3.

[53] See Daniela Petroşel (2011) 'Poetry and Technology in Marinetti's Futurist Manifestos', *Rupkatha Journal on Interdisciplinary Studies in Humanities*, 3.1, 140–6.

Marinetti says, "A new beauty is born today from the chaos of the new contradictory sensibilities that we Futurists will substitute for the former beauty, and that I call Geometric and Mechanical Splendor."[54] Cultivating this post-industrial sensibility, the poet, he declares, must set off "the enthusiastic imitation of electricity and the machine", so that he can communicate

> The ship's speed, its trajectories of fire from the height of the quarterdeck in the cool ventilation of warlike probabilities, the strange vitality of orders sent down from the admiral and suddenly become autonomous, human no longer, in the whims, impatiences, and illnesses of steel and copper. All of this radiated geometric and mechanical splendor.[55]

The rendering of the human mind and muscle but accessory to the machine is to be displayed and celebrated in poetry. The ambition is great and dear. But the kinds of things to be imported from mathematics to the craft of writing in order to effect this transition seem rather trivial. For instance, an expression like "a vast and deep boom of bells" is to be rendered in Futuristic parlance as, "bell boom breadth 20 square kilometres".[56] It is hard see how the replacement of adjectives with numerical description escapes the "imprecision and dullness" of the former:[57] the reader is enjoined to accept that the pitch and timbre of the sound, its changing volume across a 20 kilometre stretch, are unimportant—qualities secondary, in the Lockean hierarchy, to the measurable breadth.

Marinetti's sound poem "Zang Tumb Tumb" uses arithmetical signs, '+', '–', 'x' to achieve a syntheses of the visible particulars composing the imaginal tableau of the poem. Let us observe some exemplary lines: "horizon – sharp bore of the sun + 5 triangular shadows (1 kilometer wide) + 3 lozenges of rosy light + 5 fragments of hills + 30 columns of smoke – 23 flames".[58] The imagism here is admittedly striking, achieving a representational deformation and flattened intensification one might experience with analytic cubism; a stilled grouping of light and shade one finds in a Seurat painting; and even some of the vigour of Hopkins' kennings. But even if the effect is entirely successful, it is not quite a case of a crossing from mathematical method because the poem

[54] F. T. Marinetti, 'Geometric and Mechanical Splendor and the Numerical Sensibility', in *Marinetti: Selected Writings*, eds Farrar, Straus, and Giroux, translated by R. W. Flint and A. Coppotelli (New York: Farrar, Straus, and Giroux, 1972), p. 72.

[55] Marinetti, 'Geometric', p. 73. [56] Ibid., p. 74. [57] Ibid., p. 74.

[58] Filippo Tommaso Marinetti, 'Zang Tumb Tumb', *The World Magazine*, 9 August 1914. English translation of a portion of ZTT [948–1].

does not bring to bear any mathematical properties of the symbols it borrows. For instance, one of the properties of addition is that it is commutative; that is, a + b = b + a, as in 2 + 3 = 3 + 2. But the ordering is clearly important for the effect sought in the poem, as the shadow could not come before the sun; nor the rosy penumbra, before the shadow. All Marinetti really does to evoke mathematics is replace prepositions and conjunctions with mathematical operators to truncate syntax. As Empson said of his own machinery of symbols in *Complex Words*, "they are no more mathematical than road-signs".[59]

By a trivial intersection with mathematics, we mean that nothing about the nature of mathematics or its ideas and their significance is explored through art. One might say as much about the Oulipo group of writers in France in the 1960s. The Oulipo can be viewed as a reaction against the unchecked celebration of freedom and admittance of randomness in art in the twentieth century's tendency towards experimentation in free verse and deformation.[60] If their doctrine can be summarized, the Oulipo professes the seemingly contradictory view that freedom lies in limits—somewhat in the way that there can be creative chess moves only because the game has rules.[61] The Oulipo—many of whose members were mathematicians[62]—is attracted to mathematical mode only to the extent that it sees mathematics as a formal discipline governed by constraints. Patrick Bahls says,

> It is worth noting that to the Oulipans, constraint is something desirable: constraint does not confine the writer, rather, it frees him by providing structure within which it's easier to work with the resources language offers him. Put another way, constraint "foregrounds" the structures and forms that underlie the language we use every day, making the artist more aware of those conventions and more able to put them to use in his art.[63]

They thus engage with mathematics in the sense that most all poetry before the twentieth century, in being bound by metre, is 'mathematical': theirs is but a more elaborate and conscious avowal of numerical pattern. The

[59] William Empson, *Structure of Complex Words* (London: Chatto & Windus, 1951), p. 3.

[60] Natalie Berkman, *Oulipo and the Mathematics of Literature* (Austria: Peter Lang, 2022), p. 1.

[61] David Bellos (2010) 'Mathematics, Poetry, Fiction: The Adventure of the Oulipo', *BSHM Bulletin: Journal of the British Society for the History of Mathematics*, 25.2, 104–18, p. 105; see also Loveday Kempthorne, 'Relations between Modern Mathematics and Poetry: Czesław Miłosz; Zbigniew Herbert; Ion Barbu/Dan Barbilian' (PhD Dissertation: Victoria University of Wellington, 2015), ch. 2.

[62] Ibid., pp. 105–6.

[63] Patrick Bahls, 'The Mathematical Muse: Using Math to Construct Poetry', *Math Horizons*, 17.3 (2010), 14–17, p. 14.

constraints they borrow for poetry from mathematics are, ironically, quite arbitrary, often betraying the patina of gimmick. For instance, Raymond Queneau's *Cent Mille Milliards des poe'ms* (One Hundred Thousand Billion Poems) is a collection of ten 14-line sonnets.[64] The first line of each poem can be combined with the second line of any other, and so on, given the clauses and rhyme schemes of all ten poems will oblige such combinations. If we calculate the permutations, the combinations could produce up to ten^{14} poems, meriting the absurd title of the collection. Warren Motte claims that "Oulipo's interest in mathematics goes well beyond the level of mere flirtation."[65] But, as Motte himself goes on to show, the Oulipians themselves recognized they were playing games rather than contending with mathematical ideas in literary form.[66] Patrick Bahls says, *Cent Mille* "is a work of combinatorics, that just so happens to produce poems as a byproduct".[67] Like Warhol's Brillo Boxes, after making an initial point—in this case, about the potentiality of art, perhaps, or the fertility of formal constraint[68]—it is hard to see what further insight such a work can yield. Any ten sonnets that permit such or even other combinations would presumably make the same point. While such an exercise may be useful for or a precursor to machine learning algorithms attempting to produce sonnets today,[69] any value it might have is conditional upon one of 10^{14} combinations producing a good poem. We will recall our caveat in the introduction that we are using the word 'value' in the context of the agon to mean descriptive or representational power. The Oulipian movement is saturated with such game-like inducements to combination and counting, such as the S + 7 algorithm, wherein every noun in a text is replaced by the seventh word following that noun in some pre-selected dictionary.[70] The choices of the rules: of texts, nouns, the ordinal 7, and dictionary, are all arbitrary, and the resulting texts are not intended to say anything about numbers or the world.

There are, however, some instances where a poet in the modernist period attempts to clarify poetic mode using mathematical tools, which seem, at first blush, not to be trivial. A loose-knit group of modernist poets styling itself

[64] Philip Terry, ed., *The Penguin Book of Oulipo: Queneau, Perec, Calvino and the Adventure of Form* (New York: Penguin, 2019), ch. 3.

[65] Warren Motte, 'Oulipian Mathematics', in *Palgrave Handbook of Literature and Mathematics*, eds Robert Tubbs, Alice Jenkins, and Nina Engelhardt (Cham: Palgrave Macmillan, 2021), p. 228.

[66] Ibid., 229–30.

[67] Bahls, 'Mathematical Muse', p. 105.

[68] Alice Jenkins, 'Randomising Form: Stochastics and Combinatorics in Post-war Literature' in *Literature and Mathematics*, pp. 217–18.

[69] Later avatars of Oulipo apparently went the direction of text mining: Jenkins, 'Randomising Form', p. 218.

[70] Bahls, 'Mathematical Muse', p. 15.

"objectivist" emerged in America in the 1930s that sought, in fine, to recast the poem as object. The poem, for them, was not to be understood by capturing references without but by treating the poetic object as working towards itself, in some mysterious way, as its own end. Their chief theorist, Louis Zukofsky, says,

> I will tell you
> About my *poetics*—
>
> $$\int_{\textit{Speech}}^{\textit{Music}} \textit{Poetry}$$
>
> An integral
> Lower limit speech
> Upper limit music[71]

An acquaintance with basic calculus will redact the symbolism to an image of two limits or sequences of numbers ('quantities' might be preferable for the analogy) converging to two non-existent points on the number line.[72] Taking the hypothetical reality to which the quantities tend as the signification of pure sound in the upper limit, distilled from contextual distractions of daily noise, and the signification of sense in the lower limit, laden with the connotative corruptions of daily speech, we may say that the symbol idealizes the work of poetic signification as the extent to which pure meaning generated from sound stands on but overcomes the denotative limits of *parole*. My formula, 'extent to which sound overcomes sense', is meant to represent the final stage of the operation, wherein values from the upper and lower limit are plugged into the integral and subtracted (F(music) – F(speech)). Poetic method, in this model, practises a form of negative theology that reaches after pure signification lying beyond the common uses of language, assuming Schopenhauer's anagogy of the arts, with music, purged of reference to the world, seated supreme on Parnassus. But reference that lies beyond an 'outer reality' can only in actuality be reference bound within a self-referential, poetic object, however Platonized that object may then be.[73] This inclination to abstraction (whether apophatic or eidetic) readily explains Zukofsky's

[71] Louis Zukofsky, "A", intro., Barry Ahearn (New York: New Directions [1959] 2011), p. 138.

[72] The points will exist on the real number line, but not in any of the countable sets, the integers or rationals. And counting is the only way in which numbers claim existence in imagination—for the real numbers are merely logical constructions from their countable ancestors.

[73] Apart from the common-sense way in which this is the case, we will see a formalization of this argument in our discussion of Laura Riding's poetics in Chapter 4.

motivation for formulating the poetic mode in terms of integration. But the instance in the end need not be considered in the agon, not for lack of sophistication, for the terms of the analogy have distinct and clear correspondence, but because it does not say anything about mathematics, positive or negative. Zukofsky essentially employs a familiar algorithm to formalize a model of poetic meaning; and even that does not work entirely, as there is no tractable way in which one can conceptualize steady increments or decrements of meaning. What Zukofsky has in fact done is evoke a mathematical notation and its operational entailment to condense his theory of poetic meaning; this hardly suggests a consistency between the poetic and mathematical modes.

A more influential attempt at defining poetic meaning-making in mathematical terms was that of Ezra Pound in his essay "Vorticism".[74] Pound's use of mathematics is at once less trivial and less sophisticated than Zukofsky's. In the essay, he makes the rather blustering claim that "the difference between art and analytical geometry is the difference of subject matter only"[75]—or, to use our terms, poetry and mathematics differ in world-picture but not in mode. Jocelyn Rodal explains the comparison Pound seeks between poetic and algebraic equations as follows: "Representation, after all, poses one thing as another, and here Pound frames that act of representation as one of the equation."[76] However, the notions of mathematical and poetic equivalence have certain important differences not captured in her formula 'one thing as another':[77] and those differences, we will see, are central to poetry's bid for discursive authority.

We might initially reformulate Pound's claim as saying a lyric poem stands to the sensuous image it evokes just as an algebraic equation stands to the geometric curve it specifies. Let us again to that familiar x-y plane of Cartesian geometry. A point on the plane, we will recall, can then be described by its position relative to the x and y axes in the notation (x,y).

Pound says:

> Fourthly, we come to Descartian or 'analytical geometry'. Space is conceived as separated by two or by three axes (depending on whether one is treating form in one or more planes). One refers points to these axes by a series of co-ordinates. Given the idiom, one is able actually to create.

[74] Ezra Pound, 'Vorticism', *Fortnightly Review*, 46.573 (1914), 461–71.

[75] Pound, 'Vorticism', p. 468.

[76] Jocelyn Rodal, 'The Metaphor as an Equation: Ezra Pound and the Similitudes of Representation', in *Literature and Mathematics*, p. 531.

[77] Though Rodal explores a range of interesting notions of equivalence in the chapter, we are concerned specifically here with the sense of equivalence as the representational mode's value or purchase on reality.

> Thus, we learn that the equation $(x-a)^2+(y-b)^2=r^2$ governs the circle. It is the circle. It is not a particular circle, it is any circle and all circles. It is nothing that is not a circle. It is the circle free of space and time limits. It is the universal, existing in perfection, in freedom from space and time. Mathematics is dull ditchwater until one reaches analytics. But in analytics we come upon a new way of dealing with form. It is in this way that art handles life.[78]

The equation he instances specifies a circle on the Cartesian plane in the following way:

By drawing a triangle with the line joining the centre point (a,b) and any point on the circumference (x,y) as its hypotenuse, we can use the theorem of Pythagoras to calculate the radius: r^2 = base^2 + height^2. The base is simply the distance between the x-coordinates (x-a), and the height, the difference in y-coordinates (y-b), yielding the stock formula Pound uses,

$$(x-a)^2+(y-b)^2=r^2. \tag{1}$$

Given the circle's centre (a,b) in Figure 5.1 is the origin (0,0), we would have the equation

$$x^2+y^2=r^2. \tag{2}$$

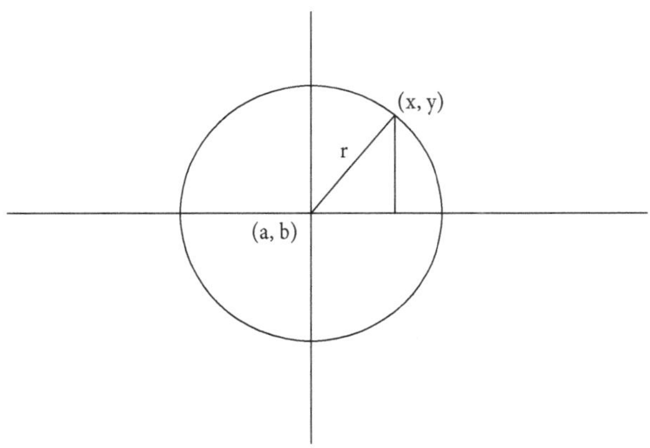

Figure 5.1 Circle on Cartesian plane with centre (a,b).
Source: Image produced by author.

[78] Pound, 'Vorticism', p. 468.

Given Pound's insistence that "by the 'image' I mean such an equation", it is essential to the analogy that equations can be rewritten in multiple ways; the above, for instance, as

$$x^2/r^2 + y^2/r^2 = 1. \tag{3}$$

Assuming Pound's analogy to hold, we can suppose that "petals on a wet, black bough" describes "the apparition of these faces in the crowd" of a metro just as the equation of a circle describes any particular circle on a Cartesian plane. The substantives in the poetic line will then serve as values of the 'constants' of the equation: as 'parameters' of the lived world, perhaps, which lend the particular 'affective shape' of the object in question. The 'variables' might then be such 'experiences' of these parameters as readers have. For competent readers, such substantives as 'petal' and 'bough', it must be insisted, will be conditioned by poetic memory: imagination will be shaped by the lyric corpus, one has to suppose, for those readers to all take in the poem in some definite way, despite considerable variation in personal experience. And lastly, the prepositions might serve as analogues, however poorly, of the operators '+', '−', and '='.[79]

Now say a thesaurus was used to concoct another haiku: 'The apparition of those faces in the depot; flowers on a wet, black branch'. If 'poetry [...] gives us equations [...] for human emotion', as Pound claims in *The Spirit of Romance*, such manipulations should certainly be possible:[80] that is, we should be able to see the two haikus as variants of some schema or template, if not as precisely as (2) and (3) are of (1). But while our facsimile has the same sentential form, it obviously goes wrong: 'flowers' and 'faces' (as opposed to 'petals' and 'faces'), for instance, do not form the instantaneous "intellectual and emotional complex" that Pound demanded of a successful image.[81]

There are some lessons in the agon to be drawn from the failure of this analogy besides Pound's ineptitude at theorizing. We argued in the first chapter that two world-pictures cannot be reduced to a common picture because they are functions of distinct sets of modal operations. In Pound's haiku, we

[79] Verbs will serve better as operators in a poem that has them. See Ann Hoff's discussion of Gertrude Stein's "Patriarchal Poetry": "Verbs tell what action is to be performed (like the actions +, −, ∗, and /), and prepositions describe the relationship between nouns (like the relations >, <, =, and ≠)" ('The Stein Differential: Gertrude Stein's mathematical aesthetic', *Mosaic: An Interdisciplinary Critical Journal*, 43.4 (2010), 1–17, p. 14).

[80] Ezra Pound, *The Spirit of Romance: An attempt to define somewhat the charm of the pre-renaissance literature of Latin Europe* (London: P. Owen [1910] 1970), p. 5.

[81] Ezra Pound, *Literary Essays of Ezra Pound*, intro. T.S. Eliot (London: Faber & Faber, 1954), p. 4.

find that words within a poem work on one another, apply perlocutionary forces on each other, in ways entirely incompatible with how mathematical objects interact in an equation. Words arrive in a poem with unique semantic orbits that often react with their rhythmic and syntactic neighbours in singular ways. The word 'apparition' transmits to the 'faces' a stark and remote character which 'petals' vividly sharpen in a way that 'flowers', or any other analogue, simply cannot. The isolate and diaphanous beauty of wet petals, along with the internally rhymed pairing of the conspicuously unpoetic and poetic words 'crowd-bough', work to *set the distance*, both physical and emotional, of the imagining eye faced with the apparition. If we momentarily label the mode of these operations, however inadequately, as 'intensification', it should become clear why arithmetic cumulations, multiplications, or combinations of discrete objects in an equation would be an altogether alien operational mode. The other point of non-correspondence between the modes is noted in passing by Michael Roberts: "for art is not like analytical geometry: there are accepted rules by which you can demonstrate the validity of your structure".[82] It may be that, as Yeats says, a real poem "comes right with a click",[83] but rightness of words in a poem is not predetermined by rules, as correctness is in mathematics.

We began in the introduction by suggesting that the modernist revolt against the mathematization of reality can be seen as a legacy of Blake's resistance to the Enlightenment: of his insistence that abstract proportions mean nothing apart from the particulars that display them. We have seen how Pound's attempt to, loosely put, 'mathematize' poetic meaning is undermined by the unique particularity of words acting in their poetic field. But beyond the inadequacies of the analogy, we find that Pound's descriptions of the poetic mode run against the grain of much of modernism. Among the copious hostilities against Pound's formula, we will focus on two poetic descendants of Blake, Auden and Yeats.

Auden declared at "A Symposium on Art and Morals at Smith College" that "certainly, art has no room for the algebraical conception of 'any'".[84] Auden resisted throughout his American years the rising dependence on science by the governing powers, both state and markets, where human beings were

[82] Michael Roberts, 'Review of a Draft of XXX Cantos by Ezra Pound', in *Adelphi* (1933), found in Michael Roberts, *Michael Roberts: Selected Poems and Prose*, ed. Frederick Grubb (Manchester: Carcanet Press, 1980), p. 85.

[83] W.B. Yeats, *Letters on Poetry from W.B. Yeats to Dorothy Wellesley* (London: Oxford University Press, 1940), p. 24.

[84] W.H. Auden, 'The Word and the Machine' (1954), in Edward Mendelson, *The Complete Works of W. H. Auden, Volume III Prose: 1949–1955* (Princeton: Princeton University Press, 2008), p. 664.

treated as replaceable numbers, crowds of "n>1",[85] to be plugged into the provisionally contrived equations of social science. It is no surprise that he rejects Pound's attempt to smuggle the 'algebraical conception of any' into artistic meaning. In "The Double Man", Auden rebukes those who would make of "works of art [...] An algebraic formula, / An abstract model of event".[86] Evoking the 'minute' and irreducible 'particulars' of Blake's *Jerusalem*, the poet says the 'particulars' of a poem "are those / That each particular artist knows, / Unique events that once took place / Within unique time and space". Although "The unique serves to typify", as algebraic equations do, the source is "still particular".[87] It is 'the apparition of *these* faces' that have occasioned the particularly successful words that are employed in typifying them.

Similarly, Yeats, in *On the Boiler*, renounces the hold that the Cartesian paradigm had on Western imagination.

> The mathematician Poincaré, according to Henry Adams, described space as the creation of our ancestors, meaning, I conclude, that mind split itself into mind and space. Space was to antiquity mind's inseparable 'other', coincident with objects, the table not the place it occupies. During the seventeenth century, it was separated from mind and objects alike, and thought of as a nothing yet a reality, the place not the table, with material objects separated from taste, smell, sound, from all the mathematicians could not measure, for its sole inhabitants, and this new matter and space, men were told, had preceded mind and would live after...yet... Berkeley...established forever the subjectivity of space. No educated man today accepts the objective matter and space of popular science, and yet deductions made by those who believed in both dominate the world, and make possible...revolutionary massacre...by substituting for the old humanity with its unique irreplaceable individual something that can be chopped and measured like a piece of cheese.[88]

He says that things in the world, to the healthy imagination of a robust people, would assert their unique modes of existence; like Olson, Yeats argues that the imposition of abstract, absolute space, the Cartesian plane, as the foundation upon which existence is anchored is an artifice of Enlightenment thinking that has served only to rob things of their identity: a circumstance

[85] W.H. Auden, *The Dyers Hand and Other Essays* (New York: Random House, 1962), p. 63.
[86] W.H. Auden, *The Double Man* (New York: Random House, 1941), p. 18. [87] Ibid.
[88] W.B. Yeats, *The Major Works*, ed. Edward Larrissy (Oxford: Oxford University Press, 2001), p. 399.

that poetry was to resist, as, for instance, in how the elements of the scene at the metro in Pound's poem jut out to the beholding reader in startling ways by enlisting the sensuous memory of detached, wet petals. Yeats extends his critique of the Cartesian artifice by tracing some of its unintended ethical and political consequences. When it comes the turn of human beings to become (x,y)'s in space or numerals in the statistics of social science, they become 'equal', not in the Enlightenment sense of having unique value and level dignity but by becoming the same: copies scuttling across the infinite plane, the "Égalité" of revolutionary zeal thus inadvertently factorizing man to an expendable. This is the mode of generalization that artistic modernism, in painting, poetry, and even prose, sought to defy by its strange transfigurations of the commonplace.

5.3 Yeats and the Modernist Project

We began this book by stating that the modernist movement can be seen as a successor to Blake insofar as it sought to unite Energy with Reason. In recovering the rhetorical styles of the poets' agon in the chapters since, we have seen that their antagonism was marshalled with a view to bringing balance between, or creating a lively tension among, the movements of mind invoked in poetry and mathematics—a dialectic that they thought would generate a truer picture of reality. We have even seen in this chapter some poets and philosophers of the period attempting to tie poetry and mathematics together with the discourse of aesthetics. We now turn to a poet that attempted in his late poetry to construct an altogether novel doctrine of aesthetics around the mathematical principle in art.

Among all the modernists, Yeats is the most conspicuous heir to Blake. Like Blake, art for Yeats was both a bringing forth and a bringing into order, and truth, we might say, emerged from the tension between proportions and particulars, pattern and energy. Like Blake, Yeats rejected the governing tenet of the mathematical world-picture, the reverence for abstraction, whether of space, time, or man. With Blake, his poetry and prose seem to declare, "I must create a system or be enslav'd by another man's."[89] The strange theory of aesthetics that emerges from Yeats's later, more doctrinaire poems, is the

[89] From 'Jerusalem', in William Blake, *The Complete Writings of William Blake: With Variant Readings*, ed. Geoffrey Keynes (London: Oxford University Press, 1966), p. 382.

last instance we will explore in this book of how the titular agon becomes manifest in modernist poetry.

Little remains to be said of "The Statues", a poem that for decades has been showered with exegesis.[90] Its interpreters, however, have not noted in the art criticism of the 1920s the prevalence of the idea that during the sixth century BC, the esoteric creed about numbers both secreted and propagated by the Pythagoreans augured a revolution in Greek sculpture; one sees it repeated across popular works of the time such as Hulme's *Speculations*, Rhys Carpenter's *The Esthetic Basis of Greek Art of the Fifth and Fourth Centuries* B.C, and even in works of history such as Cyril Robinson's *History of Greece*[91]—these texts, or the spreading influence of their ideas, likely confirmed to Yeats the fundamental aesthetic relation between art and mathematics.[92]

> Pythagoras planned it. Why did the people stare?
> His numbers, though they moved or seemed to move
> In marble or in bronze, lacked character
> But boys and girls, pale from the imagined love
> Of solitary beds, knew what they were,
> That passion could bring character enough,
> And pressed at midnight in some public place
> Live lips upon a plummet-measured face.
>
> No! Greater than Pythagoras, for the men
> That with a mallet or a chisel modelled these
> Calculations that look but casual flesh, put down

[90] See, for instance, A. Norman Jeffares, *A Commentary on the Collected Poems of W.B. Yeats* (London: Macmillan, 1975), p. 490; Hazard Adams, 'Yeatsian Art and Mathematic Form', *The Centennial Review of Arts & Science*, 4.1 (1960), 70–88; Kathleen Raine, 'Blake, Yeats and Pythagoras', in *Homage to Pythagoras: Rediscovering Sacred Science*, ed. *Christopher Bamford* (Hudson: Lindisfarne, 1994); Daniel Albright, *The Myth against Myth: A Study of Yeats's Imagination in Old Age* (Oxford: Oxford University Press, 1972), p. 129; and James Olney, *The Rhizome and the Flower: The Perennial Philosophy, Yeats and Jung* (Berkeley: University of California Press, 1980), p. 63.

[91] T.E. Hulme, *Speculations: Essays on Humanism and the Philosophy of Art*, ed. Herbert Read (London: Routledge & Kegan Paul, 1924), pp. 10–12; Rhys Carpenter, *The Esthetic Basis of the Greek Art of the Fifth and Fourth Centuries B. C.* (Bryn Mawr: Longmans, 1921); Robinson, *Greece*, p. 280.

[92] Yeats says in a letter dated 22 May 1933 to Olivia Shakespear, "My two sensations at the moment are Hulme's *Speculations* and *Lady Chatterley's Lover*" (W.B. Yeats, *The Collected Letters of W.B. Yeats, Unpublished Letters (1905–1939)*, ed. John S. Kelly (Charlottesville: InteLex Corporation, 2002), online edition)—"The Statues" is published in 1940, so Hulme had digested in Yeats for a few years by then. There is no evidence that Yeats read Carpenter, although he was often reviewed in the *North American Review* in the 1910s and 1920s (K.G.W. Cross and R.T. Dunlop, *A Bibliography of Yeats Criticism 1887–1965* (London: Macmillan, 1971), when Carpenter was regularly publishing his poems there ('Bibliography of Rhys Carpenter', *Hesperia: The Journal of the American School of Classical Studies at Athens*, 38.2 (1969), 123–32, p. 123).

All Asiatic vague immensities,
And not the banks of oars that swam upon
The many-headed foam at Salamis.
Europe put off that foam when Phidias
Gave women dreams and dreams their looking-glass.[93]

The realism of the statues overcomes the geometry that encases them, as in Ovid's Pygmalion, whose "art hides art", tempting him to kiss his own sculpture.[94] Yeats seems to have been fascinated with the story that the Greek discovery of geometric proof set off the invention of humanism in art.[95] The events clash remarkably well: according to the vague dates with which we are forced to make do in the study of ancient history, Phidias was born five years before the death of Pythagoras. Yeats, who studied Pythagorean philosophy in Burnet's *Early Greek Philosophy*, looked closely for its imprints in the plastic arts of Greece.[96]

In Doric art, as fidelity to natural form is first established as an artistic principle, realism and idealism come to mean the same thing. The sculptor wished to imbue his creation with the reality beneath the flux of subjective perception and to do so, he had to discover the eternal laws of mathematics to in turn exalt the ideal form of man. Carpenter expresses this seeming paradox as follows:

> Since all the elements of his work were ultimately man-devised and man-perfected, what possible guarantee of their objective fitness could there be, or what sanction for their claims to be the best type? That is a question which the Greek must have asked himself. He found the answer just where we to-day might wish to find it in science, though of course it was the science of his day and generation and consisted mainly of geometric theory.[97]

As with Empson's "Letter V", tension between the realism of flesh and the idealism of form is the driving energy of Yeats's poem. The stark numbers, though animated in marble, lacked 'character', the messy traits of an individual.[98]

[93] W.B. Yeats, *The Collected Poems of W.B. Yeats* (London: St. Martin's Press, 1958).
[94] Ovid, *Metamorphosis*, trans. A.S. Kline (Ann Arbor: Ann Arbor Editions, 2000), Book X, 249.
[95] Albright, *Myth*, p. 129.
[96] Olney, *Rhizome*, p. 63.
[97] Carpenter, *Esthetic Basis*, p. 120.
[98] Albright, *Myth*, p. 127.

In *On the Boiler*, Yeats says: "There are moments when I am certain that art must once again accept those Greek proportions which carry into plastic art the Pythagorean numbers, whose faces which are divine because all there is empty and measured."[99] Unlike Mackay's or Moore's, there is in Yeats's analogy between mathematical and artistic perfection a more ambivalent attitude conveyed through the alliterative and assonantal groupings of opposites—'passion, press' and 'plummet-measured', 'calculations' and 'casual flesh'. Carpenter says,

> In the statue of Herakles the right angles in which the lines and masses meet do not in themselves represent anything, any more than a geometric theorem represents actual objects (however much actual objects may exemplify and embody geometric theorems). [...] They are an abstract schema into which representational matter may be fitted, as the kneeling man is fitted into the abstract pattern of lines.[...] any one may experiment on himself to see whether he derives any emotion from contemplating such a pattern.[100]

Given the perennially powerful tendency in art to abstraction, Yeats balances warily the role of geometry in his greater system of aesthetics. "Measurement began our might: / Forms a stark Egyptian thought, / Forms that gentler Phidias wrought".[101] The rhymed coupling of the soft 'thought' and the harsh 'wrought' expresses the harmony in resisting opposites to which Blake's aphorism adverts: detached Reason requires the effort of bodily Energy to create gentle beauty. Yeats presents his aesthetics as a kind of cautionary fable: mathematics set off the greatest revolution in the history of art but mathematization, when overwrought, begins to sap the vitality it engendered. The most extreme version of this fable is the poem "Byzantium", where the mosaics "of the dancing floor / break bitter furies of complexity": all the complexity of mire and blood is washed clean in the ethereal shapes of divine geometry.[102] The geometric shapes that in Byzantine floor mosaic represented figures both human—as rectangle or line—and divine—as circle—marked to Yeats the terrible telos of the Doric elevation of geometry in art.[103] From geometry facilitating humanistic

[99] Yeats, *Works*, p. 451. [100] Carpenter, *Esthetic Basis*, p. 31.

[101] Yeats, *Collected Poems*, "Under Ben Bulben". [102] Ibid., "Byzantium".

[103] All critics have read the bird in the poem as one of the figures on the ceiling mosaic and offered an interpretation with that picture in mind (see summary of criticism in Jeffares, *Commentary*, p. 296). Dunbabin (*Mosaics*, pp. 30, 254) says that the sectile of the floor of Byzantine churches consisted of allegories told through birds and geometric forms, which agrees more closely with the birds of the poem's mosaic. Also, Otto Demus (*Byzantine Mosaic Decoration: Aspects of Monumental Art in Byzantium* (Boston: Boston Book & Art Shop, 1955), p. 24) says the floor of Byzantine churches were referred to as 'pavements', securing further our interpretation of the "Emperor's pavement".

proportions in art, the human itself—and its complexities—becomes trimmed and trapped in geometric shapes. Byzantine mosaic became in the later Yeats's doctrine an art for old men and ghosts, not those "upstanding men / that climb the streams"[104]—just as Plato and Pythagoras became figures that represented all ghostly intellectualism that withdraws man from the joy of living: for, as Socrates himself, even through his scorn, admits to Alcibiades, "the mind begins to grow critical when the bodily eye fails".[105]

This age-bound dichotomy in aesthetics extended also to music. In "Among School Children", "World-famous golden-thighed Pythagoras / Fingered upon a fiddle-stick or strings / What a star sang and careless Muses heard" [45–6]. The poet implies that Pythagoras, fabled creator of the diatonic scale, cared only for the barren notes of music, playing not for man's pleasure but a god's intrigue. Deriding his gig as mere fingering on strings, the poet relates in a sardonic tone a private concert put on by and for the heavens. Barren notes, like mosaic, are not vital art: true music, like the Greek statues, are proportions enfleshed. When Yeats uses the word 'music', in the phrase "O body swayed to music" [61], or in "Sailing to Byzantium", "caught in that sensual music" [7], he means rhythm charged with sensuous emotion.

The mathematical principle in the later Yeats's doctrine of art—which we have summed as 'proportion enfleshed'—can thus be seen as an emblem of the modernist project of resistance through reconciliation. As with Blake, it was for Yeats only amid the agon between Reason and Energy—calculation and casual flesh, harmony and music—that Nature disrobed at man's pleasure. In the final letter he ever wrote, Yeats declared, "man cannot know the truth but can only embody it".[106]

5.4 Conclusion

A poem like the "The Statues" is a typical example of how poetry became in the modernist period a medium to stage the agon between Energy and Reason. By broadening our focus in this chapter from the contest of the faculties to the clash of general principles, beauty and accuracy, say, we have caught a glimpse of the intricate interplays in the agon. Propaganda about

[104] "The Tower", in Yeats, *Collected Poems*, p. 189; the idea that geometry took over the arts of a people discomforted by existence was put forth first by Worringer, whose ideas animated Hulme's meditations on the subject (Worringer, *Abstraction*, pp. 19–20).

[105] Plato, *Symposium*, trans. Benjamin Jowett (New York: Modern Library [1871] 1996), p. 106.

[106] Richard Ellmann, *Yeats: The Man and the Masks* (New York: Macmillan, 1948), p. 285.

fidelity to reality, which is the central engine of any agon, was in this period carried out by appeal to the grasping powers of the human instrument. We have discussed in this book how players in an agon attempt to posit a pure given beneath modes of representation and show their own equipment as most capable of uncovering and agreeing with that underlying reality; but also, how across a period of discursive empire, through prolonged association of a particular mode with the posited reality, the mode itself comes to assert primacy. We said that the scientific world-picture succeeded by elevating mathematizability as a sign of an object's correspondence with the real. The agon in the early twentieth century is robust because this privileged status was being attacked from various sides; and at least intellectually, it came near to being dethroned. World-pictures were, in philosophy and poetry, shown to be merely artefacts of the instruments of measurement. The mathematical world-picture was slight because the product of superficial instruments; whereas the poetic world-picture was dense because issues from measurements taken by being in its entirety: its Energy and Reason. Man was thus called to be measure not only of all things but all of things. And the contest of the academic faculties grew into a contest of the human faculties.

This promotion of man as measurer of reality meant that an instinct for beauty was trusted wherever it bid the poet, to the Greek statues or the *Elements* of Euclid. Our discussion in this chapter of aesthetics as a principle wherein pure mathematics and poetry partook is a reminder of the strange discursive strategies and partnerships that had become necessary in the twentieth century under the prolonged imperium of the mathematical world-picture. The opposing principles of factuality and aesthetics, utility and uselessness, however, should not be mistaken as replacing the players in the agon. Principles do not compete among themselves. But neither do faculties compete apart from the individuals that people them. We must thus not view the agon between poetry and mathematics from one principal unit but rather explore the competition from its various dimensions: its practitioners, principles, methods, and instruments—just as the study of the Greek agon proceeds on the basis of principles—the Apollonian and Dionysian, say—playwrights, genres, and suchlike. We have, in this book, chronicled the poetic challenge to the mathematical world-picture by considering individual poets, taking them as representing one side of the agon. But it is hopefully clear from this chapter that this is as much a narrative choice as a focus on faculties or principles would have been. In the end, the illocutionary act of the poem is the only sure sign of an attitude, hostile or friendly, that is intended. The critical mode we have used is, therefore, among other reasons, close reading.

Coda

In that mordant and lapidary prose with which he hailed the plain speech of empirical science, Bacon introduces to the renaissance mind "the true method of experience":

> [P]er lumen iter demonstrat, incipiendo ab experientia ordinata et digesta, et minime praepostera aut erratica, atque ex ea educendo axiomata, atque ex axiomatibus constitutis rursus experimenta nova; quum nec verbum divinum in rerum massam absque ordine operatum sit. Itaque desinant homines mirari si spatium scientiarum non confectum sit, cum a via omnino aberraverint; relicta prorsus et deserta experientia, aut in ipsa (tanquam in labyrintho) se intricando, et circumcursando; cum rite institutus ordo per experientiae sylvas ad aperta axiomatum tramite constanti ducat.
>
> [It lights] the way; commencing as it does with experience duly ordered and digested, not bungling or erratic, and from it educing axioms, and from established axioms again new experiments; even as it was not without order and method that the divine word operated on the created mass. Let men therefore cease to wonder that the course of science is not yet wholly run, seeing that they have gone altogether astray, either leaving and abandoning experience entirely, or losing their way in it and wandering round and round as in a labyrinth. Whereas a method rightly ordered leads by an unbroken route through the woods of experience to the open ground of axioms.[1]

Nature would at last be comprehended by distilling life; by crumbling away the residue, the false idols, of experience. Subdued to method, channelled through instruments, experience thus condensed would escape its millennial consignment in the *labyrintho*, the running in circles, and cut a path, at last, to Alethia. More than a century on, at the eve of the French Revolution, Blake would say: "If it were not for the Poetic or Prophetic character the Philosophic & Experimental would soon be at the ratio of all things, & stand still unable

[1] Francis Bacon, *Novum Organum*, ed. Joseph Devey (New York: P.F. Collier & Son, 1902), LXXXII.

Agon: Poetry's Challenge to the Mathematization of Reality (1920s–1960s). Anirudh Sridhar, Oxford University Press.
 DOI: 10.1093/oso/9780198985457.003.0006

to do other than repeat the same dull round over again."[2] The agon had received its first reply.

Reason, aliened from body, had imprisoned phenomena in the 'ratio of all things': a new labyrinth. Blake's unusual use of 'ratio' combines the Latin and English (pre-and-post-scientific) senses, so that within the term, the 'reason' of all things becomes the 'quantitative relation' between all things. This verbal equation aptly describes Newton's definitions, considered largely, in the *Principia*:

DEFINITION I.

The quantity of matter is the measure of the same, arising from its density and bulk conjunctly.

DEFINITION II.

The quantity of motion is the measure of the same, arising from the velocity and quantity of matter conjunctly.[3]

And so he goes direly on, building ratio upon ratio, remaking phenomena brick by brick until the reason of all things is erected on the debris of quaint experience. We will recall the elementary formulae from school, where, in keeping with the first definition, we learn that Density = Mass/Volume. Andrew Motte's 1729 translation quoted above has not yet found the neutral term 'volume' and has borrowed 'bulk' from the common tongue: which, being too obtusely suggestive a word to pare of all ambiguities and render 'scientific'—an operation successfully carried out with density, velocity, and many others—is replaced after a brief career in science by 'volume'. Newton admits in his Scholium that the exercise of the *Principia* has been one of re-definition as much as anything else: "Hitherto I have laid down the definitions of such words as are less known, and explained the sense in which I would have them to be understood in the following discourse."[4] The Englishing of the *Principia*—the wresting of words from the common tongue, and so, from daily experience, to render phenomena in ratios for the privileged purview of unbodied soul—can thus be seen as a momentous assault in the agon. "Only I must observe" says the Englished *Principia*, "that the vulgar conceive those quantities under no other notions but from the relation they

[2] William Blake, *There Is No Natural Religion* (Cambridge: Harvard College Library Department of Graphic Arts [1787] 1948), p. 1.

[3] Isaac Newton, *Newton's Principia, The Mathematical Principles of Natural Philosophy* (New York: Daniel Adee [1666] 1846), p. 74.

[4] Ibid., p. 77.

bear to sensible objects. And thence arise certain prejudices, for the removing of which, it will be convenient to distinguish them into absolute and relative, true and apparent, mathematical and common."[5] The polished mirrors of Absolute Space and Absolute Time would thus excel the drag of will; and reason would mathematize vulgar experience to truth.

We will recall from the introduction to this book that Blake attempts to re-engage "Reason or the ratio of all we have already known" with will.[6] Finding the pattern and proportion in things was but Reason's role in the process of Body: a body not in passive receipt of observation but actively contending with Nature, winning through its Energy the perceptions which Reason, its soldier, would sift and order for further momentum of Life. To Blake, Bacon, and Newton, two of the three arch-fiends of his unholy trinity, had bound the 'bungling and erratic' Energy of body, leaving Reason to "the same dull round", which, "even of a Universe, would soon become a Mill with complicated wheels."[7] Like the circular syllogisms of Aristotelian science or so many concentric circles of Ptolemy blinkering our vision of the Heavens, and all other circular dogmas from which Bacon had sought to liberate the European mind, the mechanical world-picture too had finally mitigated to a labyrinth. Nature would be understood now not by enriching experience but advancing instruments that would extend the scope of the empirical purview: but this would result only in new pointer-readings being re-ratioed into more eclectic regularities.

For Blake, it is poetry that would liberate Albion from the dull round: a desire illuminated in the nitric acid that cleansed all surface but the words of his poems.

> First the notion that man has a body distinct from his soul is to be expunged: this I shall do by printing in the infernal method by corrosives, which in Hell are salutary and medicinal, melting apparent surfaces away and displaying the infinite which was hid.[8]

The words of the poem would emerge as terrible as the body of Urizon from the waters after the battle with fire.[9] If Newton's 'truth' was 'mathematical' and the 'apparent' was 'common', in Blake's trial by fire, it is the surface that proves

[5] Ibid., p. 77. [6] Blake, *Natural Religion*, p. 1. [7] Ibid., p. 1.

[8] William Blake, *The Marriage of Heaven and Hell*, ed. Geoffrey Keynes (Oxford: Oxford University Press, 1985), prov. 15.

[9] See William Blake, *The Book of the Urizon* (London: Thames and Hudson [1794] 1979), lines 58–66.

apparent, melting away to leave but words behind, poetic words of the common tongue that score the strivings of body in and against world. With the poem recovering a "sensual enjoyment" of the real, "the doors of perception", grown dusty behind the aperture of measuring instruments, "were cleansed": and "everything would appear to man as it is".[10] The poem thus became for Blake the outward sign that body, if not the measure of all things, is the best measure of things as they are, and the only possible escape, if not salvation, from the dull round of mathesis.

When we are narrating modernist poetry's resistance to the mathematization of reality, it is thus tempting to see Blake—perhaps as tenuously as Greenberg saw Kant—as the first modernist. We have seen Laura Riding reconceive the dull round as mathematical purgatory: "Numbers [...] lead only to further numbers", it is "free from any care beyond self-care", it "suspends meaning";[11] and Empson's "Apple of knowledge and forgetful mere" that "from Tantalus / Too differential bend".[12] A central theme of Empson's poems, we will recall, is that the fool who acts prevails over the ratiocinations of the alienated mind: the "Two mirrors with Infinity to dine / Drink him under the table".[13] The self-conscious mind of Empson's that stands at the diving board charting the mathematical coordinates before him begets mere reflection upon reflection, an infinite regress, a perpetual round: all of which comes to naught in the splash and eddy of activity. It is striking that such un-Blakean poets as Riding and Empson should express so consistent a note of antagonism toward the mathematical world-picture.

Blake's comparison of body and apparatus as tools for the measurement of reality is a subject that is renewed in the modernist period. In 1913, Poincaré would set the stage. The bodies that set up and read the scales

> also play the role, so to speak, of measuring instruments. These instruments, which are less precise than those of the scientist, are sufficient for everyday life, and it is with these that the child, like primitive man, has measured space or, to be more correct, has constructed a space which fulfills the needs of his daily life. Our body is our first measuring instrument.[14]

But Olson would say that the body is our first and best measuring instrument: its primitive sensation of space through depth—proprioception—contends

[10] Blake, *Natural Religion*, p. 1.
[11] Laura Riding (Jackson), *Though Gently* (Deya: Seizin, 1930), p. 3.
[12] Empson, Complete Poems, ed. Haffenden, p. 13. [13] Ibid., p. 17.
[14] Henri Poincaré, *Mathematics and Science: Last Essays*, trans. J.W.B. (New York: Dover [1913] 1963), p. 17.

with the dynamism of world in a manner that the geometric intuition of space—and its corresponding subdivision into units—only approaches by a feeble, retrospective approximation. Olson would go on to structure his metaphysics on the belief that the poem registers the body's measurement of an event in the world. The idea that the measurement of the real is registered by the breathing body and transmitted through the sounding poem is a development of Blakean doctrine in an unexpected direction. William Carlos Williams, similarly, conceptualizes the poem as a "field of action" designed to discover "a new way of measuring that will be commensurate with the social, economic world in which we are living".[15] We might in this vein also recall a theme that runs through Roberts's poem: "No analysis / can find familiar footing. Here // All reason ends; ends summer; Time / In eternal winter, watches, thrall / To faltering flesh".[16] Once reason surrenders to lived time, and time thus to flesh, as with Blake, mind is pressed into ordering sensation for further activity: "But mind in action is machine / Quarrying stern and stubborn fact; / In rhythmic pulse and discipline / It shapes the firm and final act".[17] Through the symbols of Plato and Pythagoras, lifeless, intellectual "old clothes upon old sticks to scare a bird",[18] Yeats's later poems dramatize Blake's judgement that "man", in the thrall of his measuring instruments, "has closed himself up, till he sees all things through narrow chinks of his cavern."[19] Yeats's 'poet and sculptor' took the truer measure because enfleshed it: "Measurement began our might" from the "Forms that gentler Phidias wrought".[20]

The battle over the measurement of reality between body and instrument, and the consequent one between poetry and mathematics was a central vein of the agon. Another was the contestation of the particular and the general. The binary is, however, more involved than its usual treatment would suggest. Wittgenstein remarks in his *Notebooks* that "the work of art is the object seen *sub specie aeternitatis* (under the aspect of eternity)".[21] For Wittgenstein, the work of art idealizes the object under study; but that idealization limns the unique particularity of the object into relief. But give Galileo the right conditions and he would make a feather and a boulder appear the same. We

[15] William C. Williams, *Selected Essays of William Carlos Williams* (New York: Random House [1948] 1954), p. 283.

[16] From 'Mona Lisa' (Michael Roberts, *These Our Matins* (London: E. Matthews & Marrot, 1930)), p. 13.

[17] Michael Roberts, *Collected Poems* (London: Faber & Faber, 1958), p. 47.

[18] W.B. Yeats, *The Collected Poems of W.B. Yeats* (London: St. Martin's Press, 1958), 'Among School Children'.

[19] Blake, *Natural Religion*, p. 1.

[20] Yeats, *Collected Poems*, 'Under Ben Bulben'.

[21] Ludwig Wittgenstein, *Notebooks 1914–1916*, trans. G.E.M. Anscombe (Oxford: Blackwell, 1961), [83e].

have seen Descartes performing a mathematical trick on an unsuspecting stick of wax, retrenching it of all properties until only its quantities remained. Once this process extends beyond physical objects, it leads to what Yeats described as the "substituting for the old humanity with its unique irreplaceable individual something that can be chopped and measured like a piece of cheese";[22] or Empson's humanity as a dotted line, a plain series, bulking in quantity only. The relationship between the general and the particular is thus not a mere opposition but can be described as a complex tension, even enduring mystery: for, as Whitehead wonders, "how can concrete fact exhibit entities abstract from itself and yet participated in by its own nature?"[23] In the modernist period, I believe the drama between the general and the particular is dramatized best in the ontological adventure of Empson's boy in "Letter V", caught in the poet's geometric idealizations, whose discovered face breaks free at last from the cold, imprisoning marble of mathematical purity, when the poet realizes that the particular before him "has at least the Form that has been shown".[24]

Whether or not modernist art discloses the object under the aspect of eternity, in the canvas and in verse (Wittgenstein might insist on Schoenberg too), it seems to bring the quotidian particular into strange relief. While this was laid out as a programme of Romanticism, "whereby ordinary things should be presented to the mind in an unusual aspect",[25] it is modernism that, through its many distortive ingenuities, restores to the "minute particulars", the "jewels of" Blake's "Albion",[26] their barbaric splendour. Blake says to his 'infernal trinity':

> You accumulate Particulars, & murder by analyzing, that you
> May take the aggregate; & you call the aggregate Moral Law:
> And you call that Swelld & bloated Form; a Minute Particular.
> But General Forms have their vitality in Particulars: & every
> Particular is a Man; a Divine Member[27]

[22] W.B. Yeats, *The Major Works*, ed. Edward Larrissy (Oxford: Oxford University Press, 2001), p. 399.

[23] A.N. Whitehead, *Process and Reality: An Essay in Cosmology* (Cambridge: Cambridge University Press, 1929), p. 4.

[24] Empson, *Poems*, p. 60.

[25] William Wordsworth and Samuel Taylor Coleridge, *Lyrical Ballads 1798 and 1802* (Oxford: Oxford University Press), p. 96.

[26] William Blake, 'Jerusalem', in *The Complete Writings of William Blake: With Variant Readings*, ed. Geoffrey Keynes (London: Oxford University Press, 1966), p. 428.

[27] Blake, *Complete Writings*, p. 429.

Here was first laid out Empson's vision of the ratio of things in "Doctrinal Point", the ring of pointer-readings fast enclosing man as absolute law, moral laws to govern the now false duality of choice. Against the forces of generalization, W.H. Auden, in "New Year Letter", has "Self-educated William Blake / Who threw his spectre in the lake, / Br[eak] off relations in a curse / With the Newtonian Universe".[28] Art would reject the mathematical world-picture, "an algebraic formula, / an abstract model of events / Derived from dead experience". Poetry would resist mathesis not by shrinking into an extreme nominalism, but by dramatizing the agon between the particular and general. The "particulars" of a poem would be "those / That each particular artist knows, / Unique events that once took place / Within unique time and space." And though "The unique serves to typify", it is "still particular".[29]

The art of Auden's "New Year Letter" would replace the uniformity that comes from thinking "in terms of Space and Time"[30] with experience "within unique time and space". We will note the reordering that grants time primacy. This is an artefact in some ways of a strain in fin de siècle philosophy that builds in the modernist period—from Bergson to Whitehead—towards the concept of an 'event' as the primary stuff of existence, doing away with science's view of space, absolute or otherwise, as a stable backdrop for the adventures of matter, in favour of time, as the medium through which reality unfolds. Privileging periods of time as more significant to the fact of reality as held out to experience than regions of space revealed, in fine, the poverty in the scientific account: of time as a repeated succession of uniform instants. To the reality posited by this blossoming world-picture, art that celebrated unique particulars seemed to correspond more intimately than the featureless backdrop of mathematical science. We will recall Roberts saying "poetry is a *sensuous concrete* language that forces the reader to feel the thing described as if it were actually present [...] It is a kind of algebra in which the x's and y's are not changed back into physical things till the end of the process":[31] the poem idealizes only to vivify the particular. To thus consider 'events' as the basic stuff of reality rather than mathematical time allows for a clear disambiguation of artistic idealization and mathematical generalization. ⊥ for Riding becomes a singular event like Auden's symbol of the Virgin, which stood for "analogical relations and singular events, describable only in terms of speech".[32] The poetic word recorded for Riding and Auden only those truths

[28] W.H. Auden, *New Year Letter* (London: Faber & Faber, 1941), p. 34.
[29] Auden, *New Year*, p. 20. [30] Auden, *New Year*, p. 33.
[31] Michael Roberts, *T.E. Hulme* (Manchester: Carcanet Press [1938] 1982), p. 66.
[32] W.H. Auden, *The Dyer's Hand and Other Essays* (New York: Random House, 1962), p. 137.

that breach the fabric of spacetime, the mathematical, T+, procession of uniform instants that positivists might call history, or what Auden calls the Dynamo, the "identical relations and recurrent events, describable, not in words but in terms of numbers, or rather, in algebraic terms".[33] In distinguishing the poetic from the mathematical account of things, poets were even led to speculate on parts of speech that might be more capable of expressing newly rescued events. Roberts proscribes to budding poets "the collective noun", which "has become an abstraction and its field has changed in the process. [P]oets, set out to define the abstraction, forgetting that it may not be a single positive quality at all".[34] The suspicion of the generalizing tendency of nouns led to a commandment to poets that Olson paraphrased from the Vedas: "There may be no more names than there are objects."[35]

Unlike the grand renunciations of the Romantic Rebellion,[36] however, modernist poetry, as we have seen, did not repudiate the Baconian virtues such as accuracy and the subservience of ego to nature. Their agon was pursued not from the ivory tower or pastoral estate, but through what Hill called a spiritual wrestling match,[37] using mathematical objects and theories repeatedly as metaphors or images to then toss and turn the latter's world-picture within the strictures of poetic form. Modernism's deep, and often learned, engagements with mathematics in verse—to an unprecedented degree in the history of poetry—was a sign of deep interest: but also, as we have argued, of adaptation. They were writing in a world where mathematical science had—to paraphrase Gibbon on Christianity in falling Rome—insinuated itself into the minds of men and erected the triumphant banner of number on the ruins of Albion. The mathematical world-picture's polemical bids for representational authority in the writings of Bacon, Hobbes, Spratt, and Locke in the seventeenth century enjoyed their fruition in the nineteenth and early twentieth centuries, as the lure of mathematical perfection was felt by domains not traditionally considered scientific. The *Principia Mathematica* (1910) begins declaring that an object of the treatise is "to show that, with the aid of symbolism, deductive reasoning can be extended to regions of thought not

[33] Ibid., p. 137.

[34] Michael Roberts, *Critique of Poetry* (London: Jonathan Cape, 1934), p. 63.

[35] 'Tryian Business' (Charles Olson, *The Maximus Poems*. ed. George F. Butterick (Berkeley: University of California Press, 1983), p. 40).

[36] A phrase borrowed from Kenneth Clark's *The Romantic Rebellion: Romantic versus Classical Art* (London: Harper and Row, 1973).

[37] Geoffrey Hill, "The Violent Bear it Away" (Lecture delivered at Cambridge University on the occasion of the 400th anniversary of the publication of the King James Bible, 2011).

usually supposed amenable to mathematical treatment".[38] The confidence of this utterance comes on the back of repeated successes in the agon in the previous centuries. With the combined efforts of the physiocrats of France and the cameralists of Germany, a well ordered state was reconceived as "a machine in which all the wheels and gears are precisely adjusted to one another".[39] The process was continued in England where the collective representation of human labour, once called political economy, was dissevered from the political and recast as 'economics' or the *General Mathematical Theory of Political Economy* (1962)[40]—a work of the logician William Jevons; and soon after, Jevons's collaborator, Leon Walras, would in *Elements of Pure Economics* (1874), draw out for the subject its mathematical dogma in the form of the General Equilibrium Theory.[41] The rise of measurement in other social sciences began simultaneously but took longer to dislodge their speculative origins. L.A.J. Quetelet developed from the 1820s to the '40s statistical techniques to discover the demographically 'average person'.[42] His application of statistics to the social world would prove a powerful tool of American statecraft in the nineteenth and early twentieth centuries,[43] in, say, the Department of Commerce's assemblage of data about the 'average American',[44] with which statistical outliers, 'special classes' can be identified and monitored.[45] By the 1920s, mathematization had conquered the social and began entering the last frontier of the mind. Through the works of L.L. Thurstone in the 1920s to Stanley Stevens in the 1940s, psychology[46]—mirroring political economy before it—had given way to statistical psychometry, which further ministered to the efficiency of statecraft and commerce.[47] The literary world was not

[38] A.N. Whitehead and Bertrand Russell, *Principia Mathematica, vol. 1* (Cambridge: Cambridge University Press, 1910), p. 2.

[39] J.H.G. von Justi (1761), quote found in Michael Oakeshott, *Lectures in the History of Political Thought*, eds Terry Nardin and Luke O'Sullivan (Exeter: Imprint Academic, 2006), p. 537.

[40] Irving Fisher casts Jevons's book as the start of the mathematical method in economics in *Mathematical Investigations in the Theory of Value and Prices* (Harvard College Library, 1892), Appendix III, 'The Utility and History of Mathematical Method in Economics', p. 109.

[41] Donald Gillies, 'Can Mathematics be Used Successfully in economics?', in *A Guide to What's Wrong with Economics*, ed. E. Fullbrook (London: Anthem Press, 2005).

[42] Adolphe Quetelet and Robert Knox, *A Treatise on Man and the Development of His Faculties* (Edinburgh, 1842).

[43] Kevin Donnelly, *Adolphe Quetelet, Social Physics and the Average Men of Science, 1796–1874* (Pittsburgh: University of Pittsburgh Press, 2015), p. 7.

[44] Kevin O'Keefe, *The Average American: The Extraordinary Search for the Nation's Most Ordinary Citizen* (New York: Public Affairs, 2005), p. 9.

[45] Helen Bowers, *From Lighthouses to Laserbeams: A History of the U.S. Department of Commerce, 1913–1988* (US Dept. of Commerce, Office of the Secretary, 1988), p. 81.

[46] Joel Michell, *Measurement in Psychology: A Critical History of a Methodological Concept* (Cambridge: Cambridge University Press, 1999), p. 15.

[47] Andrew J. Vinchur, *The Early Years of Industrial and Organizational Psychology* (Cambridge: Cambridge University Press, 2018), p. 73.

immune to the allure of the mathematical virtues, as we have seen in the previous chapter with Futurism and Vorticism. In England, such an attitude is betrayed in the value theory of I.A. Richards. Michael Roberts identifies and attempts to pre-empt this strain of criticism:

> Arithmetic and elementary mathematics come to be regarded as exceptionally reliable reasoning and therefore an attempt is made to force all reasoning into that mould. Mr. I.A. Richards, for example, speaks of the 'number' of impulses as if they could be counted like potatoes or 'events'. The pure mathematician, familiar with non-cumulative algebras and 'q-numbers' which cannot be expressed in terms of a finite set of numbers at all, is not liable to fall into any such error.[48]

All were thus clamouring to have their work painted 'mathematical' because, by the twentieth century, as Ernest Geller states, "when one determines whether or not something is 'scientific', one is *ipso facto* deciding whether or not it has a certain legitimate claim on our attention, and perhaps even on our credence".[49] This was the world that faced the modernists as they were taking up arms.

The resistance would thus have to be subversive, if the reader will pardon the pun. We can speculate on these strategies by observing commonalities in the mathematical notions and objects that recur in the modernist poems we have studied. Let us take matrices, quaternions, and tensors. They are all abstractions, mathematical tools, that allow complexities in the world to be elegantly expressed; for irregularities to be ironed out. These ingenious mathematical creations serve in the fabrication of the dazzling theories of modern physics. But the cultural awe in the potency of mathematics that their works inspire, particularly when their legacy extends to other, more social domains, it is feared, leads to an ominous simplification of things, "The duality of choice becomes the singularity of existence."[50] Unlike the materialism or empiricism that threatened Blake's imagination, it was what Husserl called the 'mathematisation of reality', "the cognitive process that allows us to turn reality into a 'mathematical manifold'";[51] the running of "empirical objects, events, or processes through ideal mathematical contents

[48] Michael Roberts, 'The Loneliness of Mathematics', *Adelphi*, 1.6 (1931), p. 511.

[49] Ernest Gellner, 'The Scientific Status of the Social Sciences', *International Social Science Journal*, 36.4 (1984), 567–86, p. 570.

[50] Empson, *Poems*, p. 59.

[51] Edmund Husserl, *The Crisis of European Sciences and Transcendental Phenomenology* (Evanston: Northwestern University Press, 1970), p. 23.

and to let these idealities become the prescriptive standard for what becomes actually present to us in the realm of simple life-world experience",[52] that disquieted modernist poets. Husserl's student, Alexandre Koyré notes, "[o]ne must not forget that observation and experience, in the sense of brute, common-sense experience, did not play a major role—or if it did, it was a negative one, the role of obstacle—in the foundation of modern science".[53] The modernist agon, similarly, did not much consider Bacon its antagonist because it did not see his philosophy as having truly shaped the course of the mathematical world-picture.

One might say that the replacement of Blake's materialism with his 'ratio of all things', 'mathematization', as poetry's arch antagonist, is in some ways the chief contribution of the modernist salvo in the long agon. The move allows for modernism to assimilate but extend the romantic opposition of the anima of particulars against the inertia of generalities. With mathematization as antagonist, modernist poetry attempted to directly contest for representational authority over an alleged world beneath world-pictures—leaving behind the romantic move against materialism of conjuring the universe as an infinite creative potential of the ego. The hypostatic world could now be cast as, say, more complex and elusive in ways that are better captured by poetry. Whitehead says,

> We forget how strained and paradoxical is the view of nature which modern science imposes on our thoughts. Wordsworth, to the height of genius, expresses the concrete facts of our apprehension, facts which are distorted in the scientific analysis.[54]

Whitehead saw a lyric poem as displaying faithfully the concrete facts of experience by recreating the organic patterns of an event in nature—as opposed to science which, to paraphrase Whitehead's fallacy, misplaces concreteness in the abstract.[55] Reality did not present itself to perception as discrete and independent parts but as a live complex of interrelations. Modernist poetry, in resisting mathematization, thus attempted to exhume the complex and wilful bedlam at the core of creation that was unwilling to be smoothed over by the refinements of measurement; the irregular genius of matter that

[52] Philipp Berghofer, Philip Goyal, and Harald A. Wiltsche, 'Husserl, the Mathematization of Nature, and the Informational Reconstruction of Quantum Theory', *Continental Philosophy Review*, 54 (2021), 413–36, p. 417.

[53] Alexandre Koyré, 'Galileo and Plato', *Journal of the History of Ideas*, 4.4 (1943), 400–28, p. 402.

[54] A.N. Whitehead, *Science and the Modern World* (New York: The Free Press [1925] 1967), p. 84.

[55] Ibid., p. 58.

threw off the slick sheen of mathematical description; and show the "fetid nest"[56] of life confounding what Olson called "the exact death quantity does offer";[57] the mysterious doings of "water", for instance, which was to Lawrence "H20, hydrogen two parts, oxygen one, but there is also a third thing, that makes it water and nobody knows what that is";[58] our whole "Celestial Sphere", which Empson beheld as "an acid green canvas hollow", whose "core be brown with maggots' raven".[59] J.W.N. Sullivan, in a book titled *Limitations of Science*, would say, in a manner typical of the modernist period, that

> Scientific method began by quite consciously and deliberately selecting and abstracting from the total elements of our experience. From the total wealth of impressions received from nature these men fastened upon some only as being suitable for scientific formulation. These were those elements that possess *quantitative* aspects. Between these elements mathematical relations exist.[60]

The poetic world-picture would thus capture the 'total wealth of impressions' that the mathematical, causal, scheme did not, and the modernists argued, could not, accommodate.

The causal scheme, the lingering assumption of mathematical determinism, was yet another aspect of the mathematical picture that would be contested by the modernists. Poincaré provides a sketch, tending towards caricature, of this dogma:

> We should then say that the body of laws is equivalent to a system of differential equations which link the speed of variations of the different elements of the universe to the present values of these elements. Such a system involves, as we know, an infinite number of solutions, But if we take the initial values of all the elements, that is, their values at the instant t=0, (which would correspond in ordinary language to the 'present'), the solution is completely determined, so that we can calculate the values of all the elements

[56] Charles Olson, *Collected Poems—Excluding the Maximus Poems* (Berkeley: University of California Press [1987] 1997), p. 90.

[57] 'Equal to the Real' (Charles Olson, *Collected Prose*, eds Donald Allen and Benjamin Friedlander (Berkeley: University of California Press, 1997), p. 121).

[58] From "The Third Thing" in D.H. Lawrence, *Pansies, Poems* (Fredonia: Fredonia Books [1929] 2002).

[59] From 'Value Is in Activity' in Empson, *Poems*, p. 11.

[60] J.W.N. Sullivan, *Limitations of Science* (London: Chatto & Windus, 1933), p. 197.

> at any period whatever, whether we suppose t>0, which corresponds to the 'future,' or whether we suppose t<0, which corresponds to the 'past'.[61]

Although the Newtonian world-picture was in their time coming to a close, the modernists perceived in their culture the lingering obstinacy of the premises laid out here by Poincare, even a strengthening of them, with the increasing practical successes of science.[62] Roberts laments that "[a]mong non-scientists, the commonest attitude of our time is a [...] belief that 'matter' is an ultimate reality, a neutral substance whose motions 'cause' all thought and feelings".[63] With the spreading faith in mathematical determinism, the modernists envisioned a future where human beings are wiped clean of their "complexities of mire and blood",[64] as "The effort of virtue" dissolves into "the unconsciousness of foreknowledge".[65] Where it sought to dispel determinism, modernism was free to borrow from parts of the mathematical world-picture that supported the antithesis. In classical mechanics, a precondition to predicting future states is the full knowledge of initial conditions.[66] But once the real numbers became the language of the mathematical world-picture, 'full knowledge of initial conditions' could be seen as unknowable. This might have irked the classical mechanist when working with continuous quantities, but only once the degree of physics' dependence on the formalization of the numbers was recognized in the nineteenth century would the potential impossibility of knowing the past and future be fully impressed.[67] We thus see Olson celebrating the advent of continuous mathematics, whose example, for him, would lead art, and culture with it, to escape the "rigidities of the discrete".[68] Or as Weyl notes, "A real thing can never be given adequately, its 'inner horizon' is unfolded by an infinitely continued process of ever new and more exact experiences."[69] The modernists saw the separation of cause from effect itself as an artifice of the mathematical world-picture, at odds with the world as held out to experience. Auden writes, "The intellect / That parts

[61] Henri Poincaré, *Mathematics and Science: Last Essays* trans. J.W.B. (New York: Dover [1913] 1963), p. 2.

[62] See Part 1 of Marc Lange, *Because without a Cause: Non-causal Explanations in Science and Mathematics* (Oxford: Oxford University Press, 2017).

[63] Michael Roberts, 'Science and Human Temperament', *The Adelphi*, 10.6 (1935), p. 381.

[64] Yeats, *Collected Poems*, 'Byzantium'.

[65] Empson, *Poems*, p. 59.

[66] Leonard Susskind, *Classical Mechanics: The Theoretical Minimum* (London: Penguin, 2013), p. 5.

[67] Lange, *Because*, ch. 3; Leonard Susskind, *Quantum Mechanics: The Theoretical Minimum* (London: Basic Books, 2014), p. xii.

[68] 'Equal to the Real' (Olson, *Prose*, p. 122).

[69] Hermann Weyl, *Philosophy of Mathematics and Natural Science*, transl. Olaf Helmer (Princeton: Princeton University Press, 1949), p. 41.

Cause from Effect / […] Commits a legalistic crime. / For such an unreal severance / Must falsify experience."[70] Roberts says critics must take their cue from quantum uncertainty, which in some ways brings to a definite close the Newtonian, mechanical world-picture, and use its example to dispel cultural faith in determinism. Given it is "impossible even in principle to determine the position of an ultimate particle without disturbing its velocity by an indeterminate amount, for every observation is an activity", we must "abandon 'causes' altogether."[71] He brings into his poems a rather tried romantic sentiment to celebrate the cultural liberation from determinism: "we have impulse uncontrolled / And joy not marked on causal map".[72]

Given these ends for which mathematics appears in modernist poetry, how, we might ask, would the ghost of William Blake judge when asked if modernist poetry was successful in marrying Heaven and Hell? An interesting question. Certainly, the veil of mathematical metaphor through which the verse we have read glimmered would argue the poetry for a successful marriage. But adjudicating in the context of the agon, we must consider the question more extensively: whether the modes of poetry have in the modernist period shown capable of bringing man in more intimate union with world: whether the world-picture of poetry has demonstrated itself in closer identity with a supposed world-an-sich. A strange standard of aesthetic judgement: but adjudicating so would not be to contravene Empson's dictum that poems must be judged in accordance with how they are trying to be good, with what they are trying to do. For instance, Riding's prosaic logico-philosophical poems are structured to cleanse poetic words from reference to objects in the world. Her sentences are precisely not pictures of facts. Yet the "⊥ is that which is", it "is the unmistakable".[73] Like Donne's globe-maker who takes a blank round ball and makes "that, which was nothing, All",[74] Riding's poems bring the poetic word in identity with nothing, only to make nothing (no-thing) all that there really is: for "There *is* a no-sense" [emphasis mine].[75] Nothing becomes the realm of essences, while the phenomenal world is a flux of impermanent properties. Riding's grapheme of the up tack ⊥ thus, through many re-definings and un-definings, attempts finally to impose a sense of union between word and reality. Jealously guarding the self-sufficiency of meaning, as we have also seen with Zukovsky's theory of poetry that removes words

[70] Auden, *New Year*, p. 33. [71] Roberts, 'Temperament', p. 381.
[72] Michael Roberts, *Collected Poems* (London: Faber and Faber, 1958), p. 47.
[73] Riding, *Though Gently*, p. 2.
[74] Herbert Grierson, *Metaphysical Lyrics and Poems of the Seventeenth Century* (Oxford: Oxford University Press, 1921), p. 11.
[75] Riding, *Though Gently*, p. 1.

from the corruptions of speech into the realm of pure sound, was one almost paradoxical way of defining a hypostatic world in a manner that gave poetry unique powers of access and identity. Olson redraws the compass of reality from the sufficiency of word-and-mind to that of word-and-body: "hell now / is not exterior, is not to be got out of, is / the coat of your own self".[76] And he fashions his projective verse to "turn this total thing, invert", "raise the thickets of / this wilderness" to once again overrun and overwhelm the arid plains of Absolute Space; and to turn the mathematical picture of his body, "this congerie of particles", to the "bush" of world running into and comingling with his "nerves laid open".[77] Here, again, is the Blakean attempt to confound the separation between mind and body; and word and world. Ransom says,

> We have elected to know the world through our science, and we know a great deal, but science is only the cognitive department of our animal life, and by it we know the world only as a scheme of abstract conveniences. What we cannot know constitutionally as scientists is the world which is made of whole and indefeasible objects, and this is the world which poetry recovers for us.[78]

Poetry would recover *The World's Body* not by "idealizing" but by attempting "to realise the world", for "poetry is an event in time".[79] The modernist attempt to close the distance between word and world challenged the mathematical world-picture at its foundations. We will recall Descartes pronouncing that we have knowledge only of those objects which can be clearly conceived, which participate in pure mathematics. To this, we may pose the last testament of Yeats's doctrine: that "man cannot know the truth but can only embody it".[80] The attempt to render the poetic word ingredient in the realization of reality, the poem cast as event, inviting the reader to embody world, can thus be seen as a reply to the early moves of the antagonist.

Although a deep challenge to the mathematical world-picture, this strain of modernist theory also forfeits the agon for purely representational authority: for it presents poetry as not representational art in any sense, even that of non-representational *likeness*. We might compare this species of modernist poem to the "image before the era of art", as Hans Belting phrases it, or the "the holy image", which was valued in itself apart from what it represented.[81]

[76] Olson, *Poems*, p. 158. [77] From 'In Cold Hell, in Thicket', in Olson, *Poems*, p. 156.

[78] John Crowe Ransom, *The World's Body* (New York: Charles Scribner's Sons, 1938), pp. x–xi.

[79] Ibid., p. viii.

[80] From Richard Ellmann, *Yeats: The Man and the Masks* (New York: Macmillan, 1948), p. 285.

[81] Hans Belting, *Likeness and Presence: The Image before the Era of Art*, trans. Edmund Jephcott (Chicago: University of Chicago Press, 1996), p. xxi.

The gold of the mosaic, for instance, or the stone of the idol possessed a certain *dynamis* that excited a symbolic, even ritualistic, exchange of power, which was eventually lost to the purely representational or conceptual value of renaissance painting.[82] Paint itself would lose its innate value in the fifteenth century, says Michael Baxandall. We can, for instance, trace this phenomenon from early-fifteenth-century ledgers that reserved the highest proportion of the commissioned artwork's budget for the gold and silver of the depicted armour or the ultramarine of the sky, concocted from exotic tinctures like lapis lazuli that had to be imported from the Levant.[83] The ratio of these budgets gradually changed through the century until the "skill of the artist" commanded the highest portion.[84] Paint in Western art would not again enjoy protean powers independently of the painting's fabula until, perhaps, it began to clump, drip, and assert itself along the vast canvases of Pollock: for, as Arthur Danto observes, "[t]he drips affirmed that paint has an expressive life of its own, that it is not a passive paste to be moved where the artist wants it to be moved but possesses a fluid energy over which the painter endeavors to exercise control".[85] Not unlike twentieth century painting, poetry too sought to invest its medium—diction, syntax, and rhythm—with a life of its own. Donald Davie, for instance, describes the workings of the poetic word as the "transference of power" through syntax,[86] akin to "the transferences of force from agent to object, which constitute natural phenomena".[87] One might say that modernist poetry thus re-appropriates Bacon's and the seventeenth century's slurs against the idolatry of metaphorical language. To borrow Belting's dichotomy, the trading in of "likeness" for "presence" (*bild* for *kult*), the invoking through verse of "body swayed to music",[88] is a central move in the agon. It seems a more likely endeavour than Pound's attempt to cast poetry as "a sort of inspired mathematics, which gives us equations, not for abstract figures, triangles, squares, and the like, but for the human emotions".[89] More and more into the twentieth century it seemed unlikely that poetry could—try as it may to display the total wealth of

[82] Belting, *Likeness*, p. 6.

[83] Michael Baxandall, *Painting and Experience: In Fifteenth Century Italy* (Oxford: Oxford University Press [1972] 1988), p. 11.

[84] Baxandall, *Painting*, pp. 15–17.

[85] Arthur Danto, 'Pollock and the Drip', *The Nation* (7 January 1999).

[86] Donald Davie, *Articulate Energy: An Inquiry into the Syntax of English Poetry* (London: Routledge [1958] 1976), p. 36.

[87] Ernest Fenollosa, *The Chinese Written Character as a Medium for Poetry*. ed. Ezra Pound, 1919 (London: Stanley Nott, 1936), p. 44.

[88] Yeats, *Collected Poems*, 'Among School Children'.

[89] Ezra Pound, *The Spirit of Romance: An attempt to define somewhat the charm of the pre-renaissance literature of Latin Europe* (London: P. Owen [1910] 1970), p. 5.

impressions—usurp the purely representational authority of the mathematical world-picture when the latter's victory seemed increasingly beyond impeachment.

Modernist poetry thus repeatedly attempts to win identity with world, with *ousia*, by claiming to be "part of the res itself" rather than professing "about it";[90] or as Olson says, the poetic word is "the act of the instant", not the "the act of thought *about* the instant".[91] Such a strategy, in impressing on the reader the unknowability of world (for man can embody the truth but cannot *know* it); in setting truth beyond mere description; in scoffing at the divinatory pretensions of scientific description as "one tautology";[92] can also be employed to show that poetry says just what *can* be said. If "The god approached dissolves into the air",[93] it cannot be captured by mathematical reason. The god's unknowability is all that remains to be humanly demonstrated, which it indeed is by Empson in the paradox of his opening line, which we will note, only works in the form of a poem, with its careful proleptic and retrodictive development of 'magnolias', 'grace', and 'tensors'. The poet John Ciardi would find a terse expression for this growing strain of modernist poetics: " *The way it [the poem] means is what it means.*"[94] In his essay "How does a Poem Mean?", Ciardi compares the cartoon meaning-making procedure of Mr Gradgrind's classroom numbering of phenomena on the one hand, and poetry, which is "not a meaning but an act of existence", and its mode as "enact[ing] his [the poet's] experiences rather than to talk about having them."[95] Because the medium of poetry comes to be seen as physically *alive* both verbally, through prolepsis and retrodiction and syntactically, through its 'transference of power', not only its matter but its mode is implicated in the world. That is, mode is no longer a domain anterior to world-picture—as is implicit in the common phrase, the descriptive or explanatory power of a scientific theory—but is realized as world-picture.

In judging such dear and doomed ambitions, it is important to recall that nearly all the chapters in this book end contemplating the failures of the projects set out. Empson's poetry became too involved in promoting difficulty, in a language of conscious amatory alienation, to ever set off any grand shifts in the balance of Energy and Reason. The charges of contrived difficulty and intentional alienation can just as easily be levied on Riding. And for all Olson's

[90] Wallace Stevens, *Collected Poems* (London: Faber & Faber, 1984), p. 404.

[91] 'Human Universe' in Olson, *Prose*, p. 156. [92] Empson, *Poems*, p. 59. [93] Ibid.

[94] John Ciardi, *How Does a Poem Mean?* ed. by Miller Williams (Boston: Houghton Mifflin, 1975), p. 6.

[95] Ciardi, *How*, p. 8.

acrobatics with grammar, mind and body are not equal players in the apprehension of his poetry as one often feels is the case with Yeats. Olson's affectation of grammatical pathology to shift from imagism to proprioception in verse often seems contrived or clinical. In fact, Yeats seems to have been the last poet in the English language who could meditate seriously on poetry's condition in Wordsworth's language of men. Measuring the success of modernist poetry's agon, we may thus say that this book has chronicled a magnificent failure, for the wit the poets sought to restore to the tips of the tongue, the sensuous intelligence of renaissance verse (or Melville's prose) for which modernists yearned to the last never took possession of twentieth-century mind or culture. The incommodities of modernist verse, though in many ways a necessary part of its subversive resistance, can be judged against the commodity that statistics became for the social sciences, as the kingdom of number gradually consolidated its victory in the twentieth century.[96] Without following the mathematical reasoning behind a statistical theory or tool, a social scientist can employ it to satisfy a problem.[97] But if a poem's work is in how it means, its powers are effectuated and contained in the effort it entreats. When we observe the casuistries of 'causal' modelling in economic forecasting, the fitting of data into algorithms whose predictions either support pre-existing inclinations or become self-fulfilling prophecies upon shaping policy and investment,[98] and how the conscription of artificial 'intelligence'—a phrase that must surely rank with Humpty Dumpty's 'un-birthday'—into thought, communication and prediction looks set to void entirely the "effort of virtue",[99] we are, when we recall modernist poetry's proud, dying, breaths, able to appreciate more richly our advent into the unconscious land of foreknowledge. It is worth remembering therefore such mad ambitions as to insinuate the poetic word into the *res* in an age of the household refrigerator—even if it was, with the alchemist, at last conceded that "these vessels crumble", for "none can hold / Elixir vitae we have known / Or grasp the true alchemic gold / Or touch the substance of the stone".[100]

The marriage of Heaven and Hell was modernist poetry's doctrinal point that dissolved into the air.

[96] See Arnold Zellner, *Statistics, Econometric, and Forecasting* (Cambridge: Cambridge University Press, 2004), ch. 1.

[97] From an upcoming essay by Hans Mathews, 'Inference as Commodity', in *Phalanx*.

[98] The social sciences 'remake the ground on which they build', Hans Mathews, *The Mathesis of Intelligence*, *Phalanx*, 1 (2007), 1–21; Christopher A. Sims, 'Macroeconomics and Reality', *Econometrica*, 48.1 (1980), 1–48.

[99] See Jin-Kyu Jung, Manasa Patnam, and Anna Ter-Martirosyan, 'An Algorithmic Crystal Ball: Forecasts-based on Machine Learning', International Monetary Fund, WP/18/230, 2018.

[100] Roberts, *Matins*, p. 23.

Works Cited

Adams, Barbara, 'Laura Riding's Autobiographical Poetry: "My Muse as I"', *Concerning Poetry*, 15 (1982), 71–87.

Adams, Hazard, 'Yeatsian Art and Mathematic Form', *The Centennial Review of Arts & Science*, 4.1 (1960), 70–88.

Adorno, Theodor, 'Late Style in Beethoven', in *Essays on Music*, trans. Susan Gillespie, ed. Richard Leppert (Berkeley: University of California Press, 2002): 564–568.

Adorno, Theodor, and Horkheimer, Max, *Dialectic of Enlightenment*, trans. Edmund Jephcott (Stanford: Stanford University Press, 2003).

Aggarwal, R.S., *Mathematics for Class 11* (New Delhi: Bharati Bhawan [1990] 2020).

Albright, Daniel, *The Myth against Myth: A Study of Yeats's Imagination in Old Age* (Oxford: Oxford University Press, 1972).

Alexander, Amir R., *Duel at Dawn: Heroes, Martyrs, and the Rise of Modern Mathematics* (Cambridge: Harvard University Press, 2010).

Alexandre, Koyré, 'Galileo and Plato', *Journal of the History of Ideas*, 4.4 (1943), 400–28.

Alpers, Paul, 'Empson on Pastoral', *New Literary History*, 10.1 (1978), 101–23.

Aquinas, Thomas, *Summa Theologica*, vol. 1 (London: Blackfriars, 1969).

Arendt, Hannah, *Love and Saint Augustine*, eds Joanna Vecchiarelli Scott and Judith Chelius Stark (Chicago: University of Chicago Press [1929] 1996).

Aristotle, *Categories and De Interpretatione*, trans. and ed. J.L. Ackrill (Oxford: Clarendon Press, 1963).

Armstrong, Tim, 'Poetry and Science'. *A Companion to Twentieth-century Poetry*. ed. Neil Roberts (Oxford: Blackwell Publishing, 2001).

Armstrong, Tim, '"A Transfinite Syntax": Modernism and Mathematics', *Affirmations of the Modern*, 6.1 (2019), 1–29.

Asimov, Isaac, *Prelude to Foundation* (London: Grafton Books, 1988).

Atkins, Douglas G., *Eliot and the Essay: From The Sacred Wood to Four Quartets* (Waco: Baylor University Press, 2010).

Auden, W.H., *New Year Letter* (London: Faber & Faber, 1941).

Auden, W.H., *The Double Man*. New York: Random House, 1941.

Auden, W.H., *Nones: 1947–1950* (New York: Random House, 1951).

Auden, W.H., *The Dyer's Hand and Other Essays* (New York: Random House, 1962).

Audi, Robert, ed., *The Cambridge Dictionary of Philosophy* (Cambridge: Cambridge University Press, 2015).

Auerbach, Erich, *Mimesis: Representation of Reality in Western Thought*, trans. Willard B. Trask (Princeton: Princeton University Press [1953] 2003).

Augustine, *The Confessions*, trans. by E.B. Pusey (London: Aldine Press [1907] 1957).

Auster, Paul, *Collected Prose: Autobiographical Writings, True Stories, Critical Essays, Prefaces and Collaborations with Artists* (London: Faber & Faber, 2003).

Bacon, Francis, *Novum Organum*, ed. Joseph Devey (New York: P.F. Collier & Son, 1902).

Bacon, Francis, *The Essays of Francis Bacon* (New York: Charles Scribners' Sons, 1908).

Bahls, Patrick, 'The Mathematical Muse: Using Math to Construct Poetry', *Math Horizons*, 17.3 (2010), 14–17.

Bamford, Christopher, *Homage to Pythagoras: Rediscovering Sacred Science* (Hudson: Lindisfarne, 1994).

Baumgarten, Alexander G., *Aesthetica* (Hildesheim: G. Olms [1750] 1961).

Baxandall, Michael, *Painting and Experience: In Fifteenth Century Italy* (Oxford: Oxford University Press [1972] 1988).
Bellos, David, 'Mathematics, Poetry, Fiction: The Adventure of the Oulipo', *BSHM Bulletin: Journal of the British Society for the History of Mathematics*, 25.2 (2010), 104–18.
Belting, Hans, *Likeness and Presence: The Image before the Era of Art*, trans. Edmund Jephcott (Chicago: University of Chicago Press, 1996).
Benacerraf, Paul, and Putnam, Hilary, eds, *Philosophy of Mathematics: Selected Readings* (Cambridge: Cambridge University Press [1964] 1983).
Berghofer, Philipp, Goyal, Philip, and Wiltsche, Harald A., 'Husserl, the Mathematization of Nature, and the Informational Reconstruction of Quantum Theory', *Continental Philosophy* Review, 54 ((2021), 413–36.
Bergmans, Luc, and Koetsier, Teun, eds, *Mathematics and the Divine: A Historical Study* (Amsterdam: Elsevier, 2005).
Berkeley, Bishop, *Siris: A chain of philosophical reflexions and inquiries concerning the virtues of tar water, and divers other subjects connected together and arising from one another* (Dublin, 1744).
Berkman, Natalie, *Oulipo and the Mathematics of Literature*. Austria: Peter Lang, 2022.
Berlin, Isaiah, *Roots of Romanticism*, ed. Henry Hardy (Princeton: Princeton University Press, 1999).
Bernal, J.D., *The World, the Flesh and the Devil: An Inquiry into the Future of the Three Enemies of the Rational Soul* (London: Cape, 1970).
Bernhard, Carl Gustaf, Crawford, Elisabeth, and Sorbom., Per, eds, *Science, Technology and Society in the Time of Alfred Nobel, Nobel Symposium, 52* (Oxford: Pergamon, 1982).
Bevis, Mathew, ed., *Some Versions of Empson* (Oxford: Oxford University Press, 2007).
'Bibliography of Rhys Carpenter', *Hesperia: The Journal of the American School of Classical Studies at Athens*, 38.2 (1969), 123–32.
Billiteri, Carla, *Language and the Renewal of Society in Walt Whitman, Laura (Riding) Jackson, and Charles Olson: The American Cratylus* (New York: Palgrave Macmillan, 2009).
Blake, William, *The Book of the Urizon* (London: Thames and Hudson [1794] 1979).
Blake, William, *The Complete Writings of William Blake: With Variant Readings*, ed. Geoffrey Keynes (London: Oxford University Press, 1966).
Blake, William, *The Marriage of Heaven and Hell*, ed. Geoffrey Keynes (Oxford: Oxford University Press, 1985).
Blake, William, *There Is No Natural Religion* (Cambridge: Harvard College Library Department of Graphic Arts [1787] 1948).
Bloom, Harold, *The Anxiety of Influence: A Theory of Poetry* (New York: Oxford University Press, 1973).
Bostock, David, *Philosophy of Mathematics: An Introduction* (Oxford: Wiley-Blackwell, 2009).
Boswell, Marshall, and Burn, Stephen J., eds, *A Companion to David Foster Wallace Studies* (New York: Palgrave, 2013).
Bové, Paul A., ed., *Early Postmodernism: Foundational Essays* (Durham: Duke University Press, 1995).
Bowers, Helen, *From Lighthouses to Laserbeams: A History of the U.S. Department of Commerce, 1913–1988* (US Dept. of Commerce, Office of the Secretary, 1988).
Bradbrook, M.C., and Lloyd Thomas, M.G., *Andrew Marvell* (Cambridge: Cambridge University Press, 1940).
Bradley, F.H., *Appearance and Reality: A Metaphysical Essay* (London: Swan Sonnenschein and co. [1893] 1908).
Bradshaw, Penny, '"Living at Our Full Compass": Michael Roberts and The Poetry of Mountaineering', *The Alpine Journal*, 116 (2012), 229–37.
Bram, Shachar, *Charles Olson and Alfred North Whitehead: An Essay on Poetry*. trans. Batya Stein (Lewisburg: Bucknell University Press, 2004).
Bridgman, P.W., *The Nature of Thermodynamics* (Cambridge: Harvard University Press, 1941).

Brits, Baylee, *Literary Infinities: Number and Narrative in Modern Fiction* (London: Bloomsbury, 2018).
Bronowski, Jacob, Ariotti, Piero E., and Bronowsky, Rita, eds, *The Visionary Eye: Essays in the Arts, Literature, and Science* (Cambridge: MIT Press, 1978).
Brooks, Cleanth, *The Well-wrought Urn: Studies in the Structure of Poetry* (New York: Reynal & Hitchcock, 1947).
Brooks, Cleanth, 'The Formalist Critics', *The Kenyon Review*, 13.1 (1951), 72–81.
Buchanan, Scott, *Symbolic Distance in Relation to Analogy and Fiction* (London: K. Paul, Trench, Trubner & Co, 1932).
Buchwald, Jed Z., and Fox, Robert, eds, *The Oxford Handbook of the History of Physics* (Oxford: Oxford Academic, 2013).
Burns, Allan, *Thematic Guide to American Poetry* (Westport: Greenwood Press, 2002).
Burnyeat, Myles, *Explorations in Ancient and Modern Philosophy*, vol 3, eds C. Atack, M. Schofield, and D. Sedley (Cambridge: Cambridge University Press, 2022).
Burrow, Colin, ed., *Metaphysical Poetry* (London: Penguin, 2006).
Butterick, George F., *A Guide to The Maximus Poems of Charles Olson* (Berkeley: University of California Press, 1978).
Byers, Mark, *Charles Olson and American Modernism: The Practice of the Self* (Oxford: Oxford University Press, 2018).
Byrd, Don, 'The Possibility of Measure in Olson's Maximus', *Boundary*, 2 1/2 (1973/74), 39–54.
Calvin, John, *Institutes of the Christian Religion*, trans. by Henry Beveridge. 1536 (Grand Rapids: Christian Classics Ethereal Library, 2002).
Campbell, N.R., *Physics: The Elements* (Cambridge: Cambridge University Press, 1920).
Cantor, Georg, (1883), 'Ueber unendliche, lineare Punktmannichfaltigkeiten' *Math. Ann.*, 21, 545–91.
Carlyle, Thomas, *The Collected Works, vol 3*, ed., Henry Duff Traill (London: Chapman and Hall, 1864).
Carpenter, Rhys, *The Esthetic Basis of the Greek Art of the Fifth and Fourth Centuries B.C.* (Bryn Mawr: Longmans, 1921).
Casey, John, *A Sequel to the First Six Books of the Elements of Euclid Containing an Easy Introduction to Modern Geometry with Numerous Examples* (Dublin: Hodges, Figgis, & Co., 1886).
Cassirer, Ernst, *The Philosophy of Symbolic Forms* (New Haven: Yale University Press, 1955).
Cavell, Stanley, *Must We Mean What We Say: A Book of Essays* (Cambridge: Cambridge University Press, 1978).
Chadwick, Charles, *Symbolism* (London: Routledge [1971] 2018).
Chakravartty, Anjan, 'The Structuralist Conception of Objects', *Philosophy of Science*, 70 (2003), 867–78.
Chandrashekar, S., *Truth and Beauty: Aesthetics and Motivation in Science* (Chicago: University of Chicago Press, 1987).
Childs, Donald J., *The Birth of New Criticism: Conflict and Conciliation in the Early Work of William Empson, I.A. Richards, Laura Riding, and Robert Graves* (Montreal: McGill-Queen's University Press, 2013).
Childs, Peter, *Modernism* (London: Routledge, 2000).
Christensen, Paul, *Charles Olson: Call Him Ishmael*, fore. George F. Butterick. 1975 (Austin: University of Texas Press, 1979).
Ciardi, John, *How Does a Poem Mean?* ed. Miller Williams (Boston: Houghton Mifflin, 1975).
Clark, Kenneth, *The Romantic Rebellion: Romantic versus Classical Art* (London: Harper and Row, 1973).
Cleveland, John, *The Poems of John Cleveland*, eds. Brian Morris and Eleanor Withington (Oxford: Oxford University Press, 2013).
Coffa, Alberto, *The Semantic Tradition from Kant to Carnap: to the Vienna Station*, ed. Linda Wessels (Cambridge: Cambridge University Press, 1991).

Collingwood, R.G., 'Plato's Philosophy of Art', *Mind*, 34 (1925), 154–72.
Confucius, *Book of History* (rendered and compiled by W.G. Old, 1908).
Cooke, Roger, *The History of Mathematics: A Brief Course* (New Jersey: Wiley [1997] 2013).
Coomaraswamy, Ananda, *History of Indian and Indonesian Art* (Mineola: Dover, 1927).
Cooper, Madeleine J., 'Finding a Centre: The Poetry of Charles Olson' (Dissertation: University of Nottingham, 1977).
Cornford, F.M., *From Religion to Philosophy: A Study in the Origins of Western Speculation* (New York: Longmans, Green & Co., 1912).
Cornford, F.M., *Before and after Socrates* (ebook: Pickle Partners Publishing [1932] 2016).
Cross, K.G.W., and Dunlop, R.T., *A Bibliography of Yeats Criticism 1887–1965* (London: Macmillan, 1971).
Crossland, Rachel, *Modernist Physics: Waves, Particles, and Relativities in the Writings of Virginia Woolf and D.H. Lawrence* (Oxford: Oxford University Press, 2018).
Cunningham, Valentine, *British Writers of the Thirties* (Oxford: Oxford University Press, 1988).
Dalen, Dirk van, *L.E.J. Brouwer—Topologist, Intuitionist, Philosopher: How Mathematics Is Rooted in Life* (London: Springer, 2012).
D'Ambrosio, Matteo, 'A "Poetry of Numbers"? Khlebnikov, Jakobson, Marinetti, Schwitters, Kostelanetz', *Cherchez Le Text: Proceedings of the Elo 2013 Conference.*
Danto, Arthur, 'Pollock and the Drip', *The Nation* (7 January 1999).
Darwin, Charles, *On the Origin of Species by Means of Natural Selection or The Preservation of Favoured Races in the Struggle for Life* (London: John Murray, 1859).
Dauben, Joseph Warren, *George Cantor: His Mathematics and Philosophy of the Infinite* (Princeton: Princeton University Press, 1990).
Davids, T.W. Rhys, trans. from the Pâli, *Dialogues of the Buddha (The Dîgha-Nikâya)* (London: Oxford University Press, 1899).
Davie, Donald, *Articulate Energy: An Inquiry into the Syntax of English Poetry* (London: Routledge [1958] 1976).
Debnath, Lokenath, *The Legacy of Leonard Euler: A Tricentennial Tribute* (London: Imperial College Press, 2010).
Demus, Otto, *Byzantine Mosaic Decoration: Aspects of Monumental Art in Byzantium* (Boston: Boston Book & Art Shop, 1955).
Derrida, Jacques, *On Touching—Jean-Luc Nancy*, trans., Christine Irizzary, 2000 (Stanford: Stanford University Press, 2005).
Descartes, René, *Meditations on First Philosophy*, 1641, trans., John Veitch, 1901. https://dn790005.ca.archive.org/0/items/RMCG0002/Descartes-Meditations-a1.pdf (last accessed 19 June 2025).
Descartes, René, *Principles of Philosophy*, trans. Valentine Rodger Miller and Reese P. Miller (London: Kluwer Academic Publishers, 1982).
Descartes, René, *The Discourse on Method* trans. John Veitch (London: Everyman's Library [1637] 1912).
Dicken, Paul, *A Critical Introduction to Scientific Realism* (London: Bloomsbury, 2016).
Dirac, Paul, 'Application of Quaternions to Lorentz Transformations', *Proceedings of the Royal Irish Academy. Section A: Mathematical and Physical Sciences*, 50 (1944), 261–70.
Doggett, Frank A., 'Donne's Platonism', *The Sewanee Review*, 42.3 (1934), 274–92.
Donne, John, *The Works of John Donne: Vol III.* ed. Henry Alford (London: John W. Parker, 1839).
Donnelly, Kevin, *Adolphe Quetelet, Social Physics and the Average Men of Science, 1796–1874* (Pittsburgh: University of Pittsburgh Press, 2015).
Dreyfus, Tommy, and Eisenberg, Theodore, 'On the Aesthetics of Mathematical Thought, for the Learning of Mathematics', *EPDF*, 6.1 (1986), 2–10.
Driesch, Hans, *The History and Theory of Vitalism* (London: Macmillan & Co., 1914).
Dryden, John, *Virgil's Aeneid* (New York; P.F. Collier [1697] 1909).
Dunbabin, Katherine M., *Mosaics of the Greek and Roman World* (Cambridge: Cambridge University Press, 1999).

Duncan, Joseph, *The Revival of Metaphysical Poetry* (Minneapolis: University of Minnesota Press, 1959).
Duncan-Jones, Katherine, *Shakespeare's Sonnets* (London: Thomas Nelson and Sons Ltd, 1997).
Dunne, J.W., *An Experiment with Time* (London: A. and C. Black, 1927).
Dworkin, Ronald, *Law's Empire* (Cambridge: Harvard University Press, 1986).
Eagleton, Terry, *Against the Grain: Essays 1975–1985* (London: Verso, 1986).
Eames, Rachel Fountain, *Physics and the Modernist Avant-Garde: Quantum Modernisms and Modernist Relativities* (London: Bloomsbury, 2014).
Eason, T.W., and Hamilton, R., eds. *A Portrait of Michael Roberts* (Chelsea: College of St Mark & St John, 1949).
Eddington, Arthur S., *Space, Time and Gravitation: An Outline of the General Relativity Theory* (Cambridge: Cambridge University Press, 1920).
Eddington, Arthur S., *Stars and Atoms* (Oxford: Oxford University Press 1927).
Eddington, Arthur S., *The Expanding Universe* (London: Harmondsworth, 1940).
Eddington, Arthur S., *The Nature of the Physical World*, 1928 (Cambridge: Cambridge University Press, 1948).
Einstein, Albert, *Relativity*, R.W. Lawson, 1916 (London: Methuen & Co., 1920).
Eliot, T.S., 'Ben Jonson', *TLS* (1919), 637–38.
Eliot, T.S., 'The Metaphysical Poets', *TLS*, 669–70 (1921), p. 670.
Eliot, T.S., *Inventions of the March Hare: Poems by T.S Eliot 1909–1917*, ed. Christopher Ricks (London: Faber & Faber, 1996).
Ellmann, Richard, *Yeats: The Man and the Masks* (New York: Macmillan, 1948).
Empson, William, *Some Versions of Pastoral* (London: Chatto & Windus, 1935).
Empson, William, 'Three Ethics', *Spectator* (November 1935), 912.
Empson, William, 'Chronicles: A Doctrine of Aesthetics', *Hudson Review*, 2.1 (1949), 94–7.
Empson, William, *Collected Poems* (New York: Harcourt, 1949).
Empson, William, *Seven Types of Ambiguity* (London: Chatto and Windus [1930] 1949).
Empson, William, *Structure of Complex Words* (London: Chatto & Windus, 1951).
Empson, William, 'Mr Empson and the Fire Sermon', *Essays in Criticism*, 6.4 (1956), 481–2.
Empson, William, 'Everything, Beggars, Is on Fire', *Arrows* (1957), 5.
Empson, William, *Milton's God* (London: Chatto & Windus, 1961).
Empson, William, *Using Biography* (London: Chatto & Windus, 1984).
Empson, William, *Essays on Shakespeare* (Cambridge: Cambridge University Press, 1986).
Empson, William, *Argufying: Essays on Literature and Culture*, ed. John Haffenden (London: Chatto & Windus, 1987).
Empson, William, *The Book, Film & Theatre Reviews of William Empson: Originally Printed in the Cambridge Magazine Granta, 1927–1929* (Kent: Foundling Press, 1993).
Empson, William, *Essays on Renaissance Literature, vol. 1*, ed. John Haffenden (Cambridge: Cambridge University Press, 1994).
Empson, William, *The Complete Poems*, ed. John Haffenden (London: Penguin, 2000).
Empson, William, *Selected Letters of William Empson*, ed. John Haffenden (Oxford: Oxford University Press, 2006).
Empson, William, *The Face of the Buddha*, ed. Rupert Arrowsmith (Oxford: Oxford University Press, 2016).
Engelhardt, Nina, *Modernism, Fiction and Mathematics* (Edinburgh: Edinburgh University Press, 2018).
Epple, Moritz, and Mueller, Falk, eds., *Science as Cultural Practice: Modernism in the Sciences, ca. 1900–1940* (Berlin: Akademie Verlag, 2010).
Eyers, Tom, 'The Revenge of Form: Review of C. Levine's *Forms: Whole, Rhythm, Hierarchy, Network*', *Boundary*, 2, May 9. https://www.boundary2.org/2018/05/tom-eyers-the-revenge-of-form-review-of-c-levines-forms-whole-rhythm-hierarchy-network/.
Feferman, Solomon, 'Modernism in Mathematics', *American Scientist*, 97.5 (2009), 417–20.
Fenollosa, Ernest, *The Chinese Written Character as a Medium for Poetry*. ed. Ezra Pound. 1919 (London: Stanley Nott, 1936).

Fenstad, J.E., Frolov, I.T., and Hilpinen, R., eds, *Logic, Methodology and Philosophy of Science VIII, Proceedings of the Eighth International Congress of Logic, Methodology and Philosophy of Science, Moscow* (Amsterdam: North Holland, 1987).

Ferreirós, Josè, *Labyrinth of Thought: A History of Set Theory and Its Role in Modern Mathematics* (Basel: Springer Basel AG, 2013).

Feynman, Richard P., *The Feynman Lectures on Physics, vol. 2*, eds, Robert Leighton and Matthew Sands (Boston: Addison Wesley, 1971).

Fisher, Irving, *Mathematical Investigations in the Theory of Value and Prices* (Harvard College Library, 1892).

Fisher, Tom, 'Reading Renunciation: Laura Riding's Modernism and the End of Poetry', *Journal of Modern Literature*, 33.3 (2010), 1–19.

Fitzgerald, Robert, 'Review of *The Collected Poems of Laura Riding* by Laura Riding', *Kenyon Review*, 1 (1939), 34.

Flaubert, Gustave, *Salammbo: A Realistic Romance of Ancient Carthage*, trans. J.S. Chartres (London: Gibbings & Co., 1899).

Forrest-Thomson, Veronica, *Poetic Artifice: A Theory of Twentieth-Century Poetry* (New York: St. Martin's Press, 1978).

Forster, Michael N., and Gjesdal, Kristin, eds, *The Cambridge Companion to Hermeneutics* (Cambridge: Cambridge University Press, 2019).

Fraassen, Bastian van, *The Scientific Image* (Oxford: Clarendon Press, 1980).

Frege, Gottlob, 'Ueber Sinn und Bedeutung', *The Philosophical Review*, 57 (1948), 209.

Friedman, Alan J., and Donley, Carol C., *Einstein as Myth and Muse* (Cambridge: Cambridge University Press, 1985).

Friedman, Michael, *Kant's Construction of Nature* (Cambridge: Cambridge University Press, 2015).

Fry, Edward, *Cubism*, trans. Johnathan Griffin (London: Thames and Hudson Ltd, 1966).

Frye, Northrop, *Anatomy of Criticism* (Princeton: Princeton University Press, 1957).

Fullbrook, E., ed., *A Guide to What's Wrong with Economics* (London: Anthem Press 2005).

Galilei, Galileo, *Discoveries and Opinions of Galileo*, ed. Garden City (New York: Doubleday, 1957).

Gardner, Philip, and Gardner, Averil, *The God Approached: Commentary on the Poems of William Empson* (London: Chatto & Windus, 1987).

Gay, Peter, *Modernism: The Lure of Heresy from Baudelaire to Beckett and Beyond* (New York: Vintage Books, 2007).

Gellner, Ernest, 'The Scientific Status of the Social Sciences', *International Social Science Journal*, 36.4 (1984), 567–86.

Gill, Roma, ed., *William Empson: The Man and His Works* (London: Routledge, 1974).

Gillott, Brendan C., 'Charles Olson's "Projective Verse" and the Inscription of the Breath', *Humanities*, 7.4 (2018), 1–20.

Gleizes, Albert, and Metzinger, Jean, *Du Cubisme*, trans. R. Girard & Cie (Paris: La Compagnie Française des Arts Graphiques, 1947).

Glymour, Clark, 'Realism and the Nature of Theories', in *Introduction to the Philosophy of Science* (Indianapolis: Hackett Publishing [1992] 1999), p. 114.

Goddard, L., '"True" and "Provable"', *Mind*, 67. 265 (1958), 13–31.

Gödel, Kurt, 'Über Formal UnentscheidbareSätze der Principia Mathematica und VerwandterSysteme I', *MonatsheftefürMathematik und Physik*, 38 (1931), 173–98.

Goldsmith, Maurice, Mackay, Alan, and Woudhuysen, James, eds, *Einstein: The First Hundred Years* (Oxford: Pergamon Press [1980] 2013).

Goldstein, Catherine, Gray, Jeremy, and Ritter, Jim, eds, *L' Europe mathématique: Histoires, mythes, identités* (Paris: Editions de la Maison des Sciences de l'homme, 1996).

Gombrich, Ernst, *Art and Illusion: A Study in the Psychology of Pictorial Representation* (New York: Phaidon Press, 1960).

Goodman, Nelson, *The Structure of Appearance* (Cambridge: Harvard University Press, 1951).

Goodman, Nelson, *Ways of Worldmaking* (Indianapolis: Hackett Pub., 1978).

Goodman, Sue, *Beginning Topology* (Providence: American Mathematical Society, 2009).
Gower, Barry, 'Cassirer, Schlick and "Structural" Realism: The Philosophy of the Exact Sciences in the Background to Early Logical Empiricism', *British Journal for the History of Philosophy*, 8.1 (2000), 71–106.
Grattan-Guinness, Ivor, *The Search for Mathematical Roots, 1870–1940: Logics, Set Theories and the Foundations of Mathematics from Cantor through Russell to Gödel* (Princeton: Princeton University Press, 2000).
Graves, R.P., *Life of Sir William Rowan Hamilton, vol. II* (Dublin: Hodges, Figgin and Co., 1885).
Gray, Jeremy, 'Anxiety and Abstraction in Nineteenth-Century Mathematics', *Science in Context*, 17.1–2 (2004), 23–47.
Gray, Jeremy, *Worlds out of Nothing: A Course in the History of Geometry in the 19th Century* (London: Springer, 2007).
Gray, Jeremy, *Plato's Ghost: The Modernist Transformation of Mathematics* (Princeton: Princeton University Press, 2008).
Greenberg, Clement, *Art and Culture: Critical Essays* (Boston: Beacon Press, 1961).
Grierson, Herbert, *Metaphysical Lyrics and Poems of the Seventeenth Century* (Oxford: Oxford University Press, 1921).
Haffenden, John, *William Empson: Against the Christians* (Oxford: Oxford University Press, 2006).
Haffenden, John, *William Empson: Among the Mandarins* (Oxford: Oxford University Press, 2005).
Haldane, J.B.S., *Daedalus: or Science and the Future* (London: E.P. Dutton and Co., 1924).
Hallberg, Robert Von, *Charles Olson: The Scholar's Art* (Cambridge: Harvard University Press, 1978).
Handelman, Matthew, *The Mathematical Imagination: On the Origins and Promise of Critical Theory* (New York: Fordham University Press, 2019).
Harding, Jason, *The Criterion: Cultural Politics and Periodical Networks in Inter-war Britain* (Oxford: Oxford University Press, 2002).
Hardy, G.H., *A Mathematician's Apology* (Cambridge: Cambridge University Press, 1940).
Hay, Douglas, Linebaugh, Peter, Rule, John G., Thompson, E.P., and Winslow, Cal, *Albion's Fatal Tree: Crime and Society in Eighteenth-Century England* (New York: Pantheon House, 1975).
Hazlitt, William, *Selected Essays*, ed. George Sampson (Cambridge: Cambridge University Press, 1917).
Heidegger, Martin, *Off the Beaten Track*, trans. & eds, Julian Young and Kenneth Haynes (Cambridge: Cambridge University Press, 2002).
Heisenberg, Werner, *Physics and Philosophy: The Revolution in Modern Science* (London: Allen & Unwin, 1959).
Henderson, Andrea K., *Algebraic Art: Mathematical Formalism and Victorian Culture* (Oxford: Oxford University Press, 2018).
Henderson, Linda D., *The Fourth Dimension and Non-Euclidean Geometry in Modern Art* (Cambridge: MIT Press [1985] 2013).
Heraclitus, *The Fragments of the Work of Heraclitus of Ephesus on Nature*, trans., intro. by G.T.W. Patrick (Baltimore: N. Murray, 1889).
Herbert, George, *The English Poems of George Herbert*, ed. Jacula Prudentum (London: Rivingtons, 1871).
Herd, David, *Contemporary Olson* (Manchester: Manchester University Press, 2016).
Hesse, Hermann, *The Glass Bead Game* (London: Pan Books & Cape [1949] 1987).
Hesse, Mary, *Models and Analogies in Science* (Notre Dame: University of Notre Dame Press [1966] 1970).
Hickman, Miranda, *The Geometry of Modernism: The Vorticist Idiom in Lewis, Pound, H.D., and Yeats* (Austin: University of Texas Press, 2005).
Hilbert, David, and Bernays, Paul, 'The Foundations of Mathematics', trans. Ian Mueller, *Grundlagen der Mathematik*, 1 (1934), 3.
Hill, Geoffrey, 'The Dream of Reason', *Essays in Criticism*, 14 (1964), 91–101.

Hill, Geoffrey, *Tenebrae* (London: André Deutsch, 1978).
Hill, Geoffrey, "The Violent Bear it Away" (Lecture delivered at Cambridge University on the occasion of the 400th anniversary of the publication of the King James Bible, 2011).
Hobbes, Thomas, *Leviathan or The Matter Forme, and Power of a Commonwealth Ecclesiastical and Civil*, annot. G.W.F. Hegel (Amazon Digital Services, 2019).
Hoff, Ann K., 'The Stein Differential: Gertrude Stein's Mathematical Aesthetic', *Mosaic: An Interdisciplinary Critical Journal*, 43.4 (2010), 1–17.
Holmes, John, ed., *Science in Modern Poetry: New Directions* (Liverpool: Liverpool University Press, 2012).
Howarth, Peter, *Cambridge Introduction to Modernist Poetry* (Cambridge: Cambridge University Press, 2012).
Hulme, T.E., *Speculations: Essays on Humanism and the Philosophy of Art*, ed. Herbert Read (London: Routledge & Kegan Paul, 1924).
Hume, Alexander, *Of Gods Benefites Bestowed Vpon Man* (published privately, 1599).
Hume, David, *Four Dissertations* (London: A. Millar in the Strand, 1757).
Husserl, Edmund, *The Crisis of European Sciences and Transcendental Phenomenology* (Evanston: Northwestern University Press, 1970).
Hynes, Samuel, 'Michael Roberts' Tragic View', *Contemporary Literature*, 7.4 (1971), 437–50.
Ivanova, Milena, 'Poincaré's Aesthetics of Science', *Synthese*, 194.7 (2017), 2581–94.
James, Henry, *Portrait of a Lady* (Raleigh, Freebook Publisher [1881] 2020).
Jameson, Fredric, *The Prison-house of Language: A Critical Account of Structuralism and Russian Formalism* (Princeton: Princeton University Press, 1972).
Jarrett, Joseph, *Mathematics and Late Elizabethan Drama* (Cham: Palgrave, 2019).
Jeffares, A. Norman, *A Commentary on the Collected Poems of W.B. Yeats* (London: Macmillan, 1975).
Johnson, Peter, '"Presences of the Infinite": J.M. Coetzee and Mathematics' (PhD dissertation: Royal Holloway, University of London, 2013).
Johnson, Samuel, *A Dictionary of the English Language*, eds Beth Rapp Young, Jack Lynch, William Dorner, Amy Larner Giroux, Carmen Faye Mathes, and Abigail Moreshead (London: Printed by William Strahan, [1755, 1773] 2021).
Joly, Charles Jasper, *A Manual of Quaternions* (London: Macmillan Co., 1905).
Jones, Adam (Oxford University), interviewed by Anirudh Sridhar, 03/08/2019.
Jones, William, *The Works of Sir William Jones*, ed. by Lord Teignmouth, 1807 (Cambridge: Cambridge University Press, 2013).
Jongsma, Calvin, 'Mathematization and Modern Science', in *Mathematics in a Postmodern Age: A Christian perspective*, ed. Russell W. Howell and James Bradley (Grand Rapids: Wm. B. Eerdmans, 2001).
Jonik, Michael, *Herman Melville and the Politics of the Inhuman* (Cambridge: Cambridge University Press, 2018).
Jung, Jin-Kyu, Patnam, Manasa, and Ter-Martirosyan, Anna, 'An Algorithmic Crystal Ball: Forecasts-based on Machine Learning', International Monetary Fund, WP/18/230, 2018.
Kant, Immanuel, *Kant's Prolegomena and Metaphysical Foundations of Natural Science*, ed. Ernest Belfort Bax, 1786 (London: George Bell, 1883).
Kant, Immanuel, *Critique of Judgment*, trans. Werner S. Pluhar, 1790 (Cambridge: Hackett Publishing, 1987).
Kay, Lily, *Who Wrote the Book of Life* (Stanford: Stanford University Press, 2000).
Kelvin, William T., *Popular Lectures and Addresses, vol. 1: Constitution of Matter* (Cambridge: Cambridge University Press [1889] 2011).
Kempthorne, Loveday, 'Relations between Modern Mathematics and Poetry: Czesław Miłosz; Zbigniew Herbert; Ion Barbu/Dan Barbilian' (PhD Dissertation: Victoria University of Wellington, 2015).
Kermode, Frank, *English Pastoral Poetry from the Beginnings to Marvell* (London: George G. Harrap, 1952).
Kermode, Frank, *The Romantic Image* (London: Routledge and Kegan Paul, 1957).

Kimmelman, Burt, '"Equal, That Is, to the Real Itself": The New Physics, Charles Olson, and Avant-garde Poetics', *DQR Studies in Literature*, 47 (2011), 641–67.
Kohlmann, Benjamin, *Committed Styles* (Oxford: Oxford University Press, 2014).
Lange, Marc, *Because without a Cause: Non-causal Explanations in Science and Mathematics* (Oxford: Oxford University Press, 2017).
Lange, Marc, *Philosophy of Science: An anthology* (Malden: Blackwell, 2007).
Laplace, Pierre Simon, *A Philosophical Essay on Probabilities*, trans. F.W. Truscott and F.L. Emory (New York: Dover Publications, 1951).
Lawrence, D.H., *Pansies, Poems* (Fredonia: Fredonia Books [1929] 2002).
Leavell, Linda, Miller, Cristianne, and Schulze, Robin G., *Critics and Poets on Marianne Moore: 'A Right Good Salvo of Barks'* (Lewisburg: Bucknell University Press, 2005).
Leavis, F.R., 'Cambridge Poetry', *Cambridge Review* (1929), 317–18.
Leibniz, Gottfried Wilhelm, *Discourse on Metaphysics* trans. George R. Montgomery (Chicago: The Open Court Publishing [1686] 1918).
Levi, Anthony, *Renaissance and Reformation: The Intellectual Genesis* (New Haven: Yale University Press, 2002).
Lewis, Wyndham, *Time and Western Man* (London: Chatto & Windus, 1919).
Lippert, Sarah J., *The Paragone in Nineteenth-Century Art* (Abingdon: Routledge, 2019).
Locke, John, *An Essay Concerning Human Understanding* (London: T Tegg and Son [1689] 1836).
Lord, Eric A., *Symmetry and Pattern in Projective Geometry* (London: Springer, 2013).
Louis Zukofsky, *'A'*, intro., Barry Ahearn (New York: New Directions [1959] 2011).
Mackay, A.E., *The Garden of the Gods, and Other Poems* (London: Grant Richards, 1931).
MacNeice, Louis, 'Mr Empson as Poet (review of poems)', *New Verse*, 16 (1935), 17–18.
Man, Paul de, *Blindness and Insight: Essays in the Logic of Contemporary Criticism, 2nd edn* (Minneapolis: University of Minnesota Press, 1983).
Marinetti, F.T., *Marinetti: Selected Writings*, eds Farrar, Straus, and Giroux, trans R.W. Flint and A. Coppotelli (New York: Farrar, Straus, and Giroux, 1972).
Marinetti, F.T., 'Zang Tumb Tumb', *The World Magazine*, 9 August 1914. English translation of a portion of ZTT.
Mason, H.A., 'William Empson's Verse', *Scrutiny*, 4.3 (1935), 302–4.
Masopust, Michael A., 'Laura Riding's Quarrel with Poetry', *South Central Review*, 2.1 (1985), 42–56.
Mathews, Hans, *The Mathesis of Intelligence*, *Phalanx*, 1 (2007), 1–21.
Mathews, Hans, 'Words and Picturables', *Phalanx*, 3 (2009), 1–22.
Matthiessen, F.O., *American Renaissance: Art and Expression in the Age of Emerson and Whitman* (Oxford: Oxford University Press [1941] 1968).
Maud, Ralph, *Charles Olson's Reading: A Biography* (Carbondale: Southern Illinois University Press, 1996).
McCoy, Dorothy S., *Tradition and Convention: A Study of Periphrasis in English Pastoral Poetry from 1557–1715* (The Hague: Mouton & Co, 1965).
McCrea, W.H, 'On Matrices of Quaternions and the Representation of Eddington's E-Numbers', *Proceedings of the Royal Irish Academy. Section A: Mathematical and Physical Sciences*, 45 (1938), 65–71.
McGann, Jerome, 'Laura (Riding) Jackson and the Literal Truth', *Critical Inquiry*, 18 (1992), 454–73.
Mehrtens, Herbert, *Moderne—Sprache—Mathematik: Eine Geschichte des Streits um die Grundlagen der Disziplin und des Subjekts formaler Systeme* (Frankfurt am Main: Suhrkamp, 1990).
Melville, Herman, *Moby-Dick; or, The Whale* (New York: Harper & Brothers, 1851).
Menand, Louis, 'T. S. Eliot and Modernity', *The New England Quarterly*, 69.4 (1996), 554–79.
Mendelson, Edward, *The Complete Works of W. H. Auden, Volume III Prose: 1949–1955* (Princeton: Princeton University Press, 2008).
Merill, Robert Valentine, 'Platonism in Petrarch's "Canzoniere"', *Modern Philology*, 27.2 (1929), 161–74.

Meyer, Ernst, *The Growth of Biological Thought* (Cambridge: Harvard University Press, 1982).
Meyer, Steven, ed., *The Cambridge Companion to Literature and Science* (Cambridge: Cambridge University Press, 2018), 97–113.
Michell, Joel, *Measurement in Psychology: A Critical History of a Methodological Concept* (Cambridge: Cambridge University Press, 1999).
Middleton, Peter, *Physics Envy: American Poetry and Science in the Cold War and After* (Chicago: University of Chicago Press, 2016).
Mill, J.S., *Utilitarianism* (Kitchener: Batoche Books [1861] 2001).
Millay, Edna St Vincent, *Collected Poems*, ed. Norma Millay (New York, Harper, 1956).
Moore, D.S., 'Espousing Interactions and Fielding Reactions: Addressing Laypeople's Beliefs about Genetic Determinism', *Philosophical Psychology*, 21 (2008), 331–48.
Moore, Marianne, *Collected Poems* (New York: Macmillan, 1951).
Murry, John M., 'Science and Nonsense', *Adelphi*, 7.3 (1933).
Needham, Joseph, *The Sceptical Biologist: (Ten Essays)* (London: Chatto & Windus, 1929).
Nelly, Una, *The Poet Donne: A Study in His Dialectic Method* (Cork: Cork University Press, 1969).
Newton, Isaac, *Newton's Principia, The Mathematical Principles of Natural Philosophy* (New York: Daniel Adee [1666] 1846).
Nicholls, Peter, *Modernisms: A Literary Guide* (Basingstoke: Palgrave Macmillan, 2009).
Nietszche, Friedrich, *The Birth of Tragedy: Or Hellenism and Pessimism* trans. Walter Kauffman (London: George Allen & Unwin, 1910).
Norris, Christopher, *William Empson and the Philosophy of Literary Criticism* (London: Athlone, 1978).
Norris, Christopher, and Mapp, Nigel, eds, *William Empson: The Critical Achievement* (Cambridge: Cambridge University Press, 1993).
North, Joseph, *Literary Criticism: A Concise Political History* (Cambridge: Harvard University Press, 2017).
Norton, Thomas, *The Ordinall of Alchemy*, eds, Elias Ashmole Thomas and Eric John Holmyard (London: Edward Arnold & Co, 1928).
Novikoff, Alex J., *The Medieval Culture of Disputation: Pedagogy, Practice, and Performance* (Philadelphia: University of Pennsylvania Press, 2013).
Oakeshott, Michael, *On Human Conduct* (Oxford: Oxford University Press, 1975).
Oakeshott, Michael, *Rationalism in Politics and other Essays* (Carmel: Libert Fund [1962] 1991).
Oakeshott, Michael, *Lectures in the History of Political Thought*, eds Terry Nardin and Luke O'Sullivan (Exeter: Imprint Academic, 2006).
Ogden, C.K. and Richards, I.A., *The Meaning of Meaning: A Study of the Influence of Language upon Thought and of the Science of Symbolism* (New York: Harvest, 1923).
O'Keefe, Kevin, *The Average American: The Extraordinary Search for the Nation's Most Ordinary Citizen* (New York: Public Affairs, 2005).
Olney, James, *The Rhizome and the Flower: The Perennial Philosophy, Yeats and Jung* (Berkeley: University of California Press, 1980).
Olson, Charles, 'Verse & Geometry plus E. P.', I Box 49:37, Charles Olson Research Collection, University of Connecticut Library.
Olson, Charles, *The Maximus Poems*. ed. George F. Butterick (Berkeley: University of California Press, 1983).
Olson, Charles, *Collected Poems—Excluding the Maximus Poems* (Berkeley: University of California Press [1987] 1997).
Olson, Charles, *Collected Prose*, eds Donald Allen and Benjamin Friedlander (Berkeley: University of California Press, 1997).
Olson, Charles, *Selected Letters*. ed. Ralph Maud (Berkeley: University of California Press, 2000).
Ophir, Ella Z., 'The Laura Riding Question: Modernism, Poetry, and Truth', *Modern Language Quarterly*, 66.1 (2005), 85–114.
Ovid, *Metamorphosis*, trans. A.S. Kline (Ann Arbor: Ann Arbor Editions, 2000).

Panofsky, Erwin, *Perspective as Symbolic Form*, trans. Christopher S. Wood (New York: Zone Books [1924] 1991).
Parsons, I. M., 'Three Seizins', *Spectator* 147. 5396 (November 1931), 739.
Pater, Walter, *The Renaissance* (New York: The Modern Library, 1873).
Pearson, Karl, *Grammar of Science* (Briston: Thoemmes Antiquarian Books, 1991).
Penrose, Roger, *Fashion, Faith, and Fantasy in the New Physics of the Universe* (Princeton: Princeton University Press, 2017).
Perkins, David, *A History of Modern Poetry: Modernism and After* (Cambridge: Harvard University Press, 1987).
Pestre, Dominique, *Physique et Physiciens en France 1918–1940* (Paris: Éditions des Archives Contemporaines, 1984).
Petroşel, Daniela, 'Poetry and Technology in Marinetti's Futurist Manifestos', *Rupkatha Journal on Interdisciplinary Studies in Humanities*, 3.1 (2011), 140–6.
Piaggio, H., 'The Significance and Development of Hamilton's Quaternions', *Nature*, 152 (1943), 553–55.
Plato, *The Republic of Plato*, trans. Benjamin Jowett (Oxford: Clarendon Press, 1888).
Plato, *Plato in Twelve Volumes, vol. 7*, trans. R.G. Bury (Cambridge: Harvard University Press, 1966).
Plato, *Symposium*, trans. Benjamin Jowett (New York: Modern Library [1871] 1996).
Plato, *Timaeus and Critias*, trans. Desmond Lee (London: Penguin Books, 1977).
Poincaré, Henri, *Mathematics and Science: Last Essays* trans. J.W.B. (New York: Dover [1913] 1963).
Plotnitsky, Arkady, 'Manifolds: On the Concept of Space in Riemann and Deleuze', in *Virtual Mathematics: The logic of difference*, ed. Simon B. Duffy (Manchester: Clinamen Press, 2006):187–208.
Poincaré, Henri, *Science and Method*, trans. Francis Maitland (London: Thomas Maitland and Sons [1908] 1914).
Poincaré, Henri, *Science and Hypothesis* (London: Walter Scott Publishing, 1905).
Poincaré, Henri, *The Choice of Facts*, trans. G.B. Halsted, *Monist*, 19.2 (1909), 231–239.
Poincaré, Henri, *The Value of Science: Essential Writings of Henri Poincaré*, ed. Stephen Gould (New York: Modern Library [1908] 2001).
Pope, Alexander, *Alexander Pope: Selected Poetry*, ed. Pat Rogers (Oxford: Oxford University Press, [1749] 1994).
Pope, Alexander, *The Works of the Poets of Great Britain and Ireland with Prefaces, Biographical and Critical, vol. 6*, ed. Samuel Johnson (Dublin: Pat Wogan [1711] 1804).
Porte, Rebecca, 'An Agreement with Reality: The Poetry of Logical Modernism' (PhD Dissertation: University of Michigan, 2014).
Pound, Ezra, *The Spirit of Romance: An attempt to define somewhat the charm of the pre-renaissance literature of Latin Europe* (London: P. Owen [1910] 1970).
Pound, Ezra, 'Vorticism', *Fortnightly Review*, 46.573 (1914), 461–71.
Price, Katy, 'Flame Far Too Hot: William Empson's Non-Euclidean Predicament', *Interdisciplinary Science Reviews*, 30.4 (2005), 312–22.
Prynne, J.H., *Poems* (Freemantle: Freemantle Arts Centre Press, 2005).
Prynne, J.H., and Sutherland, Keston 'Introduction to Prynne's Poems in Chinese', *The Cambridge Quarterly* 41.1 (2012), 197–207.
Quetelet, Adolphe & Knox, Robert, *A Treatise on Man and the Development of His Faculties* (Edinburgh, 1842).
Rabi, Lior (2016) 'Ortega y Gasset on Georg Cantor's Theory of Transfinite Numbers' *Kairos* (15), 46–70.
Raine, Kathleen, 'Michael Roberts and the Hero Myth', *Penguin New Writing*, 39 (1950), 84–98.
Raine, Kathleen, *The Land Unknown* (London: Hamish Hamilton, 1975).
Rainey, Lawrence, ed., *Modernism: An Anthology* (Malden: Blackwell Publishing, 2005).
Ramsey, Frank P., *The Foundations of Mathematics and Other Logical Essays*, ed. R.B. Braithwaite, intro. G.E. Moore (London: Kegan Paul, Trench, Trübner, 1931).

Ransom, John Crowe, 'Mr Empson's Muddles', *The Southern Review*, 4 (July–April 1938–9), 322–39.
Ransom, John Crowe, *The World's Body* (New York: Charles Scribner's Sons, 1938).
Read, Herbert, 'Readers and Writers', *The New Age*, 22 (December 1921), 67–8.
Read, Herbert, *Reason and Romanticism* (London: Faber & Faber, 1926).
Reynolds, Joshua, *Discourses on Art* (London: George Routledge [1778] 1909).
Richards, I.A., *Science and Poetry* (London: Kegan Paul, Trench, Trubner & Co, 1926).
Richards, I.A., 'Empson's "Poems"', *Cambridge Review*, 57.1399 (February 1936), 253.
Richards, I.A., *The Philosophy of Rhetoric* (Oxford: Oxford University Press, 1936).
Richards, I.A., 'William Empson', *Furioso*, 1/3 (1940), 44.
Richards, I.A., *Principles of Literary Criticism* (New York: Routledge [1924] 2001).
Richards, Joan L., *Mathematical Visions: The Pursuit of Geometry in Victorian England* (Boston: Academic Press, 1988).
Richeson, David, *Euler's Gem: The Polyhedron Formula and the Birth of Topology* (Princeton: Princeton University Press, 2019).
Ricks, Christopher, *True Friendship: Geoffrey Hill, Anthony Hecht, and Robert Lowell under the Sign of Eliot and Pound* (London: Yale University Press, 2010).
Riding, Laura (Jackson), 'Dimensions', *The Fugitive* (August–September 1923), 124.
Riding, Laura (Jackson), *Love as Love: Death as Death* (London: Seizin, 1928).
Riding, Laura (Jackson), *Experts are Puzzled* (London: Jonathan Cape, 1930).
Riding, Laura (Jackson), *Poems: A Joking Word* (London: Cape, 1930).
Riding, Laura (Jackson), *Though Gently* (Deya: Seizin, 1930).
Riding, Laura (Jackson), *Twenty Poems Less* (Paris: Hours Press, 1930).
Riding, Laura (Jackson), *Poet: A Lying Word* (London: A. Barker, 1933).
Riding, Laura (Jackson), *Anarchism is not Enough*, ed. Lisa Samuels (Berkeley: University of California Press [1928] 2001).
Riding, Laura (Jackson), *Essays from Epilogue: 1935–1937*, ed. Mark Jacobs (Manchester: Carcanet Press, 2001).
Riding, Laura (Jackson), *The Laura (Riding) Jackson Reader*, ed. Elizabeth Friedman (New York: Persea Books, 2005).
Riding, Laura (Jackson), *The Failure of Poetry, the Promise of Language*, ed. John Nolan (Ann Arbor: University of Michigan Press, 2007).
Riding, Laura (Jackson), *Contemporaries and Snobs*, ed. Laura Heffernan and Jane Malcolm (Tuscaloosa: University of Alabama Press [1928] 2014).
Riding, Laura (Jackson), *The Close Chaplet*, ed. Mark Jacobs (Nottingham: Nottingham University Press [1926] 2018).
Riding, Laura (Jackson), and Graves, Robert, *A Pamphlet against Anthologies* (Hertfordshire: Garden City Press, 1928).
Riding, Laura (Jackson), and Jackson, Schuyler B., *Rational Meaning: A New Foundation for the Definition of Words and Supplementary Essays*, ed. William Harmon (Charlottesville: University Press of Virginia, 1997).
Riemann, Bernhard, *On the Hypotheses which Lie as the Bases of Geometry*, ed & trans. Juergen Jost (Switzerland: Springer International [1867] 2016).
Roberts, Michael, 'Credo: A Note on Poetry and Science', *Poetry Review*, 29.3 (1928), 192–6.
Roberts, Michael, 'On Mechanical Hallelujahs, or How Not to Do It', *Poetry Review*, 19.6 (1928), 433–8.
Roberts, Michael, 'Beyond the Golden Bars', *Poetry Review*, 20.2 (1929), 109–18.
Roberts, Michael, 'Broceliande, or the Future of the Past', *Poetry Review*, 20.1 (Jan.–Feb. 1929), 43–50.
Roberts, Michael, 'Pope and English Classicism', *Poetry Review*, 21.3 (1930), 61–70.
Roberts, Michael, 'Symbolism and Romance', *Poetry Review*, 21.1 (1930), 34.
Roberts, Michael, *These Our Matins* (London: E. Matthews & Marrot, 1930).
Roberts, Michael, 'The Loneliness of Mathematics', *Adelphi*, 1.6 (1931), 510–11.
Roberts, Michael, 'Review of James Clerk Maxwell 1831–1931', *Time and Tide*, 8.5 (1932), 126–8.
Roberts, Michael, *Elizabethan Prose* (London: Jonathan Cape, 1933).

Roberts, Michael, *Critique of Poetry* (London: Jonathan Cape, 1934).
Roberts, Michael, 'Poetry and Mountains', *The Spectator* (1934), 420, 422.
Roberts, Michael, 'Thought in the Seventeenth Century', *The Philosopher*, 12.4 (1934), 170–2.
Roberts, Michael, 'Science and Human Temperament', *The Adelphi*, 10.6 (1935), 381–2.
Roberts, Michael, 'The Mathematic Way', *The London Mercury*, 32 (1935), 173–9.
Roberts, Michael, 'The Two Coleridges', *The London Mercury*, 32 (1935), 292–3.
Roberts, Michael, 'Review of William Empson, *Some Versions of Pastoral*', *The Criterion*, 15.59 (1936), 345–8.
Roberts, Michael, 'The Poetry of T.S. Eliot', *London Mercury* 34 (1936), 38–44.
Roberts, Michael (unsigned), 'Examination of Logical Positivism', *The Listener*, 17 (1937), 238.
Roberts, Michael, *The Modern Mind* (London: Faber & Faber, 1937).
Roberts, Michael, 'In Praise of Modern Poetry', *The Listener*, 29 (1938), 375–6.
Roberts, Michael, 'Poetry and the Humour of Mountaineering', *The Alpine Journal*, 52 (1940), 22–33.
Roberts, Michael, *Faber Book of Modern Verse* (London: Faber & Faber [1936] 1951).
Roberts, Michael, *Collected Poems* (London: Faber & Faber, 1958).
Roberts, Michael, *Michael Roberts: Selected Poems and Prose*, ed. Frederick Grubb (Manchester: Carcanet Press, 1980).
Roberts, Michael, *Recovery of the West* (London: Faber & Faber, 1941).
Roberts, Michael, *T.E. Hulme* (Manchester: Carcanet Press [1938] 1982).
Roberts, Michael, 'List of Verses', National Library of Scotland, acc. 13145/53.
Rodal, Jocelyn, '"A Few Sheets of Paper Covered with Arbitrary Symbols": Formalism, Modernism, Mathematics' (PhD dissertation: University of California, Berkeley).
Rodway, A. E., 'The Structure of Complex Verse: Review of *Collected Poems*', *Essays in Criticism* 6.2 (1956), 232–40.
Ruskin, John, *Modern Painters vol. 5*. 1960 (London: George Allen, 1888).
Ruskin, John, *The Two Paths: Being lectures on art and its application to decoration and manufacture delivered in 1858–59*, ed. Christine Roth (West Lafayette: Parlor Press, 2004).
Russell, Bertrand, 'On Denoting', *Mind*, 14 (1905), 479–93.
Russell, Bertrand, *Mysticism and Logic: And Other Essays* (London: Allen & Unwin, 1917).
Russell, Bertrand, *Why I Am Not a Christian* (London: George Allen & Unwin, 1957).
Russell, Bertrand, *Bertrand Russell: His Works, vol. 3: Towards the 'Principles of Mathematics' 1900–02* (New York: Routledge, 1994).
Russo, John P., 'I. A. Richards in Retrospect', *Critical Inquiry*, 8.4 (1982), 743–60.
Saussure, Ferdinand de, *Course in General Linguistics*, trans. R. Harris, 1916 (Chicago: Open Court Publishing Company, 1982).
Schiffer, Reinhold, *Oriental Panorama: British Travellers in 19th Century Turkey* (Atlanta: Radopi, 1999).
Schwartzman, Steven, *The Words of Mathematics: An Etymological Dictionary of Mathematical Terms Used in English* (Mathematical Association of America, 1996).
Sebeok, Thomas and Umiker-Sebeok, Jean, *The Semiotic Sphere* (New York: Plenum, 1986).
Seligman, Edwin R.A., and Johnson, Alvin, eds, *The Encyclopaedia of Social Science* (New York: The Macmillan Company [1933] 1962).
Sherman, Paul, *Olson's Push: Origin, Black Mountain, and Recent American Poetry* (Baton Rouge: Louisiana State University Press, 1978).
Sidney, Philip, *The Defence of Poesie* (London: Printed for William Ponsonby, 1595).
Simpson, David, 'Everything, Beggars, Is on Fire', *Arrows*, New Year Edition (1957), 5–6.
Sims, Christopher A., 'Macroeconomics and Reality', *Econometrica*, 48.1 (1980), 1–48.
Skinner, Quentin, *The Foundations of Modern Political Thought: Volume 1 The Renaissance* (Cambridge: Cambridge University Press, 1978).
Slade, Joseph W., and Lee, Judith Yaross, eds, *Beyond the Two Cultures: Essays on Science, Technology, and Literature* (Ames: Iowa State University Press, 1990).
Sloan, Thomas O., 'The Rhetoric in the Poetry of John Donne', *Studies in English Literature, 1500–1900*, 3.1 (1963), 31–44.

Snow, C.P., *The Two Cultures* (Cambridge: Cambridge University Press [1959] 1998).
Spinoza, Benedictus, *Ethics*, ed. and trans. Edwin Curley. 1677. London: Penguin, 1996.
Spratt, Thomas, *The History of the Royal Society of London for the Improving of Natural Knowledge* (London: T.R. for J. Martyn at the Bell, 1667).
Sridhar, Anirudh, 'Auden and the Infernal Sciences', *Essays in Criticism*, 71.4 (2022): 431–453.
Sridhar, Anirudh, Hosseini, Mir Ali, and Attridge, Derek, eds, *The Work of Reading: Literary Criticism in the 21st Century* (New York: Palgrave Macmillan, 2020).
Stebbing, Susan, *Philosophy and the Physicists* (London: Methuen [1937] 2018).
Sterne, Lawrence, *The Life and Opinions of Tristram Shandy, Gentleman* (London: J.M Dent and Sons [1759] 1917).
Stevens, Wallace, *Collected Poems* (London: Faber & Faber, 1984).
Stevens, Wallace, *Collected Poetry and Prose*, eds Frank Kermode and Joan Richardson (New York: Library of America, 1997).
Sullivan, J.W.N, 'The Justification of the Scientific Method', *Athenaeum*, 4644 (1919), 274–5.
Stevens, Wallace, *Limitations of Science* (London: Chatto & Windus, 1933).
Susskind, Leonard, *Classical Mechanics: The Theoretical Minimum* (London: Penguin, 2013).
Susskind, Leonard, *Quantum Mechanics: The Theoretical Minimum* (London: Basic Books, 2014).
Tao, Terrence, *Analysis I* (Delhi: Hindustan Book Agency [2006] 2022).
Temes, Peter S., '"Code of Silence" Laura (Riding) Jackson and the Refusal to Speak', *PMLA*, 109 (1994), 87–99.
Terry, Philip, ed.,*The Penguin Book of Oulipo: Queneau, Perec, Calvino and the Adventure of Form* (New York: Penguin, 2019).
Thackeray, W.M., *Vanity Fair: A Novel without a Hero* (New York: Harper & Brothers, 1849).
Tubbs, Robert, *Mathematics in Twentieth Century Literature and Art: Content, Form and Meaning* (Baltimore, Johns Hopkins University Press, 2014).
Tubbs, Robert, Jenkins, Alice, and Engelhardt, Nina, eds, *The Palgrave Handbook of Literature and Mathematics* (Cham: Palgrave Macmillan, 2021).
Turner, W.J., *The Seven Days of the Sun: A Dramatic Poem* (London: Chatto & Windus, 1925).
Tuve, Rosemond, *Elizabethan and Metaphysical Imagery* (Chicago: Chicago University Press, 1947).
Vailati, Ezio, *Leibniz and Clarke: A Study of Their Correspondence* (Oxford: Oxford University Press, 1997).
Vinchur, Andrew J., *The Early Years of Industrial and Organizational Psychology* (Cambridge: Cambridge University Press, 2018).
Voigt, W. and Young, G. Chisholm, 'Die Fundamentalen Physikalischen Eigenschaften der Krystalle', *Nature*, 58 (1898), 99.
Waldrop, Rosmarie, 'Charles Olson: Process and Relationship', *Twentieth Century Literature*, 23.4 (1977), 467–86.
Warwick, Andrew, *Masters of Theory: Cambridge and the Rise of Mathematical Physics.* Chicago: University of Chicago Press, 2003.
Weir, Alan and Zalta, Edward N., eds, 'Formalism in the Philosophy of Mathematics', *The Stanford Encyclopaedia of Philosophy* (Stanford: Metaphysics Research Lab, Stanford University, 2015).
Wesley, John, *Primitive Physic or an Easy and Natural Method of Curing Most Diseases* (privately published, 1773).
Wexler, Joyce, *Laura Riding's Pursuit of Truth* (Ohio: Ohio University Press, 1979).
Wexler, Joyce, *Laura Riding: A Bibliography* (New York: Garland and Pub, 1981).
Weyl, Hermann, *Philosophy of Mathematics and Natural Science*, transl. Olaf Helmer (Princeton: Princeton University Press, 1949).
Weyl, Hermann, *The Concept of a Riemann Surface.* trans. Gerald R. MacLane (New York: Dover [1913] 2009).
Weyl, Hermann, *Space-Time-Matter*, trans. Henry L. Brose (New York: Dover, [1918] 1952).
Whichcote, Benjamin, *Moral and Religious Aphorisms* (Norwich: Fr. Burges, 1703).

White, Hayden, *Tropics of Discourse: Essays in Cultural Criticism* (Baltimore: Johns Hopkins University Press, 1981).
Whitehead, A.N., *An Introduction to Mathematics* (London: Williams and Norgate, 1911).
Whitehead, A.N., *The Concept of Nature* (Cambridge: Cambridge University Press, 1920).
Whitehead, A.N., *Adventures of Ideas* (Cambridge: Cambridge University Press, 1933).
Whitehead, A.N., *Process and Reality: An Essay in Cosmology* (Cambridge: Cambridge University Press, 1929).
Whitehead, A.N., *Science and the Modern World* (New York: The Free Press [1925] 1967).
Whitehead, A.N., and Russell, Bertrand, *Principia Mathematica, vol. 1* (Cambridge: Cambridge University Press, 1910).
Whittington, Leah, *Renaissance Supplicants: Poetry, Antiquity, Reconciliation* (Oxford: Oxford University Press, 2016).
Whitworth, Michael, 'Physics and the Literary Community: 1905–1939' (D.Phil. thesis: Oxford University, 1994).
Whitworth, Michael, *Einstein's Wake: Relativity, Metaphor, and Modernist Literature* (Oxford: Oxford University Press, 2001).
Wiener, Norbert, *Cybernetics: Or Control and Communication in the Animal and the Machine* (Cambridge: MIT Press, 1948).
Wigner, Eugene, 'The Unreasonable Effectiveness of Mathematics in the Natural Sciences', *Communications in Pure and Applied Mathematics*, 13.1 (1960), 1–14.
Williams, William Carlos, *Selected Essays of William Carlos Williams* (New York: Random House [1948] 1954).
Willis, J.H., *William Empson* (New York: Columbia University Press, 1969).
Wittgenstein, Ludwig, *Tractatus Logico-Philosophicus* (New York: Humanities [1921] 1951).
Wittgenstein, Ludwig, *Notebooks 1914–1916*, trans. G.E.M. Anscombe (Oxford: Blackwell, 1961).
Wood, Michael, *Literature and the Taste of Knowledge* (Cambridge: Cambridge University Press, 2005).
Wood, Michael, *On Empson* (Princeton: Princeton University Press, 2017).
Wordsworth, William, *The Poems of William Wordsworth: Collected Reading Texts from the Cornell Wordsworth Series Vol 2*, ed. Jared Curtis (Tirril: Humanities-Ebooks, 2011).
Wordsworth, William, and Coleridge, Samuel Taylor, *Lyrical Ballads 1798 and 1802*, ed. Fiona Stafford (Oxford: Oxford University Press, 2013).
Worringer, Wilhelm, *Abstraction and Empathy: A Contribution to the Psychology of Style*, trans. Michael Bullock (Chicago: Ivan R. Dee [1908] 1997).
Yeats, W.B., *Letters on Poetry from W.B. Yeats to Dorothy Wellesley* (London: Oxford University Press, 1940).
Yeats, W.B., *The Collected Poems of W.B. Yeats* (London: St. Martin's Press, 1958).
Yeats, W.B., *Later Essays*, ed. William H. O'Donnell (New York: Charles Scribner's Sons, 1994).
Yeats, W.B., *The Major Works*, ed. Edward Larrissy (Oxford: Oxford University Press, 2001).
Yeats, W.B., *The Collected Letters of W.B. Yeats, Unpublished Letters (1905–1939)*, ed. John S. Kelly (Charlottesville: InteLex Corporation, 2002).
Zellner, Arnold, *Statistics, Econometric, and Forecasting* (Cambridge: Cambridge University Press, 2004).
Zukofsky, Louis, *"A", intro., Barry Ahearn* (New York: New Directions [1959] 2011).

Index